Mathematics and Visualization

Mathematics and Visualization

Series Editors:

Gerald Farin
Hans-Christian Hege
David Hoffman
Christopher R. Johnson
Konrad Polthier
Martin Rumpf

Jean-Daniel Boissonnat
Monique Teillaud

Editors

Effective Computational Geometry for Curves and Surfaces

With 120 Figures and 1 Table

 Springer

Jean-Daniel Boissonnat
Monique Teillaud
INRIA Sophia-Antipolis
2004 route des Lucioles
B.P. 93
06902 Sophia-Antipolis, France
E-mail: Jean-Daniel.Boissonnat@sophia.inria.fr
Monique.Teillaud@sophia.inria.fr

Cover Illustration:

Cover Image by Steve Oudot (INRIA, Sophia Antipolis)

The standard left trefoil knot, represented as the intersection between two algebraic surfaces that are the images through a stereographic projection of two submanifolds of the unit 3-sphere S3 – further details can be found in [1, Chap. III, Section 8.5]. This picture was obtained from a 3D model generated with the CGAL surface meshing algorithm.

[1] E. Brieskorn and H. Knörrer. Plane Algebraic Curves. Birkhäuser, Basel Boston Stuttgart, 1986.

Mathematics Subject Classification: 68U05; 65D18; 14Q05; 14Q10; 14Q20; 68N19; 68N30; 65D17; 57Q15; 57R05; 57Q55; 65D05; 57N05; 57N65; 58A05; 68W05; 68W20; 68W25; 68W40; 68W30; 33F05; 57N25; 58A10; 58A20; 58A25.

ISBN 978-3-642-06987-1   e ISBN 978-3-540-33259-6

Springer is a part of Springer Science+Business Media
springer.com
© Springer-Verlag Berlin Heidelberg 2006
Softcover reprint of the hardcover 1st edition 2006

Cover design: *design & production* GmbH, Heidelberg

Preface

Computational geometry emerged as a discipline in the seventies and has had considerable success in improving the asymptotic complexity of the solutions to basic geometric problems including constructions of data structures, convex hulls, triangulations, Voronoi diagrams and geometric arrangements as well as geometric optimisation. However, in the mid-nineties, it was recognized that the computational geometry techniques were far from satisfactory in practice and a vigorous effort has been undertaken to make computational geometry more practical. This effort led to major advances in robustness, geometric software engineering and experimental studies, and to the development of a large library of computational geometry algorithms, CGAL.

The goal of this book is to take into consideration the multidisciplinary nature of the problem and to provide solid mathematical and algorithmic foundations for effective computational geometry for curves and surfaces. This book covers two main approaches.

In a first part, we discuss *exact geometric algorithms for curves and surfaces*. We revisit two prominent data structures of computational geometry, namely *arrangements* (Chap. 1) and *Voronoi diagrams* (Chap. 2) in order to understand how these structures, which are well-known for linear objects, behave when defined on curved objects. The mathematical properties of these structures are presented together with algorithms for their construction. To ensure the effectiveness of our algorithms, the basic numerical computations that need to be performed are precisely specified, and tradeoffs are considered between the complexity of the algorithms (i.e. the number of primitive calls), and the complexity of the primitives and their numerical stability. Chap. 3 presents recent advances on *algebraic and arithmetic tools* that are keys to solve the robustness issues of geometric computations.

In a second part, we discuss mathematical and algorithmic methods for *approximating curves and surfaces*. The search for approximate representations of curved objects is motivated by the fact that algorithms for curves and surfaces are more involved, harder to ensure robustness of, and typically several orders of magnitude slower than their linear counterparts. This book provides widely applicable, fast, safe and quality-guaranteed approximations of curves and surfaces. Although these problems have received considerable attention in the past, the solutions previously proposed were mostly heuristics and limited in scope. We establish theoretical foundations to the problem and introduce two emerging new topics: *discrete differential geometry* (Chap. 4) and *computational topology* (Chap. 7). In addition, we present certified algorithms for *mesh generation* (Chap. 5) and *surface reconstruction* (Chap. 6), two problems of great practical significance.

Each chapter refers to open source software, in particular CGAL, and discusses potential applications of the presented techniques. In 1995, CGAL, the *Computational Geometry Algorithms Library*, was founded as a research project with the goal of making correct and efficient implementations for the large body of geometric algorithms developed in the field of computational geometry available for industrial applications. It has since then evolved to an open source project [2] and now is the state-of-art implementation in many areas. A short appendix (Chap. 8) on generic programming and the CGAL library is included.

This book can serve as a textbook on non-linear computational geometry. It will also be useful to engineers and researchers working in computational geometry or other fields such as structural biology, 3-dimensional medical imaging, CAD/CAM, robotics, graphics etc. Each chapter describes the state of the art algorithms as well as provides a tutorial introduction to important concepts and methods that are both well founded mathematically and efficient in practice.

This book presents recent results of the ECG project, a Shared-Cost RTD (FET Open) Project of the European Union[1] devoted to effective computational geometry for curves and surfaces. More information on ECG, including the results obtained during this project, can be found on the web site http://www-sop.inria.fr/prisme/ECG/.

We wish to thank Franz Aurenhammer, Frédéric Chazal, Éric Colin de Verdière, Tamal Dey, Ioannis Emiris, Andreas Fabri, Menelaos Karavelas, John Keyser, Edgar Ramos, Fabrice Rouillier, and many other colleagues, for their cooperation and feedback which greatly helped us to improve the quality of this book.

[1]Number IST-2000-26473

List of Contributors

Jean-Daniel Boissonnat
INRIA
BP 93
06902 Sophia Antipolis cedex
France
Jean-Daniel.Boissonnat
@sophia.inria.fr

Frédéric Cazals
INRIA
BP 93
06902 Sophia Antipolis cedex
France
Frederic.Cazals@sophia.inria.fr

David Cohen-Steiner
INRIA
BP 93
06902 Sophia Antipolis cedex
France
David.Cohen-Steiner
@sophia.inria.fr

Efraim Fogel
School of Computer Science
Tel Aviv University
Tel Aviv 69978
Israel
efif@post.tau.ac.il

Joachim Giesen
ETH Zürich
CAB G33.2, ETH Zentrum
CH-8092 Zürich
Switzerland
giesen@inf.ethz.ch

Dan Halperin
School of Computer Science
Tel Aviv University
Tel Aviv 69978
Israel
danha@tau.ac.il

Lutz Kettner
Max-Planck-Institut für Informatik
Stuhlsatzenhausweg 85
66123 Saarbrücken
Germany
kettner@mpi-inf.mpg.de

Jean-Marie Morvan
Institut Camille Jordan
Université Claude Bernard Lyon 1
43 boulevard du 11 novembre 1918
69622 Villeurbanne cedex
France
morvanjeanmarie@yahoo.fr

Bernard Mourrain
INRIA

BP 93
06902 Sophia Antipolis cedex
France
Bernard.Mourrain@sophia.inria.fr

Sylvain Pion
INRIA
BP 93
06902 Sophia Antipolis cedex
France
Sylvain.Pion@sophia.inria.fr

Günter Rote
Freie Universität Berlin
Institut für Informatik
Takustraße 9
14195 Berlin
Germany
rote@inf.fu-berlin.de

Susanne Schmitt
Max-Planck-Institut für Informatik
Stuhlsatzenhausweg 85
66123 Saarbrücken
sschmitt@mpi-inf.mpg.de

Jean-Pierre Técourt
INRIA
BP 93
06902 Sophia Antipolis cedex
France
Jean-Pierre.Tecourt
@sophia.inria.fr

Monique Teillaud
INRIA
BP 93
06902 Sophia Antipolis cedex
France
Monique.Teillaud@sophia.inria.fr

Elias Tsigaridas
Department of Informatics and
Telecommunications
National Kapodistrian University of
Athens
Panepistimiopolis 15784
Greece
et@di.uoa.gr

Gert Vegter
Institute for Mathematics and
Computer Science
University of Groningen
P.O. Box 800
9700 AV Groningen
The Netherlands
gert@cs.rug.nl

Ron Wein
School of Computer Science
Tel Aviv University
Tel Aviv 69978
Israel
wein@post.tau.ac.il

Nicola Wolpert
Max-Planck-Institut für Informatik
Stuhlsatzenhausweg 85
66123 Saarbrücken
nicola.wolpert@hft-stuttgart.de

Camille Wormser
INRIA
BP 93
06902 Sophia Antipolis cedex
France
Camille.Wormser@sophia.inria.fr

Mariette Yvinec
INRIA
BP 93
06902 Sophia Antipolis cedex
France
Mariette.Yvinec@sophia.inria.fr

Contents

2 Curved Voronoi Diagrams

3 Algebraic Issues in Computational Geometry

1

Arrangements

Efi Fogel, Dan Halperin*, Lutz Kettner, Monique Teillaud, Ron Wein, and
Nicola Wolpert

1.1 Introduction

Arrangements of geometric objects have been intensively studied in combinatorial and computational geometry for several decades. Given a finite collection S of geometric objects (such as lines, planes, or spheres) the *arrangement* $\mathcal{A}(S)$ is the *subdivision* of the space where these objects reside into cells as induced by the objects in S. Figure 1.1 illustrates a planar arrangement of circles, which consists of vertices, edges, and faces: a *vertex* is an intersection point of two (or more) circles, an *edge* is a maximal portion of a circle not containing any vertex, and a *face* is a maximal region of the plane not containing any vertex or edge. For convenience we also introduce two (artificial) vertices in each circle at the x-extreme points splitting the circle into two x-monotone arcs (thus each edge of the arrangement now has two distinct endpoints).

Arrangements are defined and have been investigated for general families of geometric objects. One of the best studied type of arrangements is that of lines in the plane. This means that arrangements may have unbounded edges

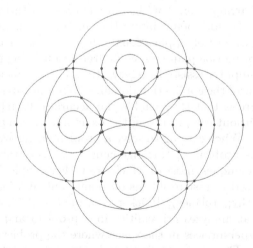

Fig. 1.1. An arrangement of 14 circles in the plane, with 53 faces (one of which is unbounded), 106 edges, and 59 vertices

---

* Chapter coordinator

and faces. Furthermore, arrangements are defined in any dimension. There are, for example, naturally defined and useful arrangements of hypersurfaces in six-dimensional space, arising in the study of rigid motion of bodies in three-dimensional space.

Written evidence of the study of arrangements goes back to the nineteenth century (see [249]). We notice three periods in the *computational* study of arrangements. From the inception of computational geometry in the seventies till the mid eighties the focus was almost exclusively on the theoretical study of arrangements of unbounded linear objects, of hyperplanes. Many of the results obtained during this period are summarized in the book by Edelsbrunner [130]. The central role of arrangements in computational geometry has fortified in the following period, from the mid-eighties to the mid-nineties, where the theoretical focus has shifted toward arrangements of curves and surfaces. Many of the results obtained in those years are summarized in the book by Sharir and Agarwal [312] and in the survey papers [15] and [196]. From the mid nineties till the time of writing this chapter, a new more practical aspect of the study of arrangements has strengthened, emphasizing implementation and usage. The goal of robustly implementing algorithms for arrangements continues to raise numerous challenging technological and scientific problems. This chapter is devoted to the developments in this applied direction, with an emphasis on curves and surfaces. It should be noted however, that the appearance of a new topic of study in arrangements never replaced previous trends, and to this very day the theoretical study of arrangements of hyperplanes or of arrangements of algebraic curves is a thriving and fruitful domain.

Besides being interesting in their own right, arrangements are useful in a variety of applications. They have been used in solving problems in robot motion planning, computer vision, GIS (geographic information systems), computer-assisted surgery, statistics, and molecular biology, to mention just a few of the application domains. What makes arrangements such a useful structure is that they enable the accurate discretization of continuous problems, without compromising the exactness or completeness of the solution.

When coming to solve a problem using arrangements, we first need to cast the problem in "arrangement terms". To this end, there is a large arsenal of techniques, most of which are rather simple. They include duality transforms (used to cast problems on point configurations as problems on arrangements of hyperplanes), Plücker coordinates [287], and the so-called locus method that analyzes criticalities in a problem and transforms the criticalities into hypersurfaces in the space where the problem is studied.

The next stage is to understand the combinatorial complexity of the relevant arrangement or of a portion thereof. Quite often, one does not need to construct the entire arrangement in order to solve a problem, and having only a substructure is sufficient (e.g., a single connected component of the ambient space often suffices in solving motion planning problems). This analysis gives a lower bound on the resources required by the algorithms and data structures that will be used in the solution.

Choosing or devising efficient algorithms and data structures to build the required (sub)arrangement is the next step of solving a problem using arrangements. This and the above two steps have been studied for decades.

Then comes the stage of effective implementation of the solution, which has various aspects to it. Sometimes asymptotically efficient algorithms are not necessarily practically the best. Precision and robustness of the solution is a central issue and raises questions such as: (i) how to cope with degeneracies, which are typically ignored in theory under the *general position* assumption; (ii) is the ready-made computer arithmetic sufficient or do we need to use more sophisticated machinery? These are the topics that this chapter focuses on.

The next section of the chapter gives a brief overview of the state-of-the-art in constructing arrangements. In Sect. 1.3 we survey the main advances in exact construction of planar arrangements, focusing mostly on the sweep-line approach, but explaining also what is needed for incremental construction. In Sect. 1.4 we review the software implementation details of these methods. The more recent work on the three-dimensional case is discussed in Sect. 1.5. Stepping away from exact computing is the topic of Sect. 1.6. A tour of implemented applications of curves and surfaces is given in Sect. 1.7. We conclude in Sect. 1.8 with suggestions for further reading and open problems.

1.2 Chronicles

In 1995, CGAL, the *Computational Geometry Algorithms Library*, was founded as a research project with the goal of making the large body of geometric algorithms developed in the field of computational geometry available for industrial applications with correct and efficient implementations [2, 222, 156]. It has since then evolved to an Open Source Project and is now representing the state-of-art in implementing computational geometry software in many areas. CGAL contains an elaborate and efficient implementation of arrangements that supports general types of curves.

The recent ECG project, which stands for *Effective Computational Geometry for Curves and Surfaces*, running from 2001 to 2004, extended the scope of implementation research towards curved objects.[1] Arrangements of curves and surfaces were an important theme in this project, and several different approaches to the topic were taken. This body of work is now collected and presented in a uniform manner in the following sections of the current chapter. In this brief section however, we present how the different results evolved.

The first branch of research, undertaken at Tel-Aviv University, Israel, is founded on the CGAL arrangement computation. Originally the CGAL arrangements package supported line segments, circular arcs and restricted types of parabolas. Wein [331] extended the CGAL implementation to ellipses and

---

[1] ⟨http://www-sop.inria.fr/prisme/ECG/⟩

arcs of conics, where the conics can be of any type. Following the newly emerging requirements from curves on the software, Fogel et al. [167, 166] improved and refined the software design of the CGAL arrangement package. This new design formed a common platform for a preliminary comparison documented in [165] of the different arrangement computation approaches described here. Recently the whole package has been revamped [333] leading to more compact, easier-to-use code, which in certain cases is much faster than the results reported in [165], sometimes by a factor of ten or more. The description in Section 1.4 pertains to this latest design.

The second branch of research, undertaken at the Max-Planck Institute of Computer Science in Saarbrücken, Germany, produces a set of C++ libraries in the project EXACUS (*Efficient and Exact Algorithms for Curves and Surfaces*) [4] as contribution in the ECG project with support for the CGAL arrangement class. The design of the EXACUS libraries is described in Berberich et al. [48]. The theory and the implementations for the different applications behind EXACUS are described in the following series of papers: Berberich et al. [49] computed arrangements of conic arcs based on the improved LEDA [251] implementation of the Bentley-Ottmann sweep-line algorithm [46]. Eigenwillig et al. [140] extended the sweep-line approach to cubic curves. A generalization of Jacobi curves for locating tangential intersections is described by Wolpert [341]. Berberich et al. [50] recently extended these techniques to special quartic curves that are projections of spatial silhouette and intersection curves of quadrics, and lifted the result back into space. The recent extension to algebraic curves of general degree by Wolpert and Seidel [310] exists currently only as a Maple prototype.

The third branch of research, undertaken at INRIA Sophia-Antipolis in France and The National University of Athens in Greece, proposes a design for a more systematic support of non-linear geometry in CGAL focusing on arrangements of curves. This implementation was limited to circular arcs and yielded preliminary results for conical arc, restricted to elliptic arcs [145]. More work on this kernel is in progress. Their approach is quite different from the approach taken in the EXACUS project in that it avoids the direct manipulation of algebraic numbers. Instead, by using algebraic tools like Sturm sequences (that are statically precomputed for small degrees for increased efficiency), they reduce operations such as comparison of algebraic numbers to computing *signs of polynomial expressions*, which can be done with exact integer arithmetic. In addition, this allows the use of efficient filtering techniques [116, 144]. This method is expected to compare favorably against the algebraic numbers used elsewhere.

Efforts towards exact and efficient implementations have been made in the libraries MAPC [226] and ESOLID [225], which deal with algebraic points and curves and with low-degree surfaces, respectively. Both libraries are not complete in the sense that they require surfaces to be in general position.

Computer algebra methods, based on exact arithmetic, guarantee correctness of their results. A particularly powerful and complete method related to our problems is *cylindrical algebraic decomposition* invented by Collins [101, 100] and subsequently implemented (with numerous refinements). Our approach to curve and curve pair analysis can be regarded as a form of cylindrical algebraic decomposition of the plane in which the lifting phase has been redesigned to take advantage of the specific geometric setting; in particular, to reduce the degree of algebraic numbers in arithmetic operations.

Much of the work described in this chapter relies on having effective algebraic software available, like algebraic number types (as the ones provided by LEDA and CORE), or other algebraic tools. Exactness and efficiency also require the use of adapted number types and filtering techniques. Developments in these areas, which are independent of computing arrangements, are discussed in Chapter 3.

1.3 Exact Construction of Planar Arrangements

Implementing geometric algorithms in a robust way is known to be notoriously difficult. The decisions made by such algorithms are based on the results of simple geometric questions, called *predicates*, solved by the evaluation of *continuous* functions subject to rounding errors (if we use the standard floating-point computer arithmetic), though the algorithms are basically of combinatorial and *discrete* nature. The *exact* evaluation of predicates has become a research topic on its own in recent years, in particular for the case of arrangements of curves.

Algorithms for computing arrangements of curves (or segments of curves) require several operations to be performed exactly. One essential predicate, for example, is the xy-comparison that takes two endpoints or intersection points of two segments and compares them lexicographically. This predicate is known to be very sensitive to numerical errors: If the relation given by the comparison test is not transitive, due to erroneous numerical computations, the algorithm may fail.

Let us assume that we are given a set $\mathcal{B}$ of planar x-monotone curves that are pairwise disjoint in their interior — that is, two curves in $\mathcal{B}$ may have a common endpoint but cannot have any other intersection points. The *doubly-connected edge list* (DCEL for short) is a data structure that allows a convenient and efficient representation of the planar subdivision $\mathcal{A}(\mathcal{B})$ induced by $\mathcal{B}$. We represent each curve using a pair of directed *halfedges*, one going from the left endpoint (lexicographically smaller) of the curve to its right endpoint, and the other (its *twin* halfedge) going in the opposite direction. The DCEL consists of three containers of records: *vertices* (associated with planar points), *halfedges*, and *faces*, where halfedges are used to separate faces and to connect vertices. We store a pointer from each halfedge to its *incident face*, which is the face lying to its left. In addition, every halfedge is followed by another halfedge sharing the same incident face, such that the destination vertex of the halfedge is the same as the origin vertex of its following halfedge.

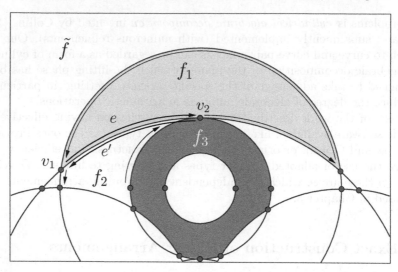

Fig. 1.2. A portion of the arrangement depicted in Fig. 1.1 with some of the DCEL records that represent it. $\tilde{f}$ is the unbounded face. The halfedge e (and its twin e') correspond to a circular arc that connects the vertices v_1 and v_2 and separates the face f_1 from f_2. The predecessors and successors of e and e' are also shown — note that e, together with its predecessor and successor halfedges form a closed chain representing the boundary of f_1 (lightly shaded). Also note that the face f_3 (darkly shaded) has a more complicated structure as it contains a hole in its interior

The halfedges are therefore connected in circular lists and form chains, such that all halfedges of a chain are incident to the same face and wind in a counterclockwise direction along its outer boundary. See Fig. 1.2 for an illustration.

The full details concerning the DCEL records are omitted here. We only mention that the DCEL representation is very useful for algorithms that require traversal of a subdivision, or for computing the overlay of two planar subdivisions. See [111, Sect. 2.2] for further details and examples.

In a scenario more general than above, we are given a set $\mathcal{C}$ of planar curves. A curve $C \in \mathcal{C}$ may not necessarily be x-monotone and it may intersect any other curve $C' \in \mathcal{C}$ in a finite number of points, or alternatively, it may (partially) overlap it. If we wish to construct a DCEL that represents $\mathcal{A}(\mathcal{C})$, the arrangement (or planar subdivision) induced by $\mathcal{C}$, we perform the following steps:

- Subdivide each $C \in \mathcal{C}$ into x-monotone segments. We denote the resulting set as $\hat{\mathcal{C}}$. We prefer that our DCEL contain only x-monotone curves as this not only makes its maintenance much simpler, but also enables us to construct a data structure on top of the DCEL, based on vertical decomposition, that enables us to answer point-location queries efficiently (see Sect. 1.3.2).

To allow for degenerate input, we only require that each curve in $\hat{\mathcal{C}}$ is weakly x-monotone, where vertical segments are also considered to be weakly x-monotone.

- Compute all the intersections between the curves in $\hat{\mathcal{C}}$, and subdivide the curves into subcurves that are pairwise disjoint in their interior. Let $\mathcal{B}$ be the resulting set.

- Construct a DCEL representing $\mathcal{A}(\mathcal{B})$. The edges in this subdivision correspond to the subcurves of $\mathcal{C}$, where we can store with each edge a pointer to the original curve $C \in \mathcal{C}$ that contributed the corresponding subcurve. In case of an overlapping edge, we may have to store multiple pointers. It immediately follows that $\mathcal{A}(\mathcal{C}) = \mathcal{A}(\mathcal{B})$.

We will discuss two different approaches for computing arrangements of curves and the implementation of the predicates they require. In Sect. 1.3.1 we present an approach based on the Bentley and Ottmann sweep-line algorithm [46]. We first describe this algorithm in its original setting for line segments. We then show how it can be adapted to arbitrary x-monotone curves and derive the necessary predicates. We explain in detail how these predicates can be implemented for conics, i.e., implicit algebraic curves of degree 2.[2] We then briefly examine the case of algebraic curves of degree 3, also known as cubic curves, and describe what difficulties emerge as the degree of the curves increases. Our experience shows that the sweep-line approach is the fastest in practice.

The second general approach for computing arrangements of curves, presented in Sect. 1.3.2, is an incremental one. The advantage of this method in comparison to the sweep-line algorithm is that it is on-line. Furthermore, as a by-product of the incremental construction one can produce an efficient point-location structure. We first explain how the arrangement is constructed by inserting the curves one by one. We then describe point location strategies, and conclude by deriving the predicates needed by the incremental approach.

1.3.1 Construction by Sweeping

The Classical Sweep-Line Algorithm

The famous *sweep-line* algorithm of Bentley and Ottmann [46] was originally formulated for sets of line segments. For completeness, we give a sketch of this algorithm, with the general position assumptions that no segment is vertical, no three segments intersect at a common point and no two segments overlap.[3]

---

[2]An *algebraic curve* of degree d is the locus of points that satisfy the equation $\sum_{i=0}^{d} \sum_{j=0}^{d-i} c_{ij} x^i y^j = 0$, where $c_{ij} \in \mathbb{R}$ for all indices i and j.

[3]The original algorithm of Bentley and Ottmann only detects and reports the intersection points between the input segments. However, it can be easily augmented to compute the arrangement.

The main idea is that the static two-dimensional problem is transformed into a dynamic one-dimensional one. An imaginary vertical line, called the *sweep line*, is swept over the plane from left to right. At each time during the sweep a subset of the input segments intersect the sweep line in a certain order. While moving the sweep line along the x-axis a change in the topology of the arrangement takes place when this ordering changes. This happens at a finite number of *event-points*: intersection points of two segments and left endpoints or right endpoints of segments.

The following invariants are maintained during the sweep:

1. All event points to the left of the sweep line (more precisely, all event points that are xy-lexicographically smaller than the current event point) have been discovered and handled.
2. The ordered sequence of segments intersecting the sweep line is stored in a dynamic structure called the *Y-structure*.
3. The event points, namely segment endpoints and all intersection points that have already been discovered but not yet handled (that is, they are to the right of the sweep line), are stored in a second dynamic structure, named the *X-structure*, in xy-lexicographic order.

We initialize the X-structure by inserting all segment endpoints into it, while the Y-structure is initially empty. We iteratively extract the lexicographically smallest event point from the X-structure. We are done if the X-structure is empty. Otherwise, we move the sweep line past this event point and update the X- and Y-structures according to the following type of the event point (in what follows, one of the segments s_a or s_b, or both, may not exist):

- If the event is the left endpoint of a segment s, we insert s into the Y-structure according to its y-order along the sweep line. Let s_a and s_b be the segments above and below s after the insertion, respectively. If s and s_a intersect, we insert their intersection point into the X-structure. We do the same for s and s_b.
- If the event is the right endpoint of a segment s, we remove s from the Y-structure. Let s_a and s_b be the segments above and below s immediately before the deletion, respectively. After the deletion s_a and s_b become adjacent in the Y-structure. If they intersect at a point with a larger x-coordinate than the current one, we insert their intersection point into the X-structure (unless it already exists there).
- If the event is the intersection point of s_1 and s_2, we swap their position in the Y-structure. Assume, without loss of generality, that after the swap s_2 lies above s_1. Let s_a be the segment above s_2 and s_b be the segment below s_1. If s_a and s_2 intersect, we insert their intersection point into the X-structure. We do the same for s_1 and s_b.

The running time of this algorithm for a set of n input segments that intersect in k points is $O((n+k)\log n)$. It is possible to guarantee $O(n)$ space complexity by removing intersection points of segments from the X-structure as soon as the segments are no longer adjacent along the sweep line.

As we have already mentioned, the sweep-line algorithm was originally formulated with some restrictions on the input segments. It can be modified in a way that it can handle any set of line-segments, containing various kinds of degeneracies — see [111, Sect. 2.1] and [251, Sect. 10.7].

Sweeping Non-Linear Curves

Already Bentley and Ottmann observed that the sweep-line algorithm can be used to handle arbitrary x-monotone curves (or x-monotone segments of arbitrary planar curves). Two implicit assumptions are made by the "classical" algorithm: a pair of segments can intersect at most once, and two segments swap their relative position when they intersect. These assumptions do not necessarily hold for general curves, but we can easily remedy the situation:

- Instead of checking whether two curves intersect, we check whether they have an intersection point to the right of the current event point p_e. If there are several intersection points lying to the right of p_e, it is sufficient, at the current event, to consider only the leftmost one. However, if all intersection points are available, we can insert them all into the X-structure.
- When we deal with an intersection event of two curves, we have to consider the *multiplicity* of the intersection point (see Chap. 3 for the exact definition of the multiplicity of an intersection point). If the multiplicity is odd, the two curves swap their relative vertical positions and we proceed as in the case of line segments. If, however, the multiplicity is even, the two curves maintain their initial positions and no new adjacencies are created in the Y-structure.

If we drop the general position assumption and allow several curves to intersect at a common point p, we have to be a bit more careful when the sweep line passes over p. For straight line segments the y-order of the intersecting segments just has to be reversed, but this, of course, is not necessarily true for arbitrary curves. The following algorithm, taken from [49], determines the y-order of k curves immediately to the right of p, whose order to the left of p is $C_1, \ldots, C_k$, in time $O(M \cdot k)$, where M is the maximal multiplicity of a pairwise intersection of the curves:

Lemma 1. *Given the y-order of the curves $C_1, \ldots, C_k$ passing through a common point p immediately to the left of p, algorithm* ORDERTORIGHT *correctly computes the y-order of the curves immediately to the right of p.*

Proof. Notice that as our x-monotone curves are defined to the left and to the right of $p = (p_x, p_y)$, we can conceptually treat them as univariate functions $y = C_i(x)$. Each C_i is developed in a Taylor series locally around p:

Algorithm 1 ORDERTORIGHT $(C_1, \ldots, C_k; p)$

1:   For each $1 \leq i < k$ do:

1.1:   Compute m_i, the multiplicity of intersection of the curves C_i and C_{i+1} at p.

2:   Let $M \longleftarrow \max \{m_1, \ldots, m_{k-1}\}$.

3:   Let $m \longleftarrow M$.

4:   While $m \geq 1$ do:

4.1:   Form maximal subsequences of curves, where two curves belong to the same subsequence, if they are not separated by a multiplicity less than m.

4.2:   Reverse the order of each subsequence.

4.3: $m \longleftarrow m - 1$.

$C_i(x) = \sum_{\nu=1}^{\infty} c_{i\nu}(x - p_x)^\nu$, with $c_{i\nu} = \frac{C_i^{(\nu)}(x_p)}{\nu!}$. Two arbitrary curves C_i and C_j intersect with multiplicity m in p iff m is the least index for which $c_{im} \neq c_{jm}$ (or equivalently, $C_i^{(m)}(p_x) \neq C_j^{(m)}(p_x)$). The y-order of C_i and C_j immediately to the left (and immediately to the right) of p is determined by the values of c_{im} and c_{jm}. This is because locally around p the low-degree terms of the Taylor expansion C_i and C_j are the dominating ones. Without loss of generality assume $c_{im} < c_{jm}$. If m is even, the y-order to the left is $C_i \prec C_j$ (C_i is below C_j), the one to the right is also $C_i \prec C_j$. If m is odd, the y-order to the left is $C_j \prec C_i$, the one to the right is $C_i \prec C_j$. Thus, for m odd the two curves change their y-order at the point p.

The algorithm above only computes the intersection multiplicities for neighboring pairs of curves with respect to their y-order to the left of p. Now let C_i and C_j, $i < j$, be two arbitrary curves with intersection multiplicity m. We claim that $m = \mu$ with $\mu = \min\{m_i, \ldots, m_{j-1}\}$. The inequality $m \geq \mu$ is trivial since all Taylor series of neighboring pairs of curves in $(C_i, \ldots, C_j)$ are identical at least up to index $\mu - 1$, and so are the ones of C_i and C_j. Thus, we assume that $m > \mu$ and let l, $i \leq l < j$, be the least index such that $\mu = m_l$. The Taylor series of C_l and C_{l+1} differ at index μ and by the choice of l, so do the series of C_i and C_{l+1}. Since C_i and C_j are identical at least up to index μ the y-order of C_i and C_{l+1} to the left of p equals the y-order of C_j and C_{l+1} there. This contradicts the given y-order $C_i \prec C_{l+1} \prec C_j$ to the left of p, and we conclude that $m = \min\{m_i, \ldots, m_{j-1}\}$. In particular, we have proved that the maximal multiplicity of intersection among all curves $C_1, \ldots, C_k$ is given by $\max \{m_1, \ldots, m_{k-1}\}$.

The y-order to the right of p of C_i and C_j differs from their y-order to the left of p iff m is odd. Observe that $m = \min\{m_i, \ldots, m_{j-1}\}$ is exactly the number of times C_i and C_j belong to the same subsequence in the algorithm, i.e. the number of times their order is reversed. We conclude the order of C_i and C_j is reversed iff C_i and C_j cross at p.

We are now ready to define the set of geometric operations (predicates and geometric constructions) needed to realize the sweep for general curves and curve segments. The first operation that must be defined in order to make the sweep applicable is the conversion of input curves, which may not necessarily be x-monotone, to a set of x-monotone curves inducing the same arrangement:

Make x-monotone: Given a curve (or a curve segment), subdivide it into maximal x-monotone segments, also referred to as *sweepable segments*.

The geometric operations then needed by the sweep-line algorithm involve points and x-monotone segments of curves:

Compare xy: Given two points p and q, compare them lexicographically. This predicate is needed to sort the event points in the X-structure.

Point position: Given an x-monotone curve segment C and a point p in the x-range of C, determine whether p is vertically above, below, or on C. In the context of the sweep algorithm, this predicate is used to insert the leftmost endpoint of a sweepable curve into the Y-structure by locating its position with respect to the existing curves in the structure. The predicate is also used by some point-location strategies (see Sect. 1.3.2).

Compare to right: Given two curves C_1 and C_2 that intersect at a given point p, determine the y-order of C_1 and C_2 immediately to the right of p. This predicate is used to insert new curves into the Y-structure, when their leftmost endpoint lies on an existing curve in the structure. (It is also used in the incremental construction algorithm to determine the location of the inserted curve when its left endpoint lies on an existing arrangement edge or coincides with an existing vertex.)

Intersections: Given two curves C_1 and C_2, compute all the intersection points of C_1 and C_2 and their multiplicities. In degenerate situations, the two curves may overlap. In this case we return the overlapping segment as their next intersection. We omit here the technical details concerning the handling of overlaps.

We next give an overview of the algebraic methods used to implement the operations above for arrangements of algebraic curves of degree 2 (*conic* curves), based on [49] and [331], and of curves of degree 3 (*cubic* curves), based on [140].

As already mentioned, it is impossible to robustly implement the sweep-line algorithm as-is using machine-precision floating-point arithmetic. Instead, we should work with a number type that can carry out mathematical operations in an exact manner. Our main task is to minimize the *set of operations* required from the number type, and we show that it is sufficient to use number types that enable the construction from an (unbounded) integer and support the arithmetic operations $\{+, -, \times, \div\}$, the square-root operation, and comparisons on these numbers, in an exact manner. Such number types are provided by the LEDA library [251] and by the CORE library [3].[4]

---

[4]The work of Emiris *et al.* [145], which is based on different methods, reducing the comparison of algebraic numbers to computing *signs of polynomial expressions*, and thus does not require the manipulation of such number types, is not described in this section. We refer the reader to Chap. 3.

Another important task is to minimize the *number* of exact numerical operations, since these operations are typically more time consuming than machine floating-point arithmetic.

In our analysis of the operations, we pay attention to special classes of real numbers that we can handle using LEDA or CORE's exact number types:

Definition 1. *A real number* $\alpha \in \mathbb{R}$ *is a* one-root number *if it can be expressed as* $q_1 + q_2 \sqrt{q_3}$ *where* $q_1, q_2, q_3 \in \mathbb{Q}$.

Any algebraic number of degree 2, namely any real-valued solution to a quadratic equation with rational coefficients $ax^2 + bx + c = 0$, is a one-root number as it has the form $\alpha = -\frac{b}{2a} \pm \frac{1}{2a} \sqrt{b^2 - 4ac}$.

Definition 2. *The field of* real-root expressions *(FRE), denoted* $\mathbb{F}$, *is the closure of the integers under the operations* $\left\{ +, -, \times, \div, \sqrt{\ } \right\}$.

$\mathbb{F}$ is a subfield of the field of real algebraic numbers and contains the set of one-root numbers. We also note that the solution of a quadratic equation whose coefficients are one-root numbers can be represented using only the $\left\{ +, -, \times, \div, \sqrt{\ } \right\}$ operations, therefore it is in $\mathbb{F}$. When we use the `leda_real` class of LEDA or the `Expr` class of CORE as our number-type, we have the important property that we can carry out exact comparisons of any two numbers in $\mathbb{F}$.

Conics

We now show how to realize the geometric operations needed by the sweep-line algorithm for sets of conic curves, or segments of such curves (we use the term *conics* for short). A *conic curve* is implicitly defined by a quadratic polynomial:

$$C(x, y) = rx^2 + sy^2 + txy + ux + vy + w \in \mathbb{R}[x, y] .$$

However, in the rest of this section we will confine ourselves to deal with curves with rational coefficients, that is, $C(x, y) \in \mathbb{Q}[x, y]$. The conic curve consists of all points $(x, y) \in \mathbb{R}^2$ in the real plane for which $C(x, y) = 0$ holds. In what follows we will always identify the curve and its defining polynomial. A *conic arc* is a segment of a conic curve, represented by a supporting conic curve, the two delimiting endpoints, and the orientation in which the two endpoints are connected (clockwise or counterclockwise).

There are three types of curved conics, namely *ellipses*, *parabolas*, and *hyperbolas*, where the sign of the expression $\Delta_C = 4rs - t^2$ characterizes the type of conic:

- $\Delta_C > 0$ is a necessary (but not sufficient, because of degeneracies — see below) condition that the conic curve C is an *ellipse* (e.g., $C(x, y) = x^2 + 2y^2 - 1$),

- $\Delta_C = 0$ is a necessary (but not sufficient) condition that the conic curve C is a *parabola* (e.g., $C(x,y) = x^2 + 4y^2 + 4xy - y$),
- $\Delta_C < 0$ is a necessary (but not sufficient) condition that the conic curve C is a *hyperbola* (e.g., $C(x,y) = x^2 - y - 1$).

There also exist some non-curved forms of conic curves. If $r = s = t = 0$, the conic C degenerates to a single line. In addition, we can encounter line-pairs that can be either intersecting (e.g., $C(x,y) = (x+y-1)(2x+y+1)$) or parallel (e.g., $C(x,y) = (x+y-1)(x+y+1)$). Conics can also be degenerate, such as the empty set (e.g., $C(x,y) = x^2 + y^2 + 1$) or a single point (e.g., $C(x,y) = x^2 + y^2$). Indeed, it is important for some applications to handle non-curved conics (see Sect. 1.7) and due to our aim of completeness we wish to handle all types of degenerate conics. For a complete implementation that can handle all, even all degenerate conics, consider for example EXACUS [4]. However, for the sake of simplicity we will not discuss all cases in full detail in this book but mainly focus on ellipses, parabolas, and hyperbolas. We just mention that each kind of conic is completely characterized by the sign of Δ_C and the number of real roots of the polynomial $p_C(x) = (tx + v)^2 - 4s(rx^2 + ux + w)$.[5] Extending the following details for curved conics to non-curved (and even degenerate) conics is not too difficult and left to the reader.

Recall that we are interested in the points $(x,y) \in \mathbb{R}^2$ with $C(x,y) = 0$. Given x_0, the points on C whose x-coordinates equal x_0 are given by solving the equation:

$$sy^2 + (tx_0 + v)y + (rx_0^2 + ux_0 + w) = 0 .$$

Let us assume that $s \neq 0$, then there may be at most two such points, with their y-coordinates given by:

$$y_{1,2}(x_0) = \frac{-(tx_0 + v) \pm \sqrt{(tx_0 + v)^2 - 4s(rx_0^2 + ux_0 + w)}}{2s} . \tag{1.1}$$

We distinguish three different cases:

1. If $(tx_0 + v)^2 - 4s(rx_0^2 + ux_0 + w) > 0$, then y_1 and y_2 are real numbers and there are two points (x_0, y_1), $(x_0, y_2) \in \mathbb{R}^2$ that lie on the curve C.
2. For $(tx_0 + v)^2 - 4s(rx_0^2 + ux_0 + w) = 0$ we have $y_1 = y_2$, thus there is a single point $(x_0, y_1) \in \mathbb{R}^2$ that lies on the curve C.
3. If $(tx_0 + v)^2 - 4s(rx_0^2 + ux_0 + w) < 0$, then $y_1, y_2 \in \mathbb{C} \setminus \mathbb{R}$ and there are no points on C whose x-coordinate equals x_0.

As both roots y_1, y_2 evolve continuously as we move x_0 along the x-axis, it follows that when case 2 above occurs, the tangent to C at (x_0, y_1) is vertical (or, if we allow line-pairs, we may have a singularity caused by the intersection point of the two lines). We are now ready to implement the first geometric operation.

---

[5]This is the number of points on C with a vertical tangent — see below. In degenerate cases, this is the number of singular points on C.

Make x-monotone: We locate the points with a vertical tangent (or at which a singularity occurs) by computing the x-values for which

$$(tx + v)^2 - 4s(rx^2 + ux + w) = 0 ,$$

which gives the following quadratic equation:

$$(t^2 - 4rs)x^2 + 2(tv - 2su) + (v^2 - 4sw) = 0 . \qquad (1.2)$$

Let x_1, x_2 be the real-valued roots of this quadratic equation. (Notice that for a parabola $t^2 - 4rs = 0$ and we have just a single point with a vertical tangent.) The y-coordinates are simply given by:

$$y_i = -\frac{tx_i + v}{2s} \qquad \text{for } i = 1, 2 .$$

The points $p_i = (x_i, y_i)$ that have vertical tangents (or at which a singularity occurs) are called *one-curve events*.[6] Having located these points, we subdivide C into maximally connected components of $\{ (x, y) \mid C(x, y) = 0 \} \setminus \{p_1, p_2\}$ which are sweepable x-monotone conic arcs.

It is important to note that as the conic coefficients are all rational, the coordinates of the one-curve events are one-root numbers. But what can we say about *two-curve events*, namely the intersection points of two conics C_1 and C_2? Using resultant calculus (see more details in Chap. 3) we can directly compute the polynomial $\xi_I(x) = \mathrm{res}_y(C_1, C_2) \in \mathbb{Q}[x]$ whose roots are exactly the x-coordinates of the intersection points of C_1 and C_2. The degree of ξ_I is at most 4. Similarly, the y-coordinates of the intersection points are the roots of the polynomial $\eta_I(y) = \mathrm{res}_x(C_1, C_2)$, such that $\deg(\eta_I) \leq 4$ — but we show that we do not have to compute them exactly.

A root α of ξ_I with multiplicity m originates either from one intersection point (α, β) of C_1 and C_2 of multiplicity m, or from two co-vertical intersection points (α, β_1) and (α, β_2) of multiplicities m_1 and m_2 respectively, where $m_1 + m_2 = m$. This especially means that a simple root of ξ_I (having multiplicity 1) is always caused by one transversal intersection of the two curves. It is easy to see that a degree-four polynomial ξ_I either has four simple roots, or all its roots are one-root numbers.

The coordinates of the intersection points can thus be represented in one of the following two ways: One-root numbers are represented explicitly, while all the simple roots α that cannot be expressed as one-root numbers are stored in their *interval representation*, namely α is represented by the tuple $\langle \xi, l, r \rangle$ where ξ is the generating polynomial ($\xi(\alpha) = 0$) and $[l, r]$ is a rational *isolating interval* ($l, r \in \mathbb{Q}$) for α. By isolating interval for α we mean that $\alpha \in [l, r]$ but $\beta \notin [l, r]$ for any other root β of ξ. Using these two representations it is clear

---

[6] The y-coordinates of a singular point are roots of the quadratic equation $(ty + u)^2 - 4r(sy^2 + vy + w) = 0$, so it is not difficult to distinguish between a vertical tangency point and a singular point.

that all algebraic numbers we obtain as the x-coordinates of either one-curve or of two-curve events are comparable. Whenever we have an interval representation $\langle \xi, l, r \rangle$ of a simple root α we know that the generating polynomial ξ is square-free. Thus, we can refine the isolating interval by iteratively halving it and examining at the signs of $\xi(l)$, $\xi(\frac{l+r}{2})$, and $\xi(r)$, until we finally obtain that $\alpha \in [l', r'] \subset [l, r]$, where $(r' - l')$ is arbitrarily small.

We are now ready to devise a framework for the implementation of the rest of the sweep-line predicates and constructions. We denote by $\hat{C}$ a sweepable x-monotone conic arc supported by the curve C, where we can write, using the notation of Equation (1.1), that either $\hat{C}(x) = y_1(x)$ or $\hat{C}(x) = y_2(x)$.

Intersections: Let C_1 and C_2 be the supporting conics of the two given x-monotone conic arcs $\hat{C}_1$ and $\hat{C}_2$, and let $\xi_I(x)$ be the resultant of C_1 and C_2 with respect to y. The only x-coordinates at which an intersection between the supporting conic curves can take place are the real roots of ξ_I. Let α be a real root of ξ_I. We show how to decide whether the two arcs $\hat{C}_1$ and $\hat{C}_2$ intersect at $x = \alpha$.

If α is a simple root of ξ_I, then the sweepable arcs $\hat{C}_1$ and $\hat{C}_2$ intersect transversally at $x = \alpha$, if they intersect there at all. Let $[\alpha_l, \alpha_r]$ be the isolating interval of α. We refine this interval until it contains no x-coordinate of any one-curve event of the supporting conics C_1 or C_2. Now our two x-monotone arcs are defined on the entire refined interval $[\alpha_l', \alpha_r']$ and they intersect at most once in this interval. It is sufficient to check whether the signs of $\hat{C}_1(\alpha_l') - \hat{C}_2(\alpha_l')$ and $\hat{C}_1(\alpha_r') - \hat{C}_2(\alpha_r')$ differ. We therefore need to compute the signs of one-root numbers in this case, since $\alpha_l', \alpha_r' \in \mathbb{Q}$. On the other hand, if α is a one-root number, we simply have to check whether the y-values $\hat{C}_1(\alpha)$ and $\hat{C}_2(\alpha)$ are equal (note that $\hat{C}_1(\alpha), \hat{C}_2(\alpha) \in \mathbb{F}$).

We still have to determine the multiplicity of the relevant intersection point(s). Let α be a root of ξ_I of multiplicity m. The case $m = 1$ is easy because we have exactly one transversal intersection point of C_1 and C_2 at $x = \alpha$. If $m > 1$, then we know a one-root expression for α. We can compute the roots of $C_1(\alpha, y) = 0$ and of $C_2(\alpha, y) = 0$ and determine whether we have a single intersection point with multiplicity m, in which case we are done, or if there are two co-vertical intersections (α, β_1) and (α, β_2) with multiplicities m_1 and m_2 respectively (where $m_1 + m_2 = m$). We assume now that we have two covertical intersection. If $m = 2$ then obviously $m_1 = m_2 = 1$ and we are done. Otherwise, we test whether $m_i \geq 2$, by checking whether the normal vectors $\nabla C_1(\alpha, \beta_i)$ and $\nabla C_2(\alpha, \beta_i)$ are parallel.[7] If $m_1, m_2 \geq 2$, then $m_1 = m_2 = 2$ as $m \leq 4$. Otherwise, $m_1 = 1$ and $m_2 = m - 1$ (or vice versa).

Compare xy: We are given two points p and q that we wish to compare lexicographically. Let us assume that p lies on the sweepable conic segment $\hat{C}_1$ and that q lies on $\hat{C}_2$.

---

[7] If $f : \mathbb{R}^2 \longrightarrow \mathbb{R}$, $\nabla f(x_0, y_0)$ denotes the vector $\left(\frac{\partial f}{\partial x}(x_0, y_0), \frac{\partial f}{\partial y}(x_0, y_0) \right)$.

We first compare the x-coordinates of p and q. If both are one-root numbers we can do this directly. If only one of the coordinates is a one-root number (say p_x), we can check whether it lies in the isolating interval of q_x; if not, we are done. Otherwise, we should check whether p_x is a root of the generating polynomial of q_x. If it is, the two x-coordinates are equal, and otherwise we can refine the isolating interval of q_x until it does not contain p_x.

The most interesting case occurs when the two coordinates are represented using their generating polynomials $\xi_I^{(p)}$ and $\xi_I^{(q)}$ with their isolating intervals. If the intervals do not overlap, we can easily compare the two values. Otherwise, we compute $g = \gcd(\xi_I^{(p)}, \xi_I^{(q)})$ and check whether $\deg(g) > 1$. If so, we check whether the gcd-polynomial has a root that equals p_x and q_x. To this end, we compute the intersection $[l, r]$ of the two isolating intervals of p and q and test whether $g(l) \cdot g(r) < 0$. If this is the case, we conclude that $p_x = q_x$. Otherwise, we conclude that $p_x \neq q_x$ and we refine the isolating intervals of the two values until they do not overlap and we can compare them unambiguously.

Let us assume that the x-coordinates of the two points both equal α and we turn to compare the y-coordinates in this case. If we store the y-coordinates in a similar manner to the x-coordinates, we can simply compare p_y and q_y as we did for the x-coordinates. However, we can implement the predicate without having to compute the y-coordinates in advance. If we know a one-root expression for α, we can simply compute $\hat{C}_1(\alpha), \hat{C}_2(\alpha) \in \mathbb{F}$ and compare their values — so assume otherwise. We compute the resultant ξ_I of the underlying conics C_1 and C_2 and compare α to its roots. If α is equal to one of the roots we proceed as described in the **Intersections** procedure. Otherwise, we refine the isolating interval of α until it contains no one-curve events of C_1 or C_2 and then use the fact that the y-order of the two segments at α is the same as the y-order at either end of the isolating interval of α.

Point position: We want to determine whether a point p in the x-range of an x-monotone conic arc $\hat{C}$ is vertically above, below, or lies on $\hat{C}$. Note that the predicate **Compare** xy described above also resolves this predicate, if we consider p and $\hat{C}(p_x)$, the point located on $\hat{C}$ having the same x-coordinate as p.

Compare to right: Given the x-monotone conic arcs $\hat{C}_1$ and $\hat{C}_2$, along with their intersection point $p = (\alpha, \beta)$, we wish to compute the y-order of the arcs immediately to the right of p. Assume that we know a rational number $r > \alpha$ such that both $\hat{C}_1$ and $\hat{C}_2$ are defined on $[\alpha, r]$ and do not intersect in the interval $(\alpha, r]$, then the y-order immediately to the right of α is the same as the y-order at r.

How can we obtain r? If α is a simple root of the resultant of C_1 and C_2, we refine its isolating interval until it contains no one-curve event points of C_1 or C_2 and take r as the right endpoint of the isolating interval. If

α is a multiple root of the resultant, we have one-root expressions for all roots and we take a rational point to the right of α and within the x-range of the two segments.

Cubics

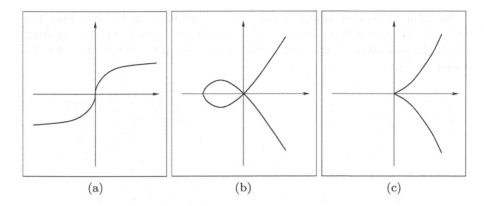

(a)   (b)   (c)

Fig. 1.3. Special points on cubic curves: (a) The curve $x = y^3$ has an inflection point with a vertical tangent at the origin; (b) $y^2 = x^3 + x^2$ intersects itself at the origin; (c) The curve $y^2 = x^3$ has a cusp at the origin

What are the difficulties in proceeding from arrangements of conics further on to arrangements of *cubic curves* (*cubics* for short), i.e., algebraic curves of degree three? The main distinction between the cases is the field of coordinates.

- As we have seen, the sweep-line algorithm works on x-monotone curve segments. Conics need to be split at their one-curve event points, and we have shown that their coordinates are in the field of real-root expressions $\mathbb{F}$. For cubics, these split points are not necessarily in $\mathbb{F}$.
- The x-monotone segments of algebraic curves have parameterizations $y(x)$, according to the Implicit Function Theorem. In the case of conics, the y-value can be expressed using one square root (see Equation (1.1)), which allows for a simple comparison of the y-coordinates of different x-monotone arcs within $\mathbb{F}$. The x-monotone segments of cubics have no parameterization as functions of x within $\mathbb{F}$.
- Conics have some properties that make them easier to handle, namely, they are always convex and never self-intersecting (with the exception of the degenerate case of a pair of intersecting lines). Cubic curves, on the other hand, can have *inflection points* where they change their convexity (especially notice that a cubic curve may have a vertical tangent at its

inflection point while still being x-monotone; see for example Fig. 1.3(a)) and *singular points* like *self-intersections* (Fig. 1.3(b)) or *cusps* (Fig. 1.3(c)).

- Two conics may have at most four intersection points and only the coordinates of transversal intersections are not in $\mathbb{F}$. No such simplification holds for cubic curves. Two cubics may have as many as nine intersection points, where the coordinate of an intersection point of multiplicity 4 (or less) is in general not in $\mathbb{F}$.

In what follows we briefly summarize the main ideas for analyzing the behavior of one and two cubic curves at any given x-coordinate $x = x_0$ from which the realization of the predicates follows. For more details we refer the reader to [140].

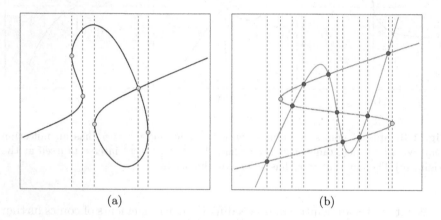

(a)   (b)

Fig. 1.4. (a) One-curve event points, and (b) two-curve event points (the one-curve events, only necessary for the dotted curve, are also marked). The vertical lines denote the division of the x-axis into maximal intervals along which the order of the branches does not change

With the goal of sweeping in mind, we have to subdivide each cubic curve C to x-monotone *branches* and to maintain their ordering at any given x-coordinate x_0. Thus we need to determine the number of branches and their relative position along the line $x = x_0$. Algebraically, this means that we are interested in the number and order of the real roots of $C(x_0, y) \in \mathbb{R}[y]$. As was the case with conic curves, these *real* roots evolve smoothly as we vary x_0, except for some special points, the *one-curve* events (see Fig. 1.4(a)). We therefore have to locate the points $p_i = (x_i, y_i)$ such that y_i is a double root of $C(x_i, y) \in \mathbb{R}[y]$. These points p_i are exactly the intersection points of $C(x, y)$ and $\frac{\partial C}{\partial y}(x, y)$, and can thus be computed using resultant calculus. By choosing an appropriate coordinate system, we avoid inflection points having a vertical tangent.

To obtain the x-coordinates of the one-curve event points of C we compute the roots of the resultant[8] $\xi_V = \operatorname{res}_y(C, \frac{\partial C}{\partial y})$. Due to degree reasons, a simple root of ξ_V is the x-coordinate of an extreme point of C having a vertical tangent, and a multiple root of ξ_V is the x-coordinate of a singular point. In general, these x-coordinates are not one-root expressions and therefore we represent them using their generating polynomial and an isolating interval.

We next analyze the behavior of C at any x-coordinate x_0. We distinguish three different cases:

1. x_0 is not a root of the resultant ξ_V: Assume x_0 is in the interval $[\alpha_i, \alpha_{i+1}]$ between two adjacent roots α_i, α_{i+1} of ξ_V. The number of branches of C is determined by substituting a rational a, where $\alpha_i < a < \alpha_{i+1}$, into C and counting the number of real roots of $C(a,y) \in \mathbb{Q}[y]$ using the univariate Sturm sequence (see Chap. 3) of $C(a,y)$. This procedure can be carried out using only rational arithmetic.

2. x_0 is a simple root of ξ_V: We want to determine which branches of C are involved in the extreme point. This can be realized using the same idea as for computing transversal intersection points between two conics. Let $[l,r]$ be the isolating interval of x_0. By isolating the real roots of the univariate polynomials $C(l,y), \frac{\partial C}{\partial y}(l,y)$ and $C(r,y), \frac{\partial C}{\partial y}(r,y)$, we compute the sequence of branches of C and $\frac{\partial C}{\partial y}$ along the lines $x = l$, slightly to the left of x_0, and along $x = r$, slightly to the right of x_0. By comparing the two sequences we can determine which branches of C are involved in the extreme point.

3. x_0 is a multiple root of ξ_V: In this case we handle a singular point and we have to determine the branches involved in each singularity, as well as the type of the singularity. In a singular point both partial derivatives of C vanish, and it is not difficult to show that a unique singular point of a cubic curve must have rational coefficients (recall that the coefficients of the cubic curve are rational). We can determine the type of a rational singularity $(\alpha, \beta) \in \mathbb{Q}^2$ by translating it to the origin and inspecting the quadratic part $ay^2 + bxy + cx^2$ of the translated polynomial $C(x+\alpha, y+\beta)$. If a singularity is not unique, we make use of the fact that this can arise only if C is a product of several *components* (say a product of three lines or a line and a conic curve) whose intersections then are the singularities of C (see, e.g., [182] for more details).

We now turn to analyze the behavior of a pair of cubic curves C_1 and C_2. For each $x_0 \in \mathbb{R}$ we want to compute a *slice* of the pair, that is, the sequence of intersections of the branches of C_1 and C_2 along the vertical line $x = x_0$. This sequence only changes at one-curve events of C_1 or C_2 and at intersection points of the two curves, referred to as the *two-curve event points*

---

[8]For conic arcs we used a slightly different technique, computing the roots of the quadratic polynomial (1.2). However, using resultant calculus we obtain an equivalent polynomial (having the same roots).

(see Fig. 1.4(b)). The *critical values* of C_1 and C_2 are the x-coordinates of either one-curve or two-curve event points. We choose the coordinate system such that no one-curve event of C_1 or C_2 has the same x-coordinate as a two-curve event.

For x_0 from an interval between two adjacent critical points, the slice is determined by substituting a rational a into C_1 and C_2, solving the univariate polynomials $C_1(a, y)$ and $C_2(a, y)$ by root isolation, and sorting the results. Using the analysis of a single curve, one can extend slicing to x-coordinates at which just a one-curve event happens. The rest of this subsection describes how to slice at the intersection points.

We use a resultant $\xi_I = \mathrm{res}_y(C_1, C_2)$, which is a polynomial of degree at most 9 in this case, to project the intersection points onto the x-axis. It is possible to select the coordinate system such that no two distinct intersection points of any pair of curves have the same x-coordinate, hence the multiplicity of a root α of ξ_I is the multiplicity of the corresponding intersection point (α, β). We next need to detect which branches of C_1 and C_2 intersect at α. We assume that no singularity is involved in the intersection, so that exactly one branch of C_1 intersects one branch of C_2.[9] We consider the following cases:

1. If the intersection multiplicity is odd, the intersecting branches can be determined by inspecting the branches of both curves slightly to the left and slightly to the right of the event point, using the rational endpoints of the interval isolating α.
2. If the intersection multiplicity is 2, there is no transposition of C_1's and C_2's branches, so there is no use comparing the branches to the left and to the right of α. However, we can consider an auxiliary curve of degree 4, the *Jacobi curve* $J = \frac{\partial C_1}{\partial x} \cdot \frac{\partial C_2}{\partial y} - \frac{\partial C_1}{\partial y} \cdot \frac{\partial C_2}{\partial x}$ of C_1 and C_2. This reduces the analysis of non-singular intersections of multiplicity 2 to the previous method for transversal intersections; see [340] and [341] for more details.
3. If α is the x-coordinate of an intersection point with an even multiplicity greater than 4, than α is a one-root number, and we can employ exact calculations in $\mathbb{F}$ to determine the branches involved in this multiple intersection.

1.3.2 Incremental Construction

In the previous section we described how it is possible to construct the planar subdivision induced by the set of arbitrary curves using the sweep-line algorithm. This construction method is very efficient, especially if the arrangement is relatively sparse (when the number of intersection points is $O(\frac{n^2}{\log n})$, where n is the number of curves).[10] However, if we use the sweep-line algorithm we must have all the n input curves in advance — which is not always the

---

[9]For intersection points that involve a singular point we refer the reader to [139].

[10]This fact follows from theoretical considerations, as the running time of the sweep-line algorithm over n curves with a total of k intersection points is $O((n +$

case, as we sometimes want to construct our arrangements on-line. In this case we should employ an incremental construction approach: We insert our curves one by one and update the arrangement accordingly. The incremental approach can also be applied to constructing dense arrangements, say with $\Theta(n^2)$ intersection points, when it becomes asymptotically faster than the sweep-line construction (see, e.g., [111, Chap. 8]).

The two construction approaches may be integrated: Suppose that we are given an initial set of curves, we can construct their planar arrangement using the sweep-line algorithm. It is then possible to insert additional curves to the resulting arrangement using the incremental insertion method, or even insert a set of new curves by "sweeping" them into an existing arrangement. All these alternatives are currently supported by the CGAL arrangement package — see Sect. 1.4. There is, however, a subtlety regarding this combined approach since, as we will describe next, incremental construction of arrangements requires answering point-location queries. We discuss this issue after presenting various point-location strategies.

The Insertion Process

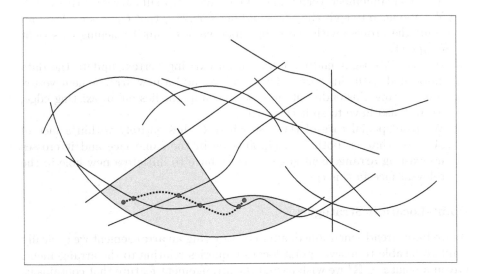

Fig. 1.5. The insertion process of a new curve (dotted) into an existing arrangement of planar curves. The zone of the new curve is lightly shaded and all the newly introduced vertices are marked

$k) \log n)$, while the incremental construction may take $O(n^2)$ time. Moreover, this was also verified by many experimental results.

Suppose we wish to insert a planar curve C_i into an existing arrangement $\mathcal{A}_{i-1}$ of the curves $C_1, \ldots, C_{i-1}$. The insertion procedure of the first curve into an empty arrangement is trivial, so we will assume that $i > 1$ and $\mathcal{A}_{i-1}$ represents the arrangement of a non-empty set of curves. We will further assume that C_i is (weakly) x-monotone — if this is not the case, we will subdivide it into several x-monotone segments and insert each segment separately.

To insert an x-monotone curve C_i we will execute the following procedure:

1. Locate C_i's left endpoint (and in case C_i is a vertical segment we start from its bottom endpoint) in $\mathcal{A}_{i-1}$ and act according to the type of the arrangement feature containing this endpoint:
   a) If the endpoint lies on an existing vertex, we have to update the data associated with this vertex.
   b) If it lies on an edge, we have to split this edge, introducing a new arrangement vertex associated with the endpoint.
   c) If the endpoint is contained in the interior of a face, we construct a new vertex associated with the endpoint within the face.
2. Traverse the *zone* of C_i, the set of arrangement faces in $\mathcal{A}_{i-1}$ that C_i crosses (see Fig. 1.5 for an illustration). Each time we discover an intersection along C_i with one of the existing arrangement elements we create a new arrangement vertex and cut C_i into two subcurves at this point. We also have to split the edges and faces of $\mathcal{A}_{i-1}$ that C_i crosses. We continue the process with the right subcurve of C_i until reaching C_i's right endpoint.
3. In case C_i's right endpoint lies on an existing vertex, update the data associated with this vertex. Otherwise we add a new arrangement vertex representing this endpoint (in case the endpoint lies on an existing edge, we will also have to split this edge).
   We take special care of the case where C_i lies entirely within a face of $\mathcal{A}_{i-1}$ — that is, both its endpoints lie in the same face and it crosses no existing arrangement edge — as we have to initiate a new hole in the relevant face in this case.

Point-Location Strategies

As we have already mentioned, after constructing an arrangement we typically want to be able to answer point-location queries relating to the arrangement: Given a point $p \in \mathbb{R}^2$ we wish to find the arrangement feature that contains it. The point can be contained in a face, but it may lie on one of the arrangement edges or coincide with an arrangement vertex. As we wish our point-location mechanism to be exact, we have to deal with the two latter cases in a robust manner. It should be noted that in the case of incremental construction the usage of point-location is inherent to the construction algorithm itself, as we start the insertion of a new curve by locating one of its endpoints in the existing arrangement. We also have to pay special attention to the type of arrangement feature we obtain as the query result.

We can benefit from refining the planar subdivision by shooting two vertical rays from each arrangement vertex, going upward and downward until hitting the next arrangement edge. In this way, we obtain a subdivision of the arrangement into simple shapes that may be regarded as "pseudo-trapezoids", as two parallel segments form their left and right edges (one of these edges can degenerate into a point), while the top and bottom edges are x-monotone segments of our input curves. It is possible to construct a search structure over these pseudo-trapezoids such that a point-location query can be answered in expected $O(\log n)$ time, where n is the complexity of the arrangement. The search structure itself consumes only linear storage. See [269] and [111, Chap. 6] for more details.

Constructing the trapezoidal-subdivision-based search structure yields efficient query times, yet the overhead involved in maintaining the search structure during the incremental construction of the arrangement may be too high for some applications. In such cases, we may consider employing a simpler point-location strategy in order to simplify the construction process. The *walk* algorithm [164] simply simulates a reverse walk along a vertical ray emanating from the query point, starting from the unbounded face of the arrangement (which represents infinity in this case) and moving toward the query point. The query time is therefore linear in the complexity of the zone of the vertical ray, and as we have to trace the zone of the inserted curve C at any case, using the walk algorithm for locating its endpoint does not increase the worst-case asymptotic complexity of the insertion procedure, as we show next. The main advantage of the walk algorithm is that it does not require any additional data structures to answer queries.

Suppose that we have a set of m curves, where each pair of curves can have at most s intersection points, for a constant s. In this case the zone complexity of a single curve is $O(\lambda_{s+2}(m))$, where $\lambda_\sigma(k)$ denotes the maximal length of a Davenport-Schinzel sequence of k elements with order σ (see [312] for more details on Davenport-Schinzel sequences; we will just mention here that for small value of σ, $\lambda_\sigma(k)$ is almost linear in k). Thus the total construction time is $O(m\lambda_{s+2}(m))$, while the complexity of the arrangement is of course $O(m^2)$.

The trapezoidal-subdivision based point-location strategy is of course more efficient, both from a theoretical and from a practical point of view (see [164] for experimental results), but, as mentioned earlier, the walk strategy does not require any auxiliary data structure on top of the arrangement. Hence, if the application at hand involves the construction of an arrangement, followed by a small number of point-location queries, it may be more efficient to construct the arrangement using the sweep-line algorithm and to use the walk strategy in order to avoid the overhead incurred by the construction of a trapezoidal-subdivision based search structure.

Another point-location strategy uses a set of "landmark" points whose location in the arrangement is known [201]. Given a query point, it uses a Kd-tree [45] to find the nearest landmark and then traverses the straight line segment connecting this landmark to the query point. Usually, the arrangement vertices are the most convenient landmarks, but other landmark sets, such as randomly sampled points or points on a grid, can be used as well. The landmarks point-location strategy offers competitive query times in comparison with the trapezoidal-based algorithm. Its main advantage is that the time needed to construct and maintain its auxiliary Kd-tree is relatively small in comparison to the running-time penalty incurred by the auxiliary data structures maintained by the trapezoidal point-location strategy.

Geometric Predicates and Constructions

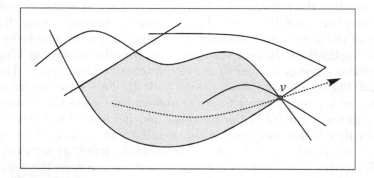

Fig. 1.6. The newly inserted curve (dashed) exits the shaded face passing through an existing arrangement vertex v

As the incremental insertion of a curve into an existing arrangement boils down to computing the zone of the curve, and as we proceed from the left endpoint of the curve to its right endpoint, the geometric predicates and constructions we use are basically the same as the ones needed in the sweep-line algorithm. However, we should take special care of the following case: Assume that the newly inserted curve passes through an existing arrangement vertex v. We therefore have to determine the position of the curve, with respect to the existing curves in the arrangement, immediately to the *left* of v (see Fig. 1.6 for an illustration). Indeed, it is possible to compare the new curve with the other curves to the right of v and then determine the order to its left using the multiplicity of the intersection. But what if one of the curves ends at v and is not defined to its right?

It seems that an additional predicate is required for the incremental construction, that is, comparing two curves to the *left* of their intersection point. However, as we strive to minimize the number of required predicates, we can implement this predicate in terms of the basic sweep-line methods.[11] Let us

---

[11]The arrangement package of CGAL lets users choose between the two alternatives: (i) supplying an additional predicate, or (ii) resorting to the basic sweep-line

assume we are given the curves C_1 and C_2 and wish to compare them to the left of their intersection point p. We can use the following procedure:

1. Compare the two left endpoints of C_1 and C_2 and let q be the (lexicographically) rightmost of the two.
2. Start computing the next intersection of the two curves from q onward to the right until reaching p. Let q' be the last intersection point of C_1 and C_2 we have discovered. It is now sufficient to compare the two curves to the *right* of q'.
3. If there are no intersection points between q and p, we simply determine the order of the curves by checking whether the endpoint q is above or below the other curve.

Constructing the point-location structure does not require any additional geometric operations, nor does the point-location query: We only have to compare the x-coordinates of two given points and to determine the position of a point relative to an x-monotone curve segment.

1.4 Software for Planar Arrangements

This section assumes some familiarity with advanced programming techniques, generic programming, and design patterns. Readers who are not interested in software issues can safely move on to the next section — the rest of the sections of this chapter do not rely on the material described in this section.

The implementation of the CGAL software packages for planar arrangements described in this section is complete and robust, as it handles all degenerate cases, and guarantees exact results. The software rigorously adapts, as does CGAL in general, the *generic programming paradigm*, briefly reviewed in Appendix 8. This approach allows for a convenient separation between the topology and the geometry of arrangements, which is a key aspect of the CGAL arrangement design, explored in Sect. 1.4.1. In this way, algorithms and data structures for arrangements can be nicely abstracted in combinatorial and topological terms, regardless of the specific geometry and algebra of the curves at hand. This abstraction constitutes the arrangement class-template. The arrangement class-template must be instantiated with a geometric *traits* class, that defines the set of geometric objects and algebraic operations on these objects required to handle a concrete type of curves, e.g., segments, polylines, or conics.

An immediate advantage of this separation is that users with limited expertise in computational geometry can employ the package with their own

---

methods (see [167] for more details). While the first option is usually more efficient, implementing the additional predicate may be a major endeavor in some cases; see for example Sect. 1.3.1.

special type of curves, provided they supply the relevant geometric traits class, which relies on (often basic) algebra. Naturally, a prospective user of the package that develops a traits class would like to face as few requirements as possible in terms of traits development. Indeed, a lot of effort in the past couple of years went into streamlining the list of operations on curves that a user has to supply as part of the traits class. The end result, the current CGAL arrangement traits, is described in Sect. 1.4.2, with ample examples of existing traits classes.

Alternative traits classes resulting from the EXACUS project, are described in Sect. 1.4.3. Sect. 1.4.4 describes an on-going effort to extend the existing CGAL kernel, which mainly deals with linear objects, to deal with curved objects. We conclude the section with notes on how to effectively use and fine-tune software for planar arrangements.

1.4.1 The CGAL Arrangements Package

CGAL, the Computational Geometry Algorithms Library, was the natural library to host a planar-arrangements package in its basic library part. Predicates and constructors from CGAL's geometric kernels are used as building blocks of the various traits classes that handle linear curves. All the traits-class components based on the kernel can be extended with minimal programming effort, as the kernel is fully adaptable and extensible [204]. The design of the Arrangements package uses many programming techniques and abstractions that C++ supports. The rest of this section introduces several of them. For the complete specifications and discussions we refer the reader to [163, 167, 333]; a comprehensive documentation of the packages with a variety of examples can be found in [2].

The Arrangement package consists of a few components. The main component is the *Arrangement_2<Traits,Dcel>*[12] class-template. It represents the planar embedding of a set of x-monotone planar curves that are pairwise disjoint in their interiors. It provides the necessary combinatorial capabilities for maintaining the planar graph, while associating geometric data with the vertices, edges, and faces of the graph. The arrangement is represented using a doubly-connected edge list (DCEL), a data structure that efficiently maintains two-dimensional subdivisions; see Sect. 1.3 for details. The *Arrangement_2* class-template includes a basic set of interface functions that access, modify, and traverse planar arrangements. For example, users can iterate over all vertices, edges, and faces of the arrangement, or insert new vertices and edges into the arrangement at a specified location (see Fig. 1.9). Additional arrangement operations that involve non-trivial geometric algorithms are supplied by free (global) functions and auxiliary classes. For example, users can construct the arrangement induced by a set of arbitrary curves from scratch, or insert these curves into an existing arrangement.

---

[12]CGAL prescribes the suffix _2 for all data structures of planar objects as a convention.

The separation between the combinatorial and geometric aspects is achieved by the decoupling of the arrangement representation from the various geometric algorithms that operate on it. As mentioned above, the insertion of a new arbitrary curve (a curve that may not necessarily be x-monotone, can intersect the existing arrangement curves, and whose insertion location is not known *a-priori*) into the arrangement is offered by free functions. When an arbitrary curve is inserted into the arrangement, it is subdivided into several x-monotone subcurves treated separately. Each x-monotone subcurve is in turn split at its intersection points with the existing arrangement features. The resulting subcurves are inserted into the arrangement using one of the special insertion-methods listed in Sect. 1.4.1. The result is a set of x-monotone and pairwise disjoint subcurves that induce a planar subdivision, equivalent to the arrangement of the original input curves. Other free functions that operate on arrangements or construct them, such as a function that computes the overlay of two arrangements, are implemented as well.

Another component of the package, named *Arrangement_with_history_2*, allows for the construction of an arrangement while maintaining its *curve history*. The class stores the input curves that induce the arrangement, where each arrangement edge stores a pointer to the input curve that induces it (or a list of pointers, in case the edge represents an overlap of several input curves). The curve history is essential in a variety of applications that use arrangements, such as robot motion-planning.

The package supports a notification mechanism implemented via the *Observer* design-pattern [175]. It defines a one-to-many dependency between objects so that if one object, referred to as the subject, changes its state then all its dependents, referred to as observers, are automatically notified and can take action accordingly. The ability to serve multiple observers plays a significant role in the arrangement package, as it satisfies many different needs with a single unified approach. One such need is to keep the *point-location* strategies (see Sect. 1.3.2), which maintain auxiliary data-structures, synchronized with the arrangement instance they relate to. Another important reason for supporting observers of arrangements is to allow users to introduce their own observers in order to maintain auxiliary application-specific data with the features (vertices, edges, or faces) of their arrangements.

We proceed with some more details about each of the components above. The *Arrangement_2* class-template must be instantiated with two classes listed below, decoupling the topological and geometric aspects of the planar subdivision; see Fig. 1.7.

- A geometric traits class tailored to handle a specific family of curves. It defines the *Point_2*, *X_monotone_curve_2*, and *Curve_2* types (see the exact details in Sect. 1.4.2) and provides a basic set of predicates and geometric constructions on objects of these types, encapsulating implementation details, such as the number type used, the coordinate representation, and the geometric or algebraic computation methods involved in the implementation of the traits-class methods.

- A DCEL class, which represents the underlying topological data structure. By default, this class is *Arr_default_dcel*, but users may extend this default DCEL class, and attach additional data with the DCEL records, or even supply their own DCEL written from scratch.

The *Arrangement_2<Traits,Dcel>* class-template defines its own vertex, halfedge (each arrangement edge is represented by a pair of halfedges with opposite orientations), and face types. These combinatorial entities have a geometric mapping, e.g., a vertex of an arrangement is associated with a *Point_2* object, and each edge is associated with an *X_monotone_curve_2* object. The *Arrangement_2* class-template maintains the incidence relation on the vertices, halfedges, and faces of the arrangement through the *Dcel* it is parameterized with. A valid arrangement has one unbounded face.[13] A *Connected Component of the Boundary* (CCB) is a cycle of at least two halfedges. Each face, except the unbounded face, has one outer CCB. Each face has a (possibly empty) set of holes referred to as the inner CCBs. In addition, a face may also contain *isolated vertices* in its interior. An empty arrangement has one unbounded face (and no halfedges nor vertices). The containment relation between a face and its holes and isolated vertices distinguishes the

---

[13] Currently, only bounded curves are supported. Arrangements of bounded curves have a single unbounded face.

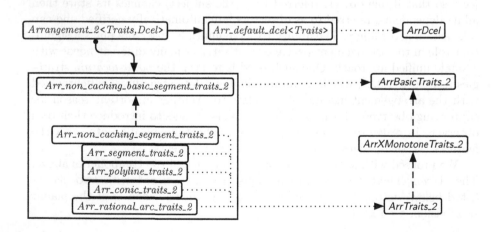

Fig. 1.7. The architecture diagram of the traits-related and DCEL-related components of the CGAL-arrangement package. Dotted lines indicate an *is-model-of* relation and dashed lines indicate a *concept refinement* or an *inheritance* relation. Solid lines indicate a *membership* relation. If the member is a pointer, the line starts with a small disk

Arrangement_2 from standard graph structures and other edge-based structures.

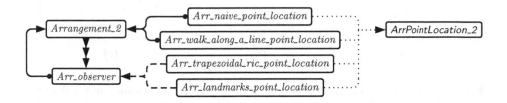

Fig. 1.8. The architecture diagram of the point-location related components of the CGAL-arrangement package. Dotted lines indicate an *is-model-of* relation, and dashed line indicate an *inheritance* relation. Solid lines indicate a *membership* relation. If the member is a pointer, the line starts with a small disk, and if the member is a container of pointers, the line is made of arrowheads

The *Arr_observer<Arrangement>* class-template must be instantiating with an arrangement class; see Fig. 1.8. It stores a pointer to an arrangement object, and is capable of receiving notifications just before a structural change occurs in the arrangement (e.g., a new vertex is created, an existing face is split into two following the insertion of a new edge, etc.) and immediately after such a change takes place. The *Arr_observer* class-template serves as a base class for other observer classes and defines a set of virtual notification functions, giving them all a default empty implementation.

The interface of *Arrangement_2* consists of various methods that enable the traversal of the arrangement. For example, the class supplies iterators for its vertices, halfedges, and faces. It is possible to visit all halfedges incident to a specific vertex or to iterate over all halfedges along the boundary of a face. The interface also supports other access methods and queries. Various input/output operations are available as well. We restrict our focus in the remainder of this section to insertion and point-location queries. A detailed review of all arrangement-related operations, can be found in the complete programming guide and reference manual [2], which also includes a large number of didactic examples.

Insertion

In geometric computing there is a major difference between algorithms that evaluate predicates only and algorithms that, in addition, construct new geometric objects. A predicate typically computes the sign of an expression used by the program control, while a construction results with a new geometric object, such as the intersection point of two segments. If we use an exact number type to ensure robustness, the newly constructed objects often have a

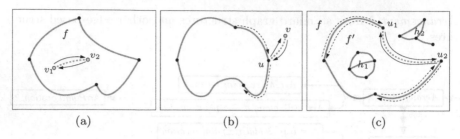

Fig. 1.9. The various insertion procedures. The inserted x-monotone curve is drawn with a light dashed line, surrounded by two solid arrows that represent the pair of twin halfedges added to the DCEL. Existing vertices are shown as black dots while new vertices are shown as light dots. Existing halfedges that are affected by the insertion operations are drawn as dashed arrows. (a) Inserting a subcurve as a new hole inside the face f. (b) Inserting a subcurve from an existing vertex u that is one of its endpoints. (c) Inserting a subcurve whose endpoints are already represented by the vertices u_1 and u_2. In this particular case, the new pair of halfedges close a new face f', where the hole h_1, which used to belong to f, now becomes an enclave in this new face

more complex representation in comparison with the input objects (e.g., the bit-length needed for their representation is often larger). Unless the overall algorithm is carefully designed to deal with these new objects, constructions will have a severe impact on the algorithm performance. To this end, we distinguish between two sets of free insertion functions. One set consists of functions that insert x-monotone curves that do not intersect in their interiors, while the other set consists of functions that insert arbitrary curves. An arrangement induced by x-monotone curves that are pairwise disjoint in their interior, can be instantiated with a limited traits class that models only a basic concept as explained in Sect. 1.4.2. Naturally, using the non-intersecting insertion-function is more efficient, since these functions avoid unnecessary computations.

An *Arrangement_2* can be built incrementally, inserting one curve at a time by traversing its *zone* (see, e.g., [196]). However, for a large number of curves that intersect rather sparsely, it is more efficient to use one of the aggregate insertion functions that insert a set of curves into an arrangement at once applying a dedicated sweep-line algorithm; see Sect. 1.3. The aggregate insertion function is more efficient in most cases, and it also requires less from the traits class in comparison with the incremental insertion operation.

The *Arrangement_2* class directly supports the basic insertion procedures for an x-monotone curve whose interior is disjoint from all existing arrangement features, where the topology information that specifies the location of the inserted x-monotone curve is also given. Note that the endpoints of such a curve may coincide with existing vertices. We distinguish between three cases (see Fig. 1.9): (i) Neither endpoint coincides with an existing vertex, so the inserted curve forms a hole inside an existing face. In this case, the users should specify the face containing the curve. (ii) Exactly one endpoint coincides with a given existing vertex, so the inserted curve forms an "antenna" rooted at this vertex. (iii) Both endpoints of the inserted x-monotone curve coincide with two given vertices.

The free insertion functions use the basic insertion functions listed above in order to insert arbitrary curves into the arrangement. For example, the incremental insertion function that accepts an x-monotone curve C starts by locating C's left endpoint in the arrangement and then traverses its zone by detecting its intersections with the existing arrangement features. C is subdivided at these intersection points into subcurves, where each subcurve is inserted using one of the special insertion functions based on the topological and geometric information available during the zone-computation algorithm.

Point Location

An important query that is often issued given an arrangement object is a point-location query (see also Sect. 1.3.2), namely identifying the arrangement cell containing a given query point q. As the arrangement representation is decoupled from geometric algorithms that operate on it, the Arrangement package provides answers to point-location queries through one of several external classes, which implement different point-location strategies available for the user to choose from; see Fig. 1.8. New strategies, perhaps with different running time and storage characteristics, can be implemented and easily plugged in.

Applying the generic programming paradigm, we define a concept named *ArrPointLocation_2* that collects the requirements for a point-location class. A model of this concept must include the definition of the *locate(q)* function that accepts an input query point q and returns an object that represents the arrangement cell that contains this point. This cell is typically a face, but in degenerate cases a query point can lie on an edge or even coincide with a vertex.

The following models of the *ArrPointLocation_2* concept are included with the arrangement package, each represents a different point-location strategy:

- *Arr_naive_point_location* locates the query point naïvely by exhaustively scanning all arrangement cells until the desired one is encountered.
- *Arr_walk_along_a_line_point_location* improves the above naïve approach by "walking" in reverse order only along the zone of the upward vertical ray emanating from the query point.
- *Arr_landmarks_point_location* uses a set of "landmark" points whose location in the arrangement is known [201]. Given a query point, it uses a nearest-neighbor search structure to find the nearest landmark and then traverses the straight line segment connecting this landmark with the query point.

- *Arr_trapezoidal_ric_point_location* implements the Randomized Incremental Construction (RIC) dynamic algorithm introduced by Mulmuley, which is based on the vertical decomposition of the arrangement into pseudo-trapezoids [270]. It requires two auxiliary data structures: a trapezoidal decomposition (of linear size) and a directed acyclic graph (DAG) used as a search structure (of expected linear size).

The first two strategies do not require any extra data The respective classes only need to store a pointer to an arrangement object, and operate directly on it when a query is issued. The remaining two strategies use an arrangement observer to maintain their auxiliary data-structures that have to be updated when the arrangement is modified; see Fig. 1.8. These strategies require preprocessing and increase the memory consumption, but they can significantly reduce the query time.

Users are free to choose the point-location strategy that best fits their application. Note that it is also possible to attach several point-location objects to a single *Arrangement_2* instance. The point-location strategy has a significant impact not only on the performance of point-location queries, but also on the performance of some of the operations that modify the arrangement. For example, when inserting a curve into a non-empty arrangement, one of the curve endpoints must be located first; see Sect. 1.3.2.

Sweep-Line Framework

The *Sweep_line_2< Traits, Visitor, Curve,EventPoint,Allocator>* class-template implements a generic sweep-line algorithm based on the algorithm of Bentley and Ottmann; see Sect. 1.3 for more details. The original algorithm is extended to support not only segments but also general curves as well as isolated points. The implementation is complete, as it handles all generic cases, such as vertical segments, multiple (more than two) curves intersecting at a single point, curves intersecting at endpoints, and overlapping curves. The class interface is carefully designed to create a general framework for implementing sweep-based algorithms.

The *Traits* template parameter provides the geometric functionality, and is tailored to handle a specific family of curves. The supplied geometric traits class must model the *ArrTraits_2* concept, also used for instantiating the *Arrangement_2* class.

The *Visitor*, *Curve*, and *EventPoint* template parameters have default values, which can be overridden to extend the sweep-line procedure. In particular, the *Visitor* class adheres to the *Visitor* design-pattern [175] and is used to implement distinct and unrelated operations within the sweep-line framework without "polluting" it. The visitor receives notifications of the events handled by the sweep-line procedure and can respond accordingly. The implementation of many sweep-based algorithms reduces to implementing an appropriate visitor class. At the same time, the sweep-line code becomes centralized, reusable and easy to maintain, since the algorithm-specific code resides in the various sweep-line visitor classes.

The arrangement package includes several sweep-line visitors, designed to efficiently perform the following operations: computing all intersection points induced by a set of input curves, constructing the arrangement of a set of input curves from scratch, inserting a set of input curves into an existing (non-empty) arrangement, performing a *batched point-location* operation given a set of query points and an arrangement, and computing the overlay of two input arrangements. Users may implement additional sweep-based algorithms by writing their own sweep-line visitors.

1.4.2 Arrangements Traits

The *Traits* parameter of the *Arrangement_2* class-template defines the abstract interface between arrangements and the primitives they use. The requirements of various components in the arrangement package create a hierarchy of traits concepts. The *ArrBasicTraits_2* concept comprises the minimal set of requirements. A model of this concept must define two types of objects, namely *X_monotone_curve_2* and *Point_2*, which is the type of the endpoints of an *X_monotone_curve_2* curve. In addition, the concept lists some predicates on these two types — a minimal set that enables the maintenance of arrangements of x-monotone curves that are pairwise disjoint in their interiors. It is possible to instantiate the *Arrangement_2* class-template with a model of this concept and construct an arrangement instance using the member functions of the class, which operate only on non-intersecting x-monotone curves. It is also possible to issue point-location queries on such arrangements, as the set of predicates the *ArrBasicTraits_2* concept comprises is sufficient for the various point-location strategies detailed in the previous section. The only exception is the "landmarks" strategy, which requires a traits class that models the refined *ArrangementLandmarkTraits_2* concept — the details of which are omitted here.

The concept *ArrXMonotoneTraits_2* refines the concept *ArrBasicTraits_2* by several construction operations, namely, computing the intersection points of two x-monotone curves and splitting an x-monotone curve at a given point in its interior. Given a model of this concept, it is possible to construct an arrangement of arbitrary x-monotone curves using one of the free insertion functions. The refined *ArrTraits_2* requires the definition of the *Curve_2*, a type that represents a general curve in the plane, and the provision of a construction operation that subdivides a given *Curve_2* object into x-monotone curves. Using a model of this concept one can construct an arrangement of arbitrary (not necessarily x-monotone) planar curves.

This level of modularity makes the package flexible and extensible. The user can choose the family of curves (and their representation) for the arrangement to handle by injecting the appropriate geometric traits class that models one of the concepts above. Users can also develop a new traits class that handles a specific family of curves of interest and a specific representation suitable for their application, as long as it conforms to the requirements of the appropriate concept. Even a simple family of curves, such as line segments, may have different representations distinguished according to the coordinate systems (e.g., Homogeneous, Cartesian), the geometric kernel (e.g., CGAL kernel, LEDA kernel, user defined), and the number type used by the arithmetic operations carried out by the predicates and constructions of the traits class. Traits classes developed for handling curves of higher degree may also differ in the underlying algebraic methods used for answering predicates and computing intersection points.

The traits class *Arr_non_caching_segment_traits_2<Kernel>* handles line segments. It is a thin layer above the parameterized kernel. It inherits the kernel types and functionality and it complements it with the necessary functors that are not directly provided by the kernel. Its implementation is simple, yet may lead to a cascaded representation of intersection points with exponentially long bit-length. The *Arr_segment_traits_2<Kernel>* traits class avoids this cascading problem by storing extra data with each segment. It achieves faster running times than the *Arr_non_caching_segment_traits_2* traits-class, when arrangements with relatively many intersection points are constructed. However, it uses more space. In both cases the kernel is parameterized with a number type, which should be exact to avoid robustness problems, although other number types could be used at the user's own risk.

The traits class *Arr_polyline_traits_2<SegmentTraits>* handles piecewise linear curves, commonly referred to as polylines. Each polyline is a chain of segments, where each two neighboring segments in the chain share a common endpoint. The traits class exploits the functionality of the parameterized *SegmentTraits* type to handle the segments that comprise the polyline curves. The type *SegmentTraits* must be instantiated with a model of the **ArrXMonotoneTraits_2** concept that handles line segments (e.g., *Arr_segment_traits_2<Kernel>* or *Arr_segment_cached_traits_2<Kernel>*), and the number type used by the instantiated segment traits should satisfy the same conditions listed above.

We next review the arrangement traits classes that handle non-linear curves. The first two traits classes are included in the public distribution of CGAL, while the others (Sect. 1.4.3) have been developed under the EXACUS project. We also review on-going work on a CGAL curved kernel (Sect. 1.4.4).

The Conic Traits-Class of CGAL

The arrangement package includes a traits class, named *Arr_conic_traits_2*, that handles finite conic arcs. A finite *conic arc* $\bar{a}$ may be either one of the following: (i) A full ellipse (which is the only type of bounded conic curve), or (ii) the 4-tuple $\langle C, p_s, p_t, o \rangle$ where p_s and p_t are the edge endpoints that must lie on the conic C, and o indicates the orientation (clockwise or counterclockwise).

The arrangement traits-class for conic arcs included with the public release of CGAL is based on the CORE library [3].CORE introduces the *root operator*, where $\mathrm{root}(p, k)$ is the kth largest real root of the polynomial $p \in \mathbb{Q}[x]$.[14] Using this operator, we obtain an explicit representation of the coordinates of the intersection points, which are roots of resultant polynomials of degree 4 at most. The usage of CORE makes the traits-class code more compact and elegant, as it implements the various geometric predicates and constructions in a straightforward manner.

Some previous versions of CGAL used to include a traits-class for conic arcs that was based on the LEDA library and its capability to carry out exact computations within the field of real-root expressions $\mathbb{F}$ (see Sect. 1.3.1). For more details on the algebraic and the numeric techniques used in the implementation of these conic-arc traits, the reader is referred to [331, 332].

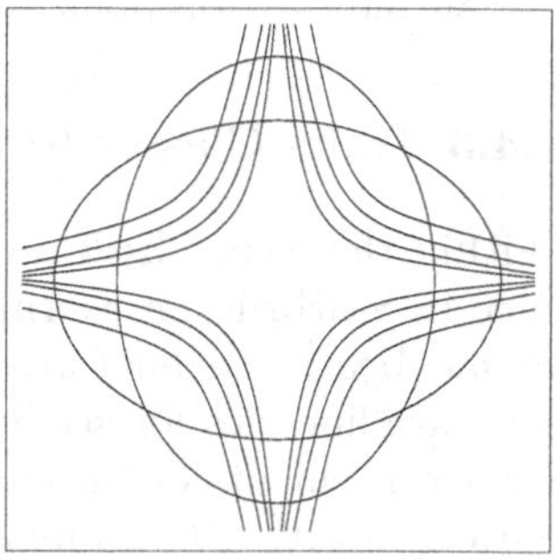

Fig. 1.10. An arrangement of conic arcs containing two canonical ellipses and 16 hyperbolic arcs

Finite Arcs of Rational Functions

Another traits class that relies on the root operator of CORE can deal with finite arcs of rational functions. Such an arc $\bar{a}$ is defined by two polynomials $p, q \in \mathbb{Q}[x]$ such that $\deg(\gcd(p, q)) = 0$ (that is, p and q have no non-trivial common divisors), and an interval $[l, r]$ such that $q(x) \neq 0$ for each $l \leq x \leq r$. The arc represents the graph of the rational function $y = \frac{p(x)}{q(x)}$ defined over $[l, r]$.

Rational functions can be used to approximate more complex planar curves (note that in particular, every univariate polynomial is a rational function), so arrangements of arcs of rational functions can be very useful in many application domains.

It should be noted that an arc of a rational function is x-monotone by definition. The implementation of the traits-class methods is straightforward. The only methods that require some attention are listed below:

Intersections: Given two arcs $\bar{a}_1 = \langle p_1, q_1, [l_1, r_1] \rangle$ and $\bar{a}_2 = \langle p_2, q_2, [l_2, r_2] \rangle$, we first have to find all x-values satisfying the equality

$$\frac{p_1(x)}{q_1(x)} = \frac{p_2(x)}{q_2(x)} .$$

We extract all the roots of the univariate polynomial $p_1 q_2 - p_2 q_1 \in \mathbb{Q}[x]$ and consider just those contained in the interval $[l_1, r_1] \cap [l_2, r_2]$, which we denote $x_1, \ldots, x_k$. These roots are the x-coordinates of the intersection points, whose corresponding y-coordinates are simply given by $y_i = \frac{p_1(x_i)}{q_1(x_i)}$.

---

[14]See [242, 285] for the theoretical background behind the implementation of the root operator.

Compare to right: Given two arcs $\bar{a}_1$ and $\bar{a}_2$ (as above) and one of their intersection points u, we begin by defining $f_1^{(0)}(x) = \frac{p_1(x)}{q_1(x)}$ and $f_2^{(0)}(x) = \frac{p_2(x)}{q_2(x)}$. We start with $m = 1$, and compute for $k = 1, 2$ the mth order derivative $f_k^{(m)} = \left(f_k^{(m-1)}\right)'$. If $f_1^{(m)}(u_x) \neq f_2^{(m)}(u_x)$, then we can determine the comparison result. Otherwise we conclude the multiplicity of u is greater than m, so we increment m and repeat the derivation process. We will need $\deg(p_1 q_2 - p_2 q_1)$ iterations at most, as this is the maximal multiplicity of an intersection point.

1.4.3 Traits Classes from Exacus

Within the recent ECG project, the Max-Planck Institute for Computer Science in Saarbrücken, Germany, initiated the project EXACUS —*Efficient and Exact Algorithms for Curves and Surfaces* [4]— where we develop several C++ software libraries for studying design and for experimenting with algorithms for arrangements of curves and surfaces. We follow the generic programming paradigm with C++ templates similar to the well established design principles in the STL [38] and in CGAL [156, 70]; cf. Chap. 8.

The EXACUS libraries are organized in a layered hierarchy with the external libraries that are (optionally) used at the bottom and the applications libraries at the top. In between, there is a small Library Support layer: the NUMERIX library contains the number types, algebra, and numerical methods, and the SWEEPX library contains the generic implementations of a sweep-line algorithm and a generic generalized polygon that supports regularized Boolean operations on regions bounded by curved arcs. In particular, the sweep-line algorithm realizes the linear-time reordering for multiple curves intersecting in a single common point [49]; see Sect. 1.3.1. Furthermore, the SWEEPX library contains *generic algebraic points and segments* (GAPS), the generic and curve-type independent part that implements the predicates and constructions for the sweep-line algorithm based upon the one-curve and two-curves analysis, as described in Sect. 1.3.1; see [48] for details.

Most notably in this context, the EXACUS libraries provide models of traits classes for the CGAL arrangement class, namely for the *ArrangementTraits_2* concept, to work with conics and conic arcs [49], cubic curves [140], and quartic curves as they result from projections of intersection and silhouette curves of quadrics in space [50]. The traits classes support all operations required for sweep and incremental construction in the CGAL arrangement. The curves and arcs may include infinite branches; they are handled completely in the traits classes and are therefore transparent for the CGAL arrangement algorithms.

We support full curves and, for conics, also arbitrary segments of curves, but both will be preprocessed into potentially smaller *sweepable segments* suitable for the algorithm. A *sweepable segment* is x-monotone, has a constant arc number in its interior (counting without multiplicities from bottom to top), and is free of one-curve events in its interior.

The implementation is not restricted to bounded arcs and curves. Endpoints of unbounded curves (for example of the hyperbola $xy - 1 = 0$) are handled symbolically and vertical line segments and vertical lines are supported as well. Two techniques are used: The x-coordinates are compactified by adding symbolic values minus- and plus-infinity to the range of values. To represent the behavior at poles, the compactified x-coordinates are then symbolically perturbed to symbolically represent "endpoints" of curves that approach a pole, and unbounded "endpoints" of vertical lines and rays. This perturbation, we call it *tendency*, can be expressed with a symbolical infinitesimal $\epsilon > 0$: curves left of the pole have endpoints with tendency $-\epsilon$, the lower endpoint of a vertical line has tendency $-\epsilon^2$, the upper endpoint of a vertical line has tendency ϵ^2, and curves right of the pole have endpoints with tendency ϵ. The result is the desired lexicographic order of event points.

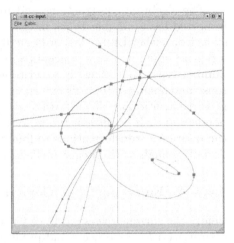

Fig. 1.11. Example arrangement of conics featuring two highly degenerate intersection points including tangential intersections

Fig. 1.12. Example arrangement of cubic curves featuring a degenerate triple intersection point

The generic parts in EXACUS are the design results of three main applications that we have realized so far: conics and conic arcs (Fig. 1.11), cubic curves (Fig. 1.12), and quartic curves as they result from projections of intersection and silhouette curves of quadrics in space (Sect. 1.5.2). The conics implementation supports line segments and imposes no restrictions on the choice of the coordinate systems. The other two applications impose restrictions on the coordinate system that are checked at runtime and that will lead to a runtime exception in case of a violation. The application would then be in charge of shearing the coordinate system and restarting the computation.

At the heart of the work is the one-curve and two-curves analysis, as described in Sect. 1.3.1. To recall it briefly, we project event points on the x-axis using resultants and isolate their roots, such that the plane is decomposed into vertical slabs with rational x-coordinates at slab boundaries. We then compute arc numbers of all arcs, of all event points, and determine the arcs getting in and out of the event points. In contrast to the cylindrical algebraic decomposition, we are able to determine those almost always by examining the intersection pattern of the curves or simple auxiliary curves at the rational x-coordinates of the slab boundaries. This keeps the polynomial degrees low for the root isolation.

The main algebraic tools that we use are reviewed in Chap. 3. Briefly, we perform root isolation using the Descartes method. It requires the polynomials to be square-free, which we have to ensure for the case analysis in the algorithms anyway. For the resultant computation we can choose between the Bézout resultant (the default) and the Sylvester matrix formulation. Arguments for these choices are given in [140].

The algebraic constructions and functions tool box in NUMERIX rely on imported basic number types integer, rationals and (optionally) real expressions from either LEDA [251] or GMP [6] and CORE [3], or EXT [307]. We define a rich set of number type concepts and supporting traits classes for these basic number types that allow us to write generic and flexible code with maximal reuse and openness towards external number type libraries. As an example, it is worth mentioning that EXACUS works transparently with polynomials independent of their coefficient types, for example rationals, and selects internally the necessary transformations to finally compute gcds or resultants efficiently on polynomials with integer coefficients.

1.4.4 An Emerging CGAL Curved Kernel

Before the current release (3.2) of CGAL, the CGAL *Kernel* provided the user mainly with *linear* objects and predicates on them. Circles and spheres were also defined but with very few functionalities. Curves have been present in the traits classes of certain specific packages of the CGAL *Basic library* for a long time: the arrangement package provides the user with a traits class for conic arcs (Sect. 1.4.2), the optimization package comes with conics and basic operations on them, needed for computing minimum enclosing ellipses, whereas the Apollonius diagram package computes the Voronoi diagram of circles.

We started to design and implement an extension of the CGAL kernel, devoted to curved objects. The first version of a 2D *Circular kernel*, defining new classes for circular arcs and points on them, as well as some basic functionalities on them, has been released in CGAL 3.2. Traits classes that serve as an interface for the CGAL arrangement class are written, based on this kernel.

For more general curves, the effort has been focused so far on defining general *concepts* (Chap. 8), so that several implementations of these concepts can be provided by several sites, in a collaboration that will allow us to compare the different methods on a large number of test cases.

Let us describe here the C++ design chosen for organizing our code, and the kind of interface provided by our kernel. As in [284, 145], our choices have been heavily inspired by the CGAL kernel design [204] which is extensible and adaptable. Indeed, one of its features is the ability to apply primitives like geometric predicates and constructions to either the geometric objects which are provided by our kernel, or to user-defined objects.

The curved kernel is parametrized by a *LinearKernel* parameter and derives from it, in order to include all needed functionality on basic geometric objects, like points, line segments, and so on.

It is clear from Sect. 1.3 that the predicates make heavy use of algebraic operations. We want to be as independent as possible from a particular implementation of the algebraic operations, so our curved kernel is parametrized by an *AlgebraicKernel* that is responsible for all algebraic computations. The declaration is the following:

```
template < typename LinearKernel, typename AlgebraicKernel >
class Curved_kernel;
```

The interface provided at the geometric level is composed of:

Types defining the objects. Some of them are inherited from the basic kernel, namely the number type (denoted as RT in the sequel), and basic geometric types. Some other types are inherited from the algebraic kernel. These types are mainly types for polynomials and roots of polynomial systems. Finally, some types are defined by the curved kernel itself, mainly types for arcs of curves and their endpoints.

Predicates and constructions defined on the above objects, as mentioned in Sect. 1.3.1.

Any implementation of the *AlgebraicKernel* template parameter of our *Curved_kernel* must be a *model* of a precise *concept*. The algebraic kernel consists of several concepts, such as bivariate polynomials and roots of systems of bivariate polynomials. These concepts must provide the operations listed below.

We only show here how the basic geometric operations for computing arrangements of algebraic curves can be translated into basic operations on algebraic numbers.

The first step consists of choosing a **representation** for the input data:
- We will consider here *implicit algebraic curves*, and focus more specifically *circles* since this corresponds to the current CGAL release. Such a curve is represented by a bivariate polynomial of total degree two. A circular *arc* is represented by a supporting circle and its two delimiting endpoints, knowing that the arc is oriented counterclockwise.
- Two kinds of points—intersection points and endpoints—can be considered in the same way, thus allowing us to have a unique representation. We use

the term *arcpoint* for either an endpoint of an arc or an intersection point. An *arcpoint* is represented by a root of a system of two bivariate polynomial equations. Finding this root reduces to the algebraic operation:

- `solve`, that computes the common roots of two given bivariate polynomials.

The basic operations necessary to implement geometric predicates and constructions are kept at a high level, not imposing any specific algebraic tool. The main operations can then be rephrased as:

- `compare`, that compares two algebraic numbers,
- `x_critical_points`, that computes the critical points of a bivariate polynomial,
- `sign_at` that computes the sign of a bivariate polynomial evaluated at the root of a system of bivariate equations.

Comparisons can be carried out *exactly* and efficiently using algebraic methods adapted to algebraic numbers of low degree [116, 144]. This is explained in Chap. 3.

It is crucial to note, however, that the concepts are general enough to allow also completely different methods, such as interval analysis, that will be used together with exact algebraic methods, for filtering purposes.

1.4.5 How To Speed Up Your Arrangement Computation in CGAL

- When the curves to be inserted into an arrangement are x-monotone and pairwise disjoint in their interior to start with, then it is more efficient (in running time) and less demanding (in traits-class functionality) to use the non-intersection insertion-functions instead of the general ones.
- The main trade-off among point-location strategies, is between time and storage. Using the naive or walk strategies, for example, takes more query time but saves storage space and requires less time for maintaining the auxiliary structures.
- If point-location queries are not performed frequently, but other modifying functions, such as removing, splitting, or merging edges are, then using a point-location strategy that does not require the maintenance of auxiliary structures, such as the naive or walk strategies, is preferable.
- When the curves to be inserted into an arrangement are available in advance (as opposed to supplied on-line), it is advised to use the more efficient aggregate (sweep-based) insertion over the incremental insertion.
- The various traits classes should be instantiated with an exact number type to ensure robustness, when the input of the operations to be carried out might be degenerate, although inexact number types could be used at the user's own risk.
- Maintaining short bit-lengths of coordinate representations may drastically decrease the time consumption of arithmetic operations on the coordinates.

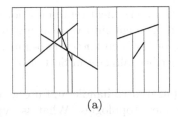 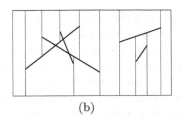

(a)   (b)

Fig. 1.13. Vertical decomposition (a) and partial decomposition (b) of an arrangement of segments [315] inside a bounding rectangle

This can be achieved by caching certain information or normalization (of rational numbers). However, both solutions should be used cautiously, as the former may lead to an undue space consumption, and indiscriminate normalization may considerably slow down the overall process.

- Geometric functions (e.g., traits methods) dominate the time consumption of most operations. Thus, calls to such function should be avoided or at least their number should be decreased, perhaps at the expense of increased combinatorial-function calls or increased space consumption. For example, repetition of geometric-function calls could be avoided by storing the results obtained by the first call, and reusing them when needed.

1.5 Exact Construction in 3-Space

Moving from two-dimensional arrangements of curves to three-dimensional arrangements of surfaces is a major endeavor. In this section we report on progress in and plans for coping with three-dimensional arrangements. We start with an efficient space-sweep algorithm for computing a useful refinement of arrangements of surfaces, and proceed with algebraic primitives and algorithms for the case of quadrics.

1.5.1 Sweeping Arrangements of Surfaces

We describe a method to effectively compute a representation of a three-dimensional arrangement of *well-behaved* surface patches[15] [315]. The method is fairly simple and efficient and has been successfully implemented for arrangements of triangles and of polyhedral surfaces. It computes the *vertical*

---

[15]For a detailed discussion of what constitute well-behaved surfaces or surface patches in the context of arrangements, see [15]. These are, for example, a collection of algebraic surface patches of bounded degree each bounded by at most some constant number of algebraic curves of bounded degree and each decomposed into a constant number of xy-monotone patches.

decomposition of the arrangement, which is a convenient representation breaking the three-dimensional cells of the arrangement into vertical prisms. It is an extension to three-dimensional space of the well-known two-dimensional trapezoidal decomposition (as briefly described already in Sect. 1.3); by vertical we mean parallel to the z-axis.

A raw arrangement is often too complicated to handle and use as it may have cells with many features and complex topologies. What is typically needed is a further refinement of the arrangement into cells, each homeomorphic to a ball and of small combinatorial complexity (that is, a small constant number of features). Additionally, we would like the refinement to be economical and not to increase the complexity of the arrangement by much. A refinement that satisfies these requirements is the so-called *vertical decomposition*. Fig. 1.13(a) depicts an arrangement of segments (in bold lines) refined by vertical decomposition: We extend a vertical line upwards and downwards from every vertex of the arrangement (either a segment endpoint or the intersection of two segments) until it hits another segment or extends to infinity. Vertical decompositions are defined for any dimension and for arrangements of any collection of "well-behaved" objects [85, 196, 312].

For simplicity of exposition we describe the three-dimensional variant for arrangements of triangles. Let $T = \{t_1, t_2, \ldots, t_n\}$ be a collection of triangles in 3-space. For a curve γ in $\mathbb{R}^3$ let $H(\gamma)$ denote the vertical wall through γ, namely the union of vertical lines intersecting γ. For an edge e of the arrangement we define the wall of the edge, denoted $W(e)$, as the union of points in $H(e)$ that can be connected to e with a vertical segment that does not cross any of the triangles in T. The vertical decomposition of $\mathcal{A}(T)$ is obtained as follows. First we erect walls from triangle boundary edges. Second, walls are erected from the intersection edges between pairs of triangles to produce a finer decomposition. Finally we refine the decomposition, in a straightforward manner, into a convex subdivision consisting of trapezoidal prisms. (See e.g. [109] for details.) We call the refined subdivision the *full* (or standard) vertical decomposition. An alternative decomposition, which induces fewer cells, called the *partial* decomposition, has also been proposed and investigated. See Fig. 1.13(b) for an illustration of the two-dimensional partial decomposition of an arrangement of segments.

De Berg et al. [109]. showed that the maximum combinatorial complexity of the vertical decomposition is the same as that of the arrangement, which is $\Theta(n^3)$, and the complexity of the vertical decomposition is sensitive to the complexity of the underlying arrangement. They gave a bound $O(n^{2+\epsilon} + K)$ where K is the complexity of the arrangement, improved by Tagansky [326] to $O(n^2 \alpha(n) \log n + K)$, where $\alpha(n)$ is the extremely slowly growing inverse of Ackermann's function. The near-quadratic overhead term is close to optimal in the worst case as there are arrangements with linear complexity whose vertical decomposition has quadratic complexity. They also gave an output-sensitive algorithm to compute the decomposition running in time $O(n^2 \log n + V \log n)$, where V is the complexity of the decomposition.

The algorithm that we sketch here is an improvement and simplification of the algorithm in [109]. It runs in time $O(n \log^2 n + V \log n)$ where V is the size of the vertical decomposition of the arrangement of n well-behaved surfaces. Thus, the algorithm has near-optimal running time. A detailed description of the algorithm and its implementation can be found in [314, 315].

We assume that the input triangles in T are in general position. For convenience, we also assume that the triangles in T are bounded inside a big simplex (four extra triangles) and we are only interested in the decomposition inside this bounding simplex. The output of the algorithm is a graph $G = (U, E)$ where each node in U describes one trapezoidal prism of the decomposition and there is an edge (u_1, u_2) in E if the two prisms corresponding to u_1 and u_2 share a vertical (artificial) wall.

The algorithm consists of one pass of a space sweep with a plane orthogonal to the x-axis moving from $x = -\infty$ to $x = \infty$. Let P_{x_1} denote the plane $x = x_1$. Let $\mathcal{A}_{x_1}$ denote the partial two-dimensional decomposition of the arrangement $\mathcal{A}(P_{x_1} \cap T)$ of segments induced on the plane P_{x_1} by intersecting it with the triangles in T and the bounding simplex and by adding vertical extensions only through segment endpoints (which looks, up to the bounding simplex, like Fig. 1.13(b)). We use $\mathcal{A}_x$ to denote this subdivision for an arbitrary x-value.

Besides the graph G in which the output is constructed, the algorithm maintains three data structures. A dynamic structure $\mathcal{F}$ that describes the subdivision $\mathcal{A}_x$, a standard event queue $\mathcal{Q}$, which maintains the events of the sweep ordered by their x-coordinate, and a dictionary $\mathcal{D}$, which connects between $\mathcal{F}$ and $\mathcal{Q}$ as we explain next.

The structure $\mathcal{F}$ supports efficient insertion or deletion of vertices, edges, and faces of the subdivision. In addition, it efficiently answers vertical ray-shooting queries. In order for the overall algorithm to be efficient, we wish to refrain from point location in the subdivision $\mathcal{A}_x$ as much as we can, since in the dynamic setting point location queries are costly. We achieve this by using the dictionary $\mathcal{D}$. Each feature in the current subdivision $\mathcal{A}_x$ is given a unique combinatorial label (we omit the straightforward details here). We keep a dictionary of all these features with cross pointers to their occurrence in $\mathcal{F}$. When we add an event that will occur later at x' to $\mathcal{Q}$ we also insert the combinatorial labels of features of $\mathcal{A}_{x'}$ that are related to the event. This way, when we come to handle the event, we could use the dictionary $\mathcal{D}$ (paying $O(\log n)$ to search in the dictionary) and with the information thus obtained we directly access $\mathcal{F}$.

If we do not compute events in advance (before the sweep starts), how do we predict all the events together with the extra labels needed? The answer lies in the observation, which is similar to the key observation in the two-dimensional Bentley-Ottmann sweep algorithm: Before an event, the involved features of the arrangement must become neighbors in $\mathcal{A}_x$. Thus it suffices to predict future events by only checking a small number of neighboring features, and repeat the test each time the neighbors of a feature change.

A one-pass sweep with a plane suffices since almost all the events can be predicted in this way, namely, every time we update the structure $\mathcal{F}$ we have to go over a constant-length checklist involving the modified features of $\mathcal{F}$. The overall cost of the prediction is $O(\log n)$ (this is a property of $\mathcal{F}$) and at most a constant number of new events is created and inserted into the queue. Some of these events may later turn out to be 'false alarms.' However, a false event never spawns a new event, the prediction of a false event can be charged to an actual event, and no actual event is charged for more than a constant number of false events. The full list of events, how they are detected and handled, is given in [314].

Most features (or events) of the arrangement $\mathcal{A}(T)$ can be fully predicted during the sweep. There is only one type of events, however, for which we cannot obtain the full event information from previous events. This is obviously the appearance of a new triangle t, namely the first time that the sweep plane hits t. This is what determines the choice of structure for $\mathcal{F}$. We use the dynamic point location structure of Goodrich and Tamassia [187], which pertains to monotone subdivisions (which in the case of triangles is even convex) and takes $O(\log n)$ per update and $O(\log^2 n)$ per point location query. Augmenting it to support vertical ray shooting in a known face in time $O(\log n)$ is trivial. We use the structure for point location exactly n times.

Now we can summarize the performance of the algorithm. The prediction work, as well as handling a single event, take $O(\log n)$ time per event for a total of $O(V \log n)$ (recall that V is the complexity of the decomposition). The extra machinery required for handling the appearance of new triangles gives rise to the overhead term $O(n \log^2 n)$ and it incurs additional work (of the point location structure, which, although we use scarcely, needs to be maintained) that is absorbed in the $O(V \log n)$ term. The storage required by the algorithm is $O(V)$—it is proportional to the complexity of the decomposition.

In the existing implementation of the algorithm the dynamic point-location structure is replaced with a naïve test that goes over all triangles to find the triangle that is immediately vertically above the query point. According to the experimental results reported in [315], this test is fast.

In the case of triangles, as mentioned above, the subdivision A_x is convex. The algorithm however does not rely on this fact in any way. For the algorithm to apply, it suffices that the subdivision A_x be y-monotone (where y is the horizontal axis on the plane P_x). The structure of Goodrich and Tamassia can handle monotone subdivisions.[16] Hence we can generalize the result to the case of well-behaved surface patches:

Theorem 1. *Given a collection S of n well-behaved surface patches in general position in three-dimensional space, the time needed to compute the full vertical decomposition of the arrangement $\mathcal{A}(S)$ is $O(n \log^2 n + V \log n)$, where V is the combinatorial complexity of the vertical decomposition.*

---

[16] M. Goodrich, Personal communication

1.5.2 Arrangements of Quadrics in 3D

Quadric surfaces, or quadrics for short, are defined as the set of roots of quadratic trivariate polynomials. For example, the ellipsoids R, G, and B in the left picture of Fig. 1.14 are defined by the following polynomials:

$$R(x, y, z) = 27x^2 + 62y^2 + 249z^2 - 10 ,$$
$$G(x, y, z) = 88x^2 + 45y^2 + 67z^2 - 66xy - 25xz + 12yz - 24x + 2y + 29z - 5 ,$$
$$B(x, y, z) = 139x^2 + 141y^2 + 71z^2 - 157xy + 97xz - 111yz - 3x - 6y - 17z - 7 .$$

On the surface of a given quadric p, the intersection curves of p with the remaining quadrics induce a two-dimensional arrangement.[17] In our example, the ellipsoid B and the ellipsoid G intersect the ellipsoid R. This leads to two intersection curves on the surface of R (the right-hand side pictures of Fig. 1.14 and Fig. 1.16). Vertices of this (sub)arrangement are common points of two intersection curves, or rather intersection points of three quadrics.

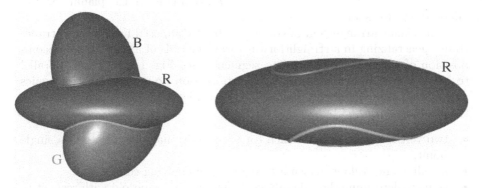

Fig. 1.14. The ellipsoid B and the ellipsoid G intersect the ellipsoid R in two spatial curves running on the surface of R

The Sweeping Approach

Sweeping a set of n quadrics $\{Q_i, i = 1, \ldots, n\}$ by a plane orthogonal to the x-axis allows to compute the so-called *vertical decomposition* of the arrangement of the quadrics (as described above in Sect. 1.5.1).

---

[17] Although we have only discussed arrangements in Euclidean space so far in the chapter, arrangements are naturally defined on curved surfaces as well. For instance, a very useful type of arrangements is defined on the surface of a sphere—see Sect. 1.7 for an application of such arrangements.

When applying this standard sweeping technique to arrangements of quadrics in $\mathbb{R}^3$, the main issue is—as usual when applying computational geometry techniques to curved objects—the implementation of the geometric primitives: predicates and constructions. Therefore, it is essential to analyze them precisely. A first study of the way the primitives needed by the sweep can be expressed in algebraic terms was conducted in [264, 265].

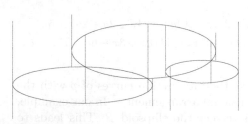

Fig. 1.15. "Trapezoidal" decomposition of conics in a plane

Let us denote by Q_i both a quadric and its equation, which is a degree 2 polynomial in the variables x, y, z. The intersection of the arrangement of quadrics with the sweeping plane in a fixed position during the sweep, gives rise to an arrangement of conics in a (y, z)-plane. The equation of a conic for a given position $x = x_0$ of the plane is $Q_i(x_0, y, z) = 0$, which is a polynomial equation in the coordinates y, z of a point in the planar section, denoted $Q_i^{x_0}$ for short.

This planar arrangement of conics can be decomposed into "curved trapezoids" generalizing in a straightforward way the cells of a trapezoidal decomposition of an arrangement of line segments (see Fig. 1.15): A *wall* parallel to the z-axis is drawn through every intersection point between two conics and every point where the tangent to the conic is parallel to the z-axis. A trapezoid is described by:

- two vertical walls, one of which (or even both) may degenerate to a single point,
- a ceiling and a floor, which are segments of evolving conics, and
- signs to determine the side (above/below) of the trapezoid with respect to each of these two conics.

When the plane is moving, the conics are deforming. The topology of the trapezoidal decomposition changes at events of the sweep. An event occurs whenever one of the two following possibilities occurs:

- either a new quadric is encountered by the sweeping plane (respectively, a quadric is left by the plane), that is, a new conic appears on (respectively, a conic disappears from) the plane
- or the description of a trapezoid is modified
  - either because its ceiling and floor intersect,
  - or because its walls coincide.

Each event corresponds to the construction of a feature of the 3D vertical decomposition.

The detection of events boils down to the manipulation of roots of systems of polynomial equations. Let us briefly illustrate this by showing an example of algebraic manipulation that is required by the algorithm.

A wall in the sweeping plane $x = x_0$ defined by the intersection of two conics $Q_i^{x_0}$ and $Q_j^{x_0}$ corresponds to a solution y of the system

$$\begin{cases} Q_i(x_0, y, z) = 0 \\ Q_j(x_0, y, z) = 0 \end{cases}$$

The worst type (in terms of algebraic degree) of event corresponds to the case when two such walls of the same trapezoid coincide, which occurs when the y-coordinate of the intersection between two conics coincides with the y-coordinate of the intersection between two other conics, which is expressed as:

$$x_0 \text{ such that } \exists y, \exists(z_1, z_2), \begin{cases} Q_i(x_0, y, z_1) = 0 \\ Q_j(x_0, y, z_1) = 0 \end{cases} \text{ and } \begin{cases} Q_k(x_0, y, z_2) = 0 \\ Q_l(x_0, y, z_2) = 0 \end{cases}$$

A solution x_0 lies in an algebraic extension of degree at most 16. The sweep requires that events be sorted, which implies that we must compare *exactly* two events, or equivalently, determine the sign of the difference of the corresponding two algebraic numbers. In the worst case, we are interested in comparing algebraic numbers of degree 16 belonging to independent algebraic extensions of the initial field. So, the algorithm is highly demanding in terms of algebraic manipulations.

The solution proposed in [265] uses algebraic tools like Descartes' rule, Sturm sequences, and rational univariate representation, described in Chap. 3. The practicality of this solution is yet to be proven.

Future work on replacing the vertical decomposition by a partial decomposition is likely to improve the behavior of the algorithm:

• Whereas the complexity of the arrangement is $O(n^3)$, the complexity V of the vertical decomposition is known to be bounded by $O(n^3 \beta(n))$ where $\beta(n) = \lambda_s(n)/n = 2^{\alpha(n)^{16}}$ [85]. .

Note that, though the size of this decomposition is slightly larger than the size of the arrangement, it is much smaller than the size of Collins' cylindrical algebraic decomposition [101].

It is well known that in practice the number of cells in the vertical decomposition can be much bigger than the number of cells in the arrangement. It was shown experimentally, in the case of triangles in 3D, that the number of cells in a partial decomposition can be smaller [315].

• Another important motivation is that it could lead to smaller degree predicates. In fact we have seen that the highest degree predicates we get come from the comparison of the y-coordinates of intersections of conics in the sweeping plane. Decompositions where regions would not be defined by intersections of conics would decrease the degree.

The Projection Approach

An alternative approach to constructing arrangements of quadric surfaces in three-dimensional space is based on projection [178]. We sketch the main ideas of the approach below and refer the reader to detailed description of theoretical and implementation aspects of the method [340], [47].

Independent of the special information about the arrangement of the quadrics one may be interested in, for example, the topological description of a cell or of the whole arrangement, the basic computation that has to be carried out in nearly all cases is: For each quadric p, locate and sort all vertices along the intersection curves on the surface of p.

Our approach for solving this problem operates similar to Collins' cylindrical algebraic decomposition. By projection, it reduces the three-dimensional problem to the one of computing planar arrangements of algebraic curves. We project for each quadric p all its intersection curves with the other quadrics and additionally its silhouette onto the plane. This projection step applied to the quadrics Fig. 1.14 results in the two-dimensional arrangement shown in Fig. 1.16.

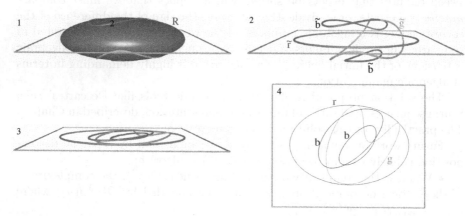

Fig. 1.16. Project the intersection curves $\tilde{b}$ and $\tilde{g}$ of the ellipsoid R with the ellipsoid B and G, respectively, together with the silhouette $\tilde{r}$ of R into the plane. This leads to the planar curves b, g, and r

We have to compute the planar arrangement resulting from the projection. All curves of the planar arrangement turn out to be defined by polynomials of degree at most 4. For example, the curve g is the set of roots of the polynomial

$$408332484x^4 + 51939673y^4 - 664779204x^3y - 24101506y^3x$$
$$+564185724x^2y^2 - 250019406x^3 + 17767644y^3$$
$$+221120964x^2y - 123026916y^2x + 16691919x^2 + 4764152y^2$$
$$+14441004xy + 10482900x + 2305740y - 1763465.$$

The reduction is algebraically optimal in the sense that it does not affect the algebraic degree of the curves we consider. But due to the projection, the curves in the planar arrangement can have six singular points and two curves can intersect in up to 16 points. The most difficult problem we face stems from the high degree of the algebraic numbers that arise in the computation.

For computing the resulting planar arrangements we again must be able to perform the analysis of a single curve and of a pair of curves. The extreme points of one curve f are, as in the case for cubics, computed by comparing the order of f and f_y slightly to the left and slightly to the right of the extreme points. For locating the singular points we make use of the fact that we consider projected quadric intersection curves. One can prove that at most two singular points result from the projection in the sense that two non-intersecting branches of the spatial intersection curve are projected on top of each other. For example, the intersection curve $\tilde{b}$ in Fig. 1.16 consists of two non-intersecting loops. They are projected on top of each other causing two self-intersections. We can compute the coordinates of these singular points as one-root numbers. In most cases one can express the coordinates of the remaining singular points as roots of quadratic rational polynomials. Only in the case that the spatial intersection curve consists of four lines do the coordinates require a second square root.

As described earlier in the section about planar arrangements of conics and cubics, transversal intersection points of two curves are easy to compute by determining the sequence of hits slightly to the left and slightly to the right of the intersection points. Non-singular intersections of multiplicity 2 are computed using the additional Jacobi curve already mentioned in the section about cubics. We know that the Jacobi curve cuts transversally through both involved curves. This fact enables us to reduce the problem of detecting tangential intersections of multiplicity 2 to the one of locating transversal intersections. For all remaining non-singular intersections of multiplicity > 2 one can prove that their coordinates are one-root numbers and can thus be solved directly.

Parameterizing the Intersection of Two Quadrics

A parallel and complementary work was conducted outside ECG, on the conversion of object representation from Constructive Solid Geometry (CSG) to Boundary Representation (BRep), motivated by modeling for rendering. A fundamental step of this conversion is the computation of the intersection of two primitive volumes. Dupont *et al* presented a robust and optimal algorithm for the computation of an exact parametric form of the intersection of two quadrics [128]. Their method is based on the projective formalism, techniques of linear algebra and number theory, and new theorems characterizing the rationality of the intersection. These theoretical results are major in the sense that the output solution is a rational parameterization whenever one exists and the coefficients are algebraic numbers with at most one extra square root.

Furthermore, for each geometric type of intersection, the number of square roots in the coefficients is always minimal in the worst case. This method is usable in practice (as opposed to the approach of Levin used before [241]).

The algorithm was improved later to minimize the size (i.e., the number of digits) of the integer coefficients that appear in the parameterizations. It was implemented in C++ [11] and its practical performance was analyzed [238].

1.6 Controlled Perturbation: Fixed-Precision Approximation of Arrangements

The approaches taken in recent years to cope with precision and robustness problems in geometric computing can be roughly categorized in one of the following two archethemes: (i) exact computing and (ii) fixed-precision approximation. The former mimics the real RAM model for certain primitives, whereas the latter adjusts the algorithmic solutions to the standard computer limited-precision arithmetic.

The approach favored by the CGAL project and related projects, as reflected also in most of this chapter, is exact computing. Exact geometric computing has many advantages, among them is solving the precision and robustness problem, giving the ultimate true results, and enabling the transcription of the geometric algorithms in the literature if the input is in general position.

However, exact computing has some disadvantages. In spite of the constant progress, it is still slower than machine arithmetic especially when the objects are non-linear or higher dimensional (beyond curves in the plane). Also, exact computing does not solve the degeneracy problem, and if we anticipate input that is not in general position, as is often the case in practice, degeneracies require special (and often very tedious) treatment.[18]

Here we outline a method that has been employed for the robust implementation of arrangements of curves and surfaces while using floating-point arithmetic. The scheme seeks to perturb the input objects slightly such that after the perturbation their arrangement is degeneracy free and all the predicates that arise in the construction of the arrangement can be accurately computed with the given machine precision. Controlled perturbation has been successfully applied to arrangements of circles [198], spheres [200, 153, 154], polygons [279], polyhedral surfaces [292], and more recently to Delaunay triangulations [172].

One can view the goal of controlled perturbation as follows. We look to move the input objects slightly from their original placement such that when constructing the arrangement of the perturbed objects while using a fixed precision floating-point filter (see, e.g., [251, 346]), the filter will always succeed and we will never need to resort to higher precision or exact computation.

---

[18] In CGAL though, the implementation of algorithms for two-dimensional arrangements do not assume general position and handle degeneracies.

We give an overview of the controlled-perturbation scheme. Although the ideas that we present here could have been described in a more general setting, and have been applied more generally, we concentrate, for ease of exposition, on arrangements of circles.

The input to the perturbation algorithm is a collection $\mathcal{C} = \{C_1, \ldots, C_n\}$ of circles, each circle C_i is given by the Cartesian coordinates of its center (x_i, y_i) and its radius r_i; we assume that all the input parameters are representable as floating-point numbers with a given precision. The input consists of three additional parameters: (i) the machine precision p, namely the length of the mantissa in the floating-point representation, (ii) an upper bound on the absolute value of each input number x_i, y_i, and r_i, and (iii) Δ — the maximum perturbation size allowed.[19]

For an input circle C_i, the perturbation algorithm will output a copy C_i' with the same radius but with its center possibly perturbed. We define $\mathcal{C}_j$ as the collection of circles $\{C_1, \ldots, C_j\}$, and $\mathcal{C}_j'$ as the collection of circles $\{C_1', \ldots, C_j'\}$. The perturbation scheme transforms the set $\mathcal{C} = \mathcal{C}_n$ into the set $\mathcal{C}' = \mathcal{C}_n'$.

We carry out the perturbation in an incremental fashion, and if there is a potential degeneracy while adding the current circle, we perturb it so that no degeneracies occur. Once the j-th step of the procedure is completed, we do not move the circles in $\mathcal{C}_j'$ again. We next describe the two key parameters that govern the perturbation scheme—the resolution bound and the perturbation bound.

Resolution Bound

A degeneracy occurs when a predicate evaluates to zero. The goal of the perturbation is to cause all the values of all the predicate expressions (that arise during the construction of the arrangement of the circles) to become "significantly non-zero", namely to be sufficiently far away from zero so that our limited precision arithmetic could enable us to safely determine whether they are positive or negative.

The degeneracies that arise in arrangements of circles have a natural geometric characterization as *incidences*. For example, in *outer tangency* (Fig. 1.17), two circles intersect in a single point. We transform the requirement that the predicates will evaluate to sufficiently-far-from-zero values into a geometric distance requirement.

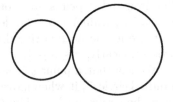

Fig. 1.17. Outer tangency — two circles (bounding interior-disjoint disks) intersect in a single point

An outer tangency between C_1 and C_2 occurs when

$$\sqrt{(x_1 - x_2)^2 + (y_1 - y_2)^2} = r_1 + r_2 .$$

---

[19]The exact size of Δ depends on the specific application of the perturbed arrangement.

We look for a distance $\varepsilon > 0$ such that if we move one circle relative to the other ε away from the degenerate configuration, we could safely determine the sign of the predicate with our limited precision arithmetic, that is, we look for a relocation (x_2', y_2') of the center of C_2 such that

$$|\sqrt{(x_1 - x_2')^2 + (y_1 - y_2')^2} - (r_1 + r_2)| \geq \varepsilon.$$

This is a crucial aspect of the scheme—the transformation of the non-degeneracy requirement into a separation distance. We call the bound on the minimum required separation distance the *resolution bound* [20] and denote it by ε. If the separation distance is less than ε, then there is a *potential degeneracy*. The bound ε depends on the size of the input numbers (center coordinates and radii) and the machine precision. It is independent of the number n of input circles.

Deriving a good resolution bound is a non-trivial task [198]. Notice however that in order to carry out the scheme successfully, one does not need to know the resolution bound. This bound is only needed in order to derive the perturbation bound (see next paragraph) which gives a guarantee on the maximum perturbation magnitude, as may be required by some applications. It also allows to determine what precision is necessary in order to accommodate a given maximum on the allowed perturbation. If a user is not concerned about the perturbation magnitude, then the scheme requires very little analysis.

Perturbation Bound

Suppose that ε is indeed the resolution bound for all the possible degeneracies in the case of an arrangement of circles for a given machine precision. When we consider the current circle C_i to be added, it could induce many degeneracies with the circles in C_{i-1}'. Just moving it by ε away from one degeneracy may cause it to come closer to other degeneracies. This is why we use a second bound δ, the *perturbation bound* — the maximum distance by which we perturb the center of any of the circles away from its original placement. The bound δ depends on ε, on the maximum radius of a circle in C, and on a *density* parameter ρ. The density parameter bounds the number of circles that are in the neighborhood of any given circle or may effect it during the process; clearly, $\rho \leq n - 1$.

We say that a point q is a valid placement for the center of the currently handled circle C_i if, when moved to q, this circle will not induce any degeneracy with the circles in C_{i-1}'. The bound δ is computed such that inside the disk D_δ of radius δ centered at the original center of C_i, at least half the points (constituting half of the area of D_δ) are valid placements for the circle. This means that if we choose a point uniformly at random inside D_δ to relocate the center of the current circle, it is a valid placement with probability at least $\frac{1}{2}$.

---

[20]It would have also been suitable to call it a *separation bound*, but we use *resolution bound* to avoid confusion with separation bounds of exact algebraic computing.

After the perturbation, the arrangement $\mathcal{A}(\mathcal{C}')$ is degeneracy free. Moreover, $\mathcal{A}(\mathcal{C}')$ can be robustly constructed with the given machine precision. The perturbation algorithm should not be confused with the actual construction of the arrangement. It is only a preprocessing stage. However, it is convenient to combine the perturbation with an incremental construction of the arrangement, or more generally, the scheme can be conveniently and efficiently interwoven with geometric Randomized Incremental Construction (RIC) algorithms [172]

The full technical details of the method are given for circles in [198]. In all applications to date, even in fairly involved applications such as dynamic maintenance of molecular surfaces under conformational changes [154], the perturbation consumes a very small fraction of the running time of the construction algorithm.

1.7 Applications

Arrangements of curved objects arise in a variety of applications from different fields, such as robotics, computer-aided design (CAD) and computer-aided manufacturing (CAM), graphics and molecular modeling. While textbook algorithms sometimes supply elegant algorithms for solving such problems, implementing reliable and efficient software solutions pose a real challenge to any computer programmer, due to the high complexity of the algebra involved. In this section we review some of these applications and describe the state-of-the-art software solutions. Most of these solutions were implemented using the geometric and the algebraic infrastructure described in the previous sections.

1.7.1 Boolean Operations on Generalized Polygons

We can describe point sets in the plane bounded by piecewise arcs of curves with a suitably labeled arrangement of (arcs of) curves. Therefore, we extend the arrangement data structure with a Boolean *selection mark* for all arrangement features (vertices, edges, and faces). The selection mark is *true* if the item is part of the represented point set and *false* otherwise.

We describe how to implement standard Boolean set-operations, such as *union, intersection, complement*, and *difference*, as well as topological operations, such as *interior, closure*, and *boundary*. A particular relevant operation is the *regularization* of a point set A, denoted $A^\star$, which is defined as the closure of the interior of the point set A. A point set A is called *regular* if it is equal to its regularization, i.e., $A = A^\star$. By regularizing the result of Boolean set-operations we obtain the *regularized Boolean set-operations*. Regularization is particularly relevant for solid modeling because regular sets are closed under regularized Boolean operations and because regularization eliminates lower dimensional features, thus simplifying and restricting the representation to physically meaningful solids. A simplified representation does not have to

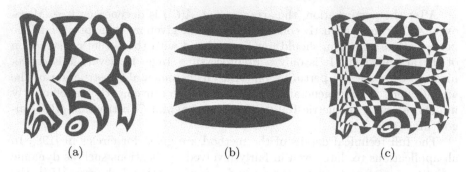

Fig. 1.18. Two generalized polygons (a) and (b), bounded by piecewise arcs of conic curves, and their symmetric difference (c)

store selection marks; they are implicitly always set for vertices and edges, and for faces they are deduced from a suitably chosen orientation condition on their boundaries.

The implementation of all operations can be broken into three distinct steps — *refinement, selection,* and *simplification* — although in practice the first two (if not all three) can be performed in a single step. The *refinement* step is ignored for unary operations.

Refinement: Given two input arrangements, we compute an arrangement that is a common refinement of the two, also referred to as the *overlay* of the two arrangements. This is done using the sweep-line algorithm presented in Sect. 1.3. We also take advantage of the fact that the two input arrangements are represented using two DCEL structures, which contain only x-monotone curves that do not intersect in their interior. We therefore never have to compute intersection between two curves that originate from the same DCEL.[21]

The result of the refinement is an arrangement including proper faces. For each feature in the result, we have a unique *support feature* in each input arrangement, in particular, for each result face we know the corresponding face in each input arrangement.

Selection: We determine the value for each selection mark in the refined arrangement:

- For binary point-set operations, we evaluate the corresponding Boolean function on the selection marks of the two support features.
- For a complement operation, we negate all selection marks.

---

[21] A variant of this problem has been studied under the name of *red-blue segment intersection* for the linear case, but we do not know of extensions for curves. For example, the algorithm in [248] (and also others) makes explicit use of the fact that the edges intersect at most once.

- For an interior operation, we set the selection marks of all edges and vertices to false.
- For the closure operation, we set the selection mark of all edges and vertices incident to a marked face and of all vertices incident to a marked edge to true.
- For the boundary operation, we compute the closure and then set all face selection marks to false.

Simplification: Refinement and selection might result in a redundant representation of the point set. For example, the union of two half-discs still contains the dividing diameter, which is redundant for the representation of a full disc. Simplification is important as it enables a unique representation of a point set and also reduces the memory requirements by eliminating redundant features. Moreover, it simplifies the selection step; in this step we assume that input arrangements are simplified, otherwise the selection rules would become more involved.

The simplification applies the following three rules, in that order, until no rule applies and the unique representation is reached. These rules are based on a complete classification of local neighborhoods in the point set:

- If the selection marks of an edge and its two incident faces have the same value, then the edge is redundant and is removed from the arrangement, merging the two faces in the process.
- If a vertex is incident to exactly two edges that have the same supporting curve and all have the same selection-mark value, then the vertex is redundant and is removed from the arrangement, such that the two edges are merged to represent a single curve (an exception to this rule are vertices that divide curves to form x-monotone sub-curves).
- If an isolated vertex has the same selection mark as its surrounding face, then this vertex is redundant and is removed from the arrangement.

Clearly, the refinement step is computationally the most demanding of the three and it dominates the run-time of the overall algorithm. However, when the intermediate arrangement consists of many redundant features, the simplification step may also be non-negligible.

The representation of point sets starting from linear half-spaces as primitives have been implemented in the full generality described here as CGAL packages: Seel [308], [309] implemented the *planar Nef polyhedra* package while Granados et al. [189] and Hachenberger et al. [195] describe the implementation of *3D Nef polyhedra*, based on the work of [274], [53] and [297]. Regular point sets with regularized Boolean set-operations have been implemented under the name of *generalized polygons* [251, Sect. 10.8] in LEDA. This implementation has been extended to handle regions bounded by conic arcs in the EXACUS project [49], see Fig. 1.18 for an example.

The Boolean set-operation package included in CGAL Version 3.2 [2] implements *regularized* Boolean set-operations on point sets given as collections of generalized polygons. A generalized polygon is defined by a circular sequence of edges realized as arbitrary directed x-monotone curves, with the property that the target point of each x-monotone curve equals the source point of its successor. Each type of x-monotone curve, as defined by one of the traits classes described in Sect. 1.4.2 (namely a conic arc, a segment of a rational function, etc.), may be used to represent a polygon edge.

An important feature of the Boolean set-operation package is that it supports aggregated operations in an efficient manner. Suppose we are given a set of generalized polygons $P_1, \ldots, P_N$ and we wish to compute $\bigcup_{k=1}^{N} P_k$. We do so using a divide-and-conquer approach — namely, we divide the polygon set into two sets of approximately the same size, recursively compute the aggregated union of each subset and finally compute the binary union of the two results. Similarly, we can define and implement an aggregated intersection and symmetric difference operations.

In the implementation of the divide-and-conquer algorithm we use the two following observations in order to optimize the process and reduce its running time:

- Simplifying the intermediate arrangements that represent the partial results in the various recursion steps is too costly. We therefore avoid the simplification steps along the divide-and-conquer process. Only the final result of the overall operation is simplified.
- At each of the recursion steps we have to compute the union of two sets by overlaying two arrangements. This is done by the sweep-line algorithm, whose first step is initializing its event queue (the X-structure) by sorting all curve endpoints. We take advantage of the fact that these endpoint are exactly the the arrangement vertices, and since each intermediate arrangement has been constructed using the sweep-line algorithm in the previous recursion steps, we know that their vertices were already created in ascending xy-lexicographic order. We can therefore apply a linear-time merger of the two vertex lists, achieving a considerable reduction in the number of geometric comparisons we perform, therefore reducing the running time.

Figure 1.19 shows two VLSI models that represent electronic circuits, the components of which have been dilated by a small radius r. The models therefore contain a large number of general polygons (representing dilated segments or dilated polygons) and circles (representing dilated points or dilated circles). We compute the union of the dilated components, using the aggregated union operation provided by the CGAL package. We use a specialized traits class for handling line segments and circular arc with rational coefficient [337]. This traits class uses the fact that the coordinates of all intersection points it handles are one-root numbers (see Sect. 1.3.1) to evaluate all the predicates involving curves and points using exact rational arithmetic. This property makes it highly efficient. The result, which represents the forbidden locations for a circular tool-tip of radius r within the model, can be expressed as a set

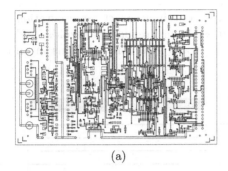
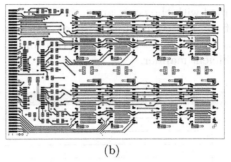

(a)   (b)

Fig. 1.19. Computing the aggregated union of generalized polygons that originate from industrial VLSI models. (a) The union of a set of 2593 generalized polygons and 645 circles, which comprises 13067 vertices, 13067 edges and 624 faces. It is computed in less than 2 seconds on a 3 GHz Pentium IV machine with 2 Gb of RAM. (b) The union of a set of 22406 generalized polygons and 294 circles, which comprises 14614 vertices, 14614 edges and 357 faces and takes 20 seconds to compute

of disjoint general polygons that contain holes. Note that in both cases the union is computed in just a few seconds in an *exact* manner (see Figure 1.19).

1.7.2 Motion Planning for Discs

The simplest variant of the motion-planning problem for a disc is as follows: We are given a disc-shaped robot with radius r, moving within a bounded polygonal region cluttered with polygonal obstacles. Given a start and a goal configuration in the plane (specified, for example, by the coordinates of the center of the disc) determine whether there exists a collision-free motion path for the robot between the two end configurations, and if so, compute such a path.

As this variant of the motion-planning problem has two degrees of freedom, the forbidden configuration space can be explicitly constructed by computing the union of the Minkowski sums of each obstacle with the disc (see, e.g., [40]). Each Minkowski sum is obtained by "inflating" the corresponding polygon by r, resulting in a shape bounded by line segments and circular arcs. The free configuration space can then be decomposed into pseudo-trapezoidal cells, and it is possible to define a roadmap over those cells that captures the connectivity of the free space (see, e.g., [237]). When the start and goal configurations, both free, are specified as a query, it suffices to locate the two corresponding cells and check whether they can be connected in the roadmap.[22]

Leiserowitz and Hirsch [240] implemented the motion-planning algorithm described above. They used CGAL's arrangement package (Sect. 1.4.1), com-

---

[22] Another approach for solving the motion-planning problem for a disc, presented by Ó'Dúnlaing and Yap [275], involves the construction of the Voronoi diagram of the obstacle edges.

bined with the conic arc traits (Sect. 1.4.2) to obtain an exact representation of the free configuration space, to decompose it into pseudo-trapezoids, and to answer point-location queries for the start and goal configurations in an efficient manner.

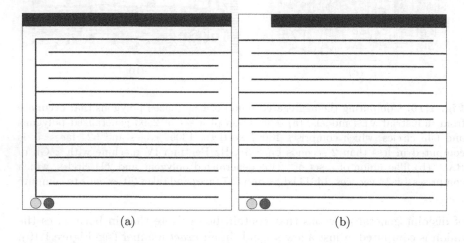

Fig. 1.20. Two scenes for motion planning of two disc robots: (a) *Annulus*, and (b) *Maze*

Hirsch and Halperin [209] suggest an approach called *hybrid motion-planning* for coordinating the motion of two disc robots D_1, D_2 moving in the plane among polygonal obstacles. The configuration space in this case is 4-dimensional and it is rather difficult to construct explicitly. Instead, the 2-dimensional free space of each robot is constructed independently and decomposed into pseudo-trapezoidal cells using the implementation of [240]. Let $c_1^{(1)}, \ldots, c_{m_1}^{(1)}$ and $c_1^{(2)}, \ldots, c_{m_2}^{(2)}$ be the free cells obtained for the two robots respectively, then the 4-dimensional free space is clearly a subset of $\bigcup_{i=1}^{m_1} \bigcup_{j=2}^{m_2} c_i^{(1)} \times c_j^{(2)}$. We now examine the cells obtained by this Cartesian product:

- Each 4-dimensional cell $c_{ij} = c_i^{(1)} \times c_j^{(2)}$ obtained from two 2-dimensional cells that are sufficiently distant from one another, namely when $(c_i^{(1)} \oplus D_2) \cap (c_j^{(2)} \oplus D_1) = \emptyset$, is entirely free: If we locate D_1 in $c_i^{(1)}$ and D_2 in $c_j^{(2)}$ each robot does not collide with any obstacles and the two robots cannot collide with one another.
- The other 4-dimensional cells may contain both free and forbidden configurations, which need to be distinguished. The computation of the free configurations within each cell is approximated using a probabilistic roadmap (see, e.g., [220]), whose construction relies on simple local planners.

The free cells and the local roadmaps are then stitched together to form a global roadmap that captures the connectivity of the entire 4-dimensional free space.

The hybrid motion planner, which combines exact arrangement computation with probabilistic techniques, has several advantages over the prevalent probabilistic approaches. First, it uses exact methods wherever possible and thus it is less sensitive to the existence of narrow (and even tight) passages. Secondly, only a small portion of the configuration space is computed using probabilistic methods, giving the ability to concentrate more computational efforts on these parts and to sample them more densely.

Fig. 1.20 shows two motion-planning problems that the hybrid planner successfully solves. In both scenes the two robots have to switch places. In the *Annulus* scene one of the robots has to go all the way through the maze, while the other should only slightly move; this solution is found deterministically, as there is almost no need to coordinate the two robots. In the *Maze* scene the two robots have to go all the way up the maze, switch places at the wide part at the top-left corner of the maze, and then go all the way down. The probabilistic methods are mostly used only at the top part of the maze, where the motion of the two discs should be carefully coordinated. These examples show that the introduction of exact two-dimensional arrangements can greatly enhance the capabilities of practical motion planners to solve difficult (tight) problems.

1.7.3 Lower Envelopes for Path Verification in Multi-Axis NC-Machining

In a typical multi-axis NC-machining collision-avoidance problem we are given a rotating milling-cutter (also called a *tool*), whose profile — with respect to its axis of symmetry — is typically piecewise linear or circular, moving in space among polyhedral solids bounded by triangular facets. These triangles model the workpiece sculptured by the tool as well as other static parts of the NC-machine [277]. Our goal is to verify a given motion path for the tool between two given configurations,[23] so it can move near the workpiece without damaging it or any of the other static parts of the machine.

The path-verification problem was addressed by many researchers (see, e.g., [348, 211]) but most proposed algorithms are based on various approximation schemes or pose certain restrictions on the shape of the tool. However, two recent papers by the same group of authors offer a novel approach for path verification, based on exact geometric computations. In [213] it is shown how to answer a single tool–model interference query, when the tool remains static at a given configuration, by computing the *lower envelope* of algebraic arcs

---

[23]In the NC-machining literature, a configuration is often referred to as a *contact location (CL) point*, as the tool tip is typically in contact with the workpiece, and this is the only type of contact that we allow.

of degree $2.$[24] We assume, without loss of generality, that the tool tip is positioned at the origin with the z-axis being its axis of symmetry. We now wish to locate all the closest model triangles to the z-axis. This is performed, taking advantage of the symmetry of the rotating tool, by radially projecting the relevant model triangles around the z-axis onto the yz-plane, i.e., by applying a transformation $\mathcal{R} : \mathbb{R}^3 \to \mathbb{R}^2$ such that $(\hat{x}, \hat{y}) = \mathcal{R}(x, y, z) = (z, \sqrt{x^2 + y^2})$. The image of a triangle under $\mathcal{R}$ is the *trace* that the triangle etches on the yz-plane (more precisely, on the half-plane $y > 0$) when rotated around the z-axis. It is possible to show that the trace of the triangle is determined by the projected curves that result from applying $\mathcal{R}$ on the triangle edges — these are finite arcs of canonical hyperbolas — plus (possibly) an additional line segment in the $\hat{x}\hat{y}$-plane.

It is now possible to compute the lower envelope of the set of hyperbolic arcs and line segments obtained from the triangles. This is carried out using an extension of CGAL's arrangement package (see Sect. 1.4.1) that computes the lower envelope of any set of curves in a robust manner, using the divide-and-conquer approach, combined with the traits class designed for conic arcs (see Sect. 1.4.2). Finally, it is possible to perform a simultaneous traversal over the lower envelope and the tool's profile along the $\hat{x}$-axis (which corresponds to the original z-axis) and compare the two entities. If at some point the profile lies above the lower envelope, we conclude that there is a collision between the tool and the model and identify the triangles that are intersected by the tool; see the example depicted in Fig. 1.21 for an illustration. If we have n triangles, the total running time is $O(2^{\alpha(n)} n \log n + m)$ where m is the number of segments in the tool's profile, and α denotes the inverse Ackermann function.

The lower envelope approach is generalized in [335], where it is used to implement a *continuous* path verification scheme for multi-axis NC-machining, that is, to detect collisions between the model and the tool while the latter continuously moves along some given path. The tool-path is approximated by a sequence of sub-paths of pure translational motions interleaved with pure rotational motions, guaranteeing that the approximation error is bounded by some prescribed ε. Each sub-path is separately verified.

It is possible to apply the appropriate transformation, with respect to each sub-path, that ensures that the translation (or rotation) in the sub-path is taking place on the xz-plane. We now assume that the tool is fixed, such that the tip of the tool is always positioned at the origin with the z-axis being its axis of symmetry, and the model is moving. The radial projection of any segment (triangle edge) in the model around the z-axis is therefore continuously changing. Thus it creates a surface patch σ of a terrain $\hat{y} = \hat{y}(\hat{x}, \tau)$ in the $\hat{x}\hat{y}\tau$-space, where τ parametrizes the motion in the corresponding

---

[24]Given a set of planar curves, we can regard each x-monotone curve as the graph of a continuous univariate function defined on an interval of the x-axis, such that the lower envelope of the set is the point-wise minimum of these functions.

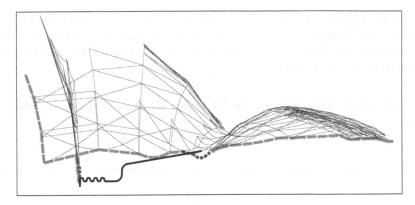

Fig. 1.21. The profile of a complex tool containing 5000 segments (drawn in a thick dark line) as it interferes with the lower envelope of a set of about 800 hyperbolic arcs and line segments obtained by radially projecting the model triangles around the tool's symmetry axis

sub-path. The key observation made here is that we are only interested in the "lowest" points in this surface patch, the ones closest to the z-axis. To this end, we compute the *silhouette*[25] of the patch σ, which is a planar curve in the $\hat{x}\hat{y}$-plane, defined as

$$\mathrm{sil}_\sigma(\hat{x}) = \inf_\tau \hat{y}(\hat{x}, \tau) \ . \tag{1.3}$$

It is now possible to compute the lower envelope of the silhouette curves of all relevant triangle edges and to proceed as we did in the discrete case. However, we now have to inflate the profile of the tool by ε before comparing it with the lower envelope, in order to account for the approximation error caused by our path decomposition scheme.

In [335] it is shown that the silhouette of a line segment (representing a triangle edge) in case of a pure translational motion is comprised of at most five hyperbolic arcs, so in this case the continuous collision check is asymptotically as fast as the discrete query using the exact-computation mechanisms. Things get more complicated for rotational motions, but in this case one can obtain a good approximation of the silhouette curves using rational arcs of a relatively low degree.

The collision-detection algorithms were implemented in the IRIT modeling environment [7], while the lower-envelope calculations and the comparison of the envelope with the tool profile were carried out with a recent extension to the CGAL arrangement package [334]. It should be noted that the hyperbolic arcs we encounter are portions of canonical hyperbolas with the $\hat{x}$-axis being their major axis. Thus, the intersection points between each pair of hyperbolic

---

[25]The silhouette is often referred to as the *envelope*. To avoid confusion with lower envelopes of finite sets of curves, we stick with the term silhouette.

arcs can actually be computed by solving only quadratic equations. This fact
not only simplifies the code that handles the geometric constructions and
predicates for our planar curves, but also helps in significantly reducing the
running time of the algorithm.

1.7.4 Maximal Axis-Symmetric Polygon Contained in a Simple Polygon

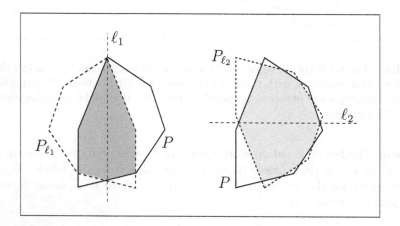

Fig. 1.22. The maximal contained polygons in P with respect to two given axes ℓ_1
and ℓ_2, obtained by the intersection of P and the mirrored polygons P_{ℓ_1} and P_{ℓ_2},
respectively. Note that the two symmetric polygons $S_{\ell_1} = P \cap P_{\ell_1}$ and $S_{\ell_2} = P \cap P_{\ell_2}$
have different signatures

We are given a simple polygon P. What is the axially-symmetric polygon
with maximal area that is contained in P? Given any axis of symmetry ℓ it
is possible to compute P_ℓ, the mirrored version of P with respect to this axis,
and obtain the maximal contained polygon $S_\ell = P \cap P_\ell$ with respect to the
given axis. Furthermore, the *signature* of S_ℓ is defined as the order of original
polygon edges and reflected edges as they occur along S_ℓ's boundary.

The number of possible axis positions is obviously infinite, yet the number
of distinct signatures of S_ℓ is restricted combinatorially by the number of
intersections between edges of the original polygon and its mirrored version.
In the example shown in Fig. 1.22, S_{ℓ_1} and S_{ℓ_2} have different signatures.

Barequet and Rogol [43] use this observation to solve the optimization
problem. They find the polygon of maximum area with a given signature, for
all possible signatures, and choose the one with maximal area from this finite
set of polygons.

The problem is transformed into the dual plane, which is subdivided into
a finite number of cells, such that two points that belong to the same cell

correspond to lines ℓ_1 and ℓ_2 such that S_{ℓ_1} and S_{ℓ_2} have the same signature. It turns out that such a subdivision is induced by the arrangement of a set of lines and hyperbolic arcs. Barequet and Rogol used CGAL's arrangement package to compute the subdivision and then used heuristic optimization techniques to find the polygon with maximal area within each cell. For a convex polygon, the complexity of the arrangement is $\Theta(n^3)$ in the worst case, and the total running time of the algorithm is $O(n^3(\log n + T(n)))$, where $T(n)$ is the average running time of a single optimization step. Further details can also be found in [296].

1.7.5 Molecular Surfaces

A common approach to representing the three-dimensional geometric structure of a molecule is to represent each of its atoms by a "hard" ball or sphere. In certain applications it is also assumed that the relative displacement of the spheres is fixed. Based on the application at hand, there are recommended values for the radius of each atom sphere and for the distance between the centers of every pair of spheres. In this model, the spheres are allowed to interpenetrate one another, therefore it is sometimes referred to as the "fused spheres" model (see Fig. 1.23). The envelope surface of the fused spheres may be regarded as a formal *molecular surface*. It is evident that various properties of

Fig. 1.23. The hard-sphere model of the molecule *crambin* (a protein comprised of 46 amino acids) with 327 atom spheres. The figure was produced using the software described in [199]

molecules are disregarded in this simple model. However, in spite of its approximate nature, it has proved useful in many practical applications. There are several closely related types of molecular surfaces. One is the so-called *solvent accessible surface* in which each atom sphere is expanded by a fixed radius r to reflect an approximation of a water molecule (of that radius) rolling over the molecule. From a geometric-computing point of view, the two types of surfaces are the same and they both amount to computing the boundary of the union of balls. There is yet another commonly used type of surfaces, so-called *smooth molecular surfaces*,[26] proposed by Richards [294] and implemented by Connolly [103], which can be derived from the union. For more

---

[26]The so-called smooth molecular surfaces are possibly self intersecting; see, e.g., [303].

background material and references, see for example the surveys [253, 104]. For a discussion of the relation between the three types of surfaces from a computational point of view see [199].

Let $\mathcal{M} = \{B_1, \ldots, B_n\}$ be the balls describing the atoms of a molecule. The goal is therefore to produce a useful representation of the boundary of the set $\mathcal{M}$, so that it can be later used to compute the surface area, to determine which atoms contribute to the surface area and how much they contribute, to detect voids, and more. The approach suggested in [199] and robustly implemented in [200] proceeds as follows: (1) For each ball in $\mathcal{M}$ identify the other balls of $\mathcal{M}$ intersecting it. (2) For each ball compute its (potentially null) contribution to the union boundary. (3) Transform the local information into global structures describing the required connected component of the union boundary of the balls in $\mathcal{M}$.

We focus here on step 2, which is carried out by computing the *spherical arrangement* describing the intersection of the boundary sphere S_i of a ball B_i in $\mathcal{M}$ with other balls in $\mathcal{M}$. This is an arrangement of *little* circles (namely, not necessarily great circles) on S_i; see Fig. 1.24(a) for an illustration. Attempts to compute such spherical arrangements using standard machine floating-point arithmetic have failed. To avoid this problem, it is possible to resort to controlled perturbation (see Sect. 1.6) [200, 152]. Molecular surfaces are suited for controlled perturbation since the model parameters are approximated to start with and the amount of perturbation introduced by the scheme is orders of magnitude smaller compared to variance of these parameters in the biochemistry literature.

Another noteworthy aspect of the implementation in [200] is that it uses a partial decomposition of the arrangement (Fig. 1.24(c)), which is an effective coarsening of the trapezoidal decomposition (Fig. 1.24(b)) as explained in Sect. 1.5, or more precisely of its variant on the sphere. The usage of a partial decomposition over a full trapezoidal decomposition leads to great savings in running time.

The underlying algorithmic ideas together with an analysis of the geometric properties of the hard-sphere model of molecules appear in [199]. The details of controlled perturbation applied to molecular surfaces are explained in [200]. Recent improvements leading to significant speedups are reported in [152]. The scheme has also been recently extended to the case of dynamic maintenance of molecular surfaces when the molecules undergo conformational changes [153, 154].

1.7.6 Additional Applications

The Pursuit-Evasion Problem

Gerkey et al. [180] studied the *pursuit-evasion* problem, in which one or more searchers move in a given polygonal environment (say in one floor of a museum) and detect evaders in this environment. They give a complete algorithm

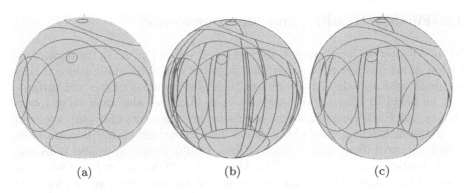

(a)   (b)   (c)

Fig. 1.24. A spherical arrangement (a), its full trapezoidal decomposition (b), and its partial trapezoidal decomposition (c)

for the case of a single robot with a limited field of view of φ radians, called a φ-searcher, based on the arrangement of curves that represent the visibility constraints induced by the environment and the searchers' field of view. Each obstacle edge defines a critical curve that is the locus of all points that "see" it at an angle φ, namely a pair of circular arcs. The algorithm for a single robot can also be generalized for multiple searches (albeit at a loss of completeness). See [179] for further details and on-line examples.

Shortest Path with Clearance

Wein et al. [336] devise a new structure for finding the shortest path for a point robot moving in the plane among polygonal obstacles between a source and a goal configuration, while trying to guarantee that the clearance between the robot and the obstacles is at least c. The main idea is to "inflate" each obstacle by a radius c (see also Sect. 1.7.2), and compute the visibility diagram of the dilated obstacles. A visibility edge is a bitangent to rounded corners of the dilated obstacle. When one encounters a region where it is impossible to guarantee a distance of at least c from the obstacles, which is characterized by an overlap between the dilated obstacles, the Voronoi diagram of the original obstacles is computed and combined into the visibility diagram, representing a path with maximal clearance in this region. The combined diagram therefore contains line segments, circular arcs, and parabolic arcs.[27] It is constructed using the conic-arc traits of CGAL's arrangement package.

---

[27]The Voronoi diagram of polygons is a collection of line segments and parabolic arcs — a parabolic arc is the locus of points equidistant from a polygon vertex and an edge of another polygon.

1.8 Further Reading and Open problems

In this chapter we have concentrated on recent developments in the implementation of arrangements of curves and surfaces and their applications. Many combinatorial and algorithmic results for arrangements of curves and surfaces can be found in Sharir and Agarwal's book [312], see also their more recent survey [15]. Earlier combinatorial results for arrangements of hyperplanes can be found in Edelsbrunner's book [130]. The recent survey by Halperin [196] summarizes combinatorial and algorithmic results as well as applications and implementation issues. The book by Matoušek [249] has several chapters dedicated to arrangements of hyperplanes and of surfaces, with an emphasis on combinatorics.

CGAL and LEDA are discussed in the survey by Kettner and Näher [222]. The CGAL website www.cgal.org has a wealth of information on the library including large-scale detailed documentation.

To conclude the chapter, we outline several open problems related to the effective handling of arrangements.

- The chapter has largely focused on the two-dimensional case. A major current direction is to extend the implementation to three and higher dimensional arrangements.
- The issue of effective decompositions has been mentioned in passing in Sect. 1.5. Devising effective decompositions for arrangement of curves and surfaces is a problem whose good solution could have significant implications on the usefulness of arrangements in practice.
- So far, most of the effort in implementing arrangements has been carried out using exact arithmetic. Fixed precision methods for arrangements could make them more widely used. One of the most challenging problems is to develop efficient and consistent rounding schemes for arrangements. Such schemes are not only needed to avoid using exact computing. The contrary is true. With the advance of exact computing we notice that quite often the exact numerical output is too big to be useful, while in cases involving curves and surfaces finite numerical output simply cannot be given.
- Most of the existing implementations, with few exceptions, compute entire arrangements. Efficiently extracting substructures (such as envelopes) in arrangements is one of the immediate targets of the practical work on arrangements.

2

Curved Voronoi Diagrams

Jean-Daniel Boissonnat[*], Camille Wormser, and Mariette Yvinec

Abstract

Voronoi diagrams are fundamental data structures that have been extensively studied in Computational Geometry. A Voronoi diagram can be defined as the minimization diagram of a finite set of continuous functions. Usually, each of those functions is interpreted as the distance function to an object. The associated Voronoi diagram subdivides the embedding space into regions, each region consisting of the points that are closer to a given object than to the others. We may define many variants of Voronoi diagrams depending on the class of objects, the distance functions and the embedding space. Affine diagrams, i.e. diagrams whose cells are convex polytopes, are well understood. Their properties can be deduced from the properties of polytopes and they can be constructed efficiently. The situation is very different for Voronoi diagrams with curved regions. Curved Voronoi diagrams arise in various contexts where the objects are not punctual or the distance is not the Euclidean distance. We survey the main results on curved Voronoi diagrams. We describe in some detail two general mechanisms to obtain effective algorithms for some classes of curved Voronoi diagrams. The first one consists in linearizing the diagram and applies, in particular, to diagrams whose bisectors are algebraic hypersurfaces. The second one is a randomized incremental paradigm that can construct affine and several planar non-affine diagrams.

We finally introduce the concept of Medial Axis which generalizes the concept of Voronoi diagram to infinite sets. Interestingly, it is possible to efficiently construct a certified approximation of the medial axis of a bounded set from the Voronoi diagram of a sample of points on the boundary of the set.

---

[*] Chapter coordinator

2.1 Introduction

Voronoi diagrams are fundamental data structures that have been extensively studied in Computational Geometry. Given n objects, the associated Voronoi diagram subdivides $\mathbb{R}^d$ into regions, each region consisting of the points that are closer to a given object than to any other object. We may define many variants of Voronoi diagrams depending on the class of objects, the distance function and the embedding space. Although Voronoi diagrams are most often defined in a metric setting, they can be defined in a more abstract way. In Sect. 2.2, we define them as minimization diagrams of any finite set of continuous functions without referring to a set of objects.

Given a finite set of objects and associated distance functions, we call bisector the locus of the points that are at equal distance from two objects. Voronoi diagrams can be classified according to the nature of the bisectors of the pairs of objects, called the bisectors of the diagram for short. An important class of Voronoi diagrams is the class of affine diagrams, whose bisectors are hyperplanes. Euclidean Voronoi diagrams of finite point sets are affine diagrams. Other examples of affine diagrams are the so-called power (or Laguerre) diagrams, where the objects are no longer points but hyperspheres and the Euclidean distance is replaced by the power of a point to a hypersphere. In Sect. 2.3, we recall well-known facts about affine diagrams. In particular, we characterize affine diagrams and establish a connection between affine diagrams and polytopes. As a consequence, we obtain tight combinatorial bounds and efficient algorithms. We also obtain a dual structure that is a triangulation under a general position assumption.

Non-affine diagrams are by far less well understood. Non-affine diagrams are obtained if one changes the distance function: additively and multiplicatively weighted distances are typical examples. Such diagrams allow to model growing processes and have important applications in biology, ecology, chemistry and other fields (see Sect. 2.9). Euclidean Voronoi diagrams of non-punctual objects are also non-affine diagrams. They are of particular interest in robotics, CAD and molecular biology. Even for the simplest diagrams, e.g. Euclidean Voronoi diagrams of lines, triangles or spheres in 3-space, obtaining tight combinatorial bounds, efficient algorithms and effective implementations are difficult research questions.

A first class of non-affine diagrams to be discussed in Sect. 2.4 is the case of diagrams whose bisectors are algebraic hypersurfaces. We first consider the case of Möbius diagrams whose bisectors are hyperspheres and the case of anisotropic diagrams whose bisectors are quadratic hypersurfaces (see Sect. 2.4.2). The related case of Apollonius (or Johnson-Mehl) diagrams is also described in Sect. 2.4.

The key to obtaining effective algorithms for computing those non-affine diagrams is a linearization procedure that reduces the construction of a non-affine diagram to intersecting an affine diagram with a manifold in some higher dimensional space. This mechanism is studied in full generality in Sect. 2.5.

In this section, we introduce abstract diagrams, which are diagrams defined in terms of their bisectors. By imposing suitable conditions on these bisectors, any abstract diagram can be built as the minimization diagram of some distance functions, thus showing that the class of abstract diagrams is the same as the class of Voronoi diagrams. Furthermore, the linearization technique introduced in Sect. 2.5 allows to prove that if the bisectors of a diagram belong to a certain class of bisectors, the distance functions defining the diagram can be chosen among a precise class of functions. For instance, affine diagrams are identified with power diagrams, spherical diagrams are identified with Möbius diagrams, and quadratic diagrams with anisotropic diagrams.

In Sect. 2.6, we introduce the incremental paradigm for constructing various diagrams. Under some topological conditions to be satisfied by the diagram, the incremental construction is efficient. The algorithm can be further improved by using a randomized data structure called the Voronoi hierarchy that allows fast localization of new objects. We then obtain fast randomized incremental algorithms for affine diagrams in any dimension and several non-affine diagrams in the plane. Going beyond those simple cases is difficult. As mentioned above, tight combinatorial bounds and efficient algorithms are lacking even for simple cases. Moreover, the numerical issues are delicate and robust implementations are still far ahead of the state of the art. This motivates the quest for approximate solutions.

In Sect. 2.7, we introduce the concept of Medial Axis of a bounded set Ω, which can be seen as an extension of the notion of Voronoi diagram to infinite sets. Interestingly, it is possible to construct certified approximations of the medial axis of quite general sets efficiently. One approach to be described consists in sampling the boundary of Ω and then computing an appropriate subset of the Voronoi diagram of the sample that approximates the medial axis. Hence the problem of approximating the medial axis of Ω boils down to sampling the boundary of Ω, a problem that is closely related to mesh generation (see Chap. 5).

Sect. 2.8 is devoted to the main CGAL software packages for computing Voronoi diagrams. Sect. 2.9 discusses some applications of curved Voronoi diagrams.

This chapter focuses on curved Voronoi diagrams defined in $\mathbb{R}^d$ and aims at providing useful background and effective algorithms. Additional material can be found in surveys on Voronoi diagrams [276, 37] and in text books on Computational Geometry [110, 67]. This chapter does not consider Voronoi diagrams defined in more general spaces. Voronoi diagrams can be defined in hyperbolic geometry without much difficulty [60, 67]. In the Poincaré model of hyperbolic geometry, the bisectors are hyperspheres and hyperbolic diagrams of finite point sets are a special case of Möbius diagrams. Computing Voronoi diagrams on Riemannian manifolds is much more involved and very few is known about such diagrams and their construction [239].

Notation: We identify a point $x \in \mathbb{R}^d$ and the vector of its coordinates. We note $x \cdot y$ the dot product of x and y, $x^2 = x \cdot x = \|x\|^2$ the squared Euclidean norm of x, and $\|x - y\|$ the Euclidean distance between points x and y.

We call *hypersurface* a manifold of codimension 1. Examples to be used in this chapter are hyperplanes, hyperspheres and quadratic hypersurfaces.

2.2 Lower Envelopes and Minimization Diagrams

Let $\mathcal{F} = \{f_1, \ldots, f_n\}$ be a set of d-variate continuous functions defined over $\mathbb{R}^d$. The *lower envelope* of $\mathcal{F}$ is defined as

$$\mathcal{F}^- = \min_{1 \le i \le n} f_i.$$

From $\mathcal{F}$ and $\mathcal{F}^-$, we define a natural partition of $\mathbb{R}^d$ called the *minimization diagram* of $\mathcal{F}$. For a point $x \in \mathbb{R}^d$, we define the *index set* $I(x)$ of x as the set of all indices i such that $\mathcal{F}^-(x) = f_i(x)$. An equivalence relation noted $\equiv$ can then be defined between two points of $\mathbb{R}^d$ if they have the same index set:

$$x \equiv y \quad \Leftrightarrow \quad I(x) = I(y).$$

The equivalence classes $\mathbb{R}^d / \equiv$ are relatively open sets that cover $\mathbb{R}^d$. Their closures are called the *faces* of the minimization diagram of $\mathcal{F}$ (see Fig. 2.1). The index set of a face is defined as the largest subset of indices common to all the points of the face. Conversely, the face of index set I is the set of all points x such that $I \subset I(x)$.

Observe that the faces of this diagram are not necessarily contractible nor even connected. In particular, a 0-dimensional face may consist of several distinct points.

Lower envelopes and minimization diagrams have been well studied. We recall an important result due to Sharir [311] which provides an almost optimal result when the f_i are supposed to be multivariate polynomials of constant maximum degree.

Theorem 1 (Sharir). *The number of faces of the minimization diagram of a set $\mathcal{F}$ of n multivariate polynomials of constant maximum degree η is $O(n^{d+\varepsilon})$ for any $\varepsilon > 0$, where the constant of proportionality depends on ε, d and η. The vertices, edges and 2-faces of the diagram can be computed in randomized expected time $O(n^{d+\varepsilon})$ for any $\varepsilon > 0$.*

This general result is close to optimal in the worst-case (see Exercise 2). It has been improved in some special cases. For more information and other related results, one should consult the book by Sharir and Agarwal [312].

Voronoi diagrams, in their general setting, are just minimization diagrams of a finite set of continuous functions. This general definition encompasses

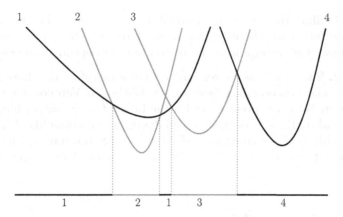

Fig. 2.1. The lower envelope of a set of univariate functions. The minimization diagram is drawn on the horizontal line with the corresponding indices. The face of index $\{1\}$ consists of two components

the more traditional definition of Voronoi diagrams where the functions are defined as distance functions to a finite set of objects. Consider a set of objects $\mathcal{O} = \{o_1, \ldots, o_n\}$. To each object o_i is attached a continuous function δ_i that measures the distance from a point x of $\mathbb{R}^d$ to o_i. In the simplest case, $\mathcal{O}$ is a finite set of points and $\delta_i(x)$ is the Euclidean distance from x to o_i. The Voronoi diagram of $\mathcal{O}$ is defined as the minimization diagram of $\Delta = \{\delta_1, \ldots, \delta_n\}$. The concept of Voronoi diagram has been generalized and various other diagrams have been defined by considering more general objects and other distance functions. Distance is then not to be taken with too much rigor. The function δ_i is only supposed to be continuous.

Theorem 1 provides very general bounds on the complexity of Voronoi diagrams. However, this result calls for improvement. First, in some special cases, much better bounds can be obtained by other approaches to be discussed later in this chapter. In particular, we will see that the most popular Euclidean Voronoi diagram of points has a much smaller combinatorial complexity than the one given in the theorem.

A second issue is the algorithmic complexity. The algorithm mentioned in the theorem fails to provide a complete description of the diagram since only faces of dimensions up to 2 are computed.

Moreover, the implementation of such an algorithm remains a critical issue. As evidenced in Chap. 1, computing lower envelopes of algebraic functions is a formidable task, even in the simplest cases, e.g. quadratic bi-variate functions. We do not know of any implementation for higher degrees and dimensions.

The main goal of the following sections is to present effective algorithms for a variety of Voronoi diagrams for which some additional structure can be exhibited.

Exercise 1. Show that the combinatorial complexity of the lower envelope of n univariate functions whose graphs intersect pairwise in at most two points is $O(n)$. Show that the envelope can be computed in optimal time $\Theta(n \log n)$.

Exercise 2. Show that the convex hull of n ellipsoids of $\mathbb{R}^d$ may have $\Omega(n^{d-1})$ faces. Since the non-bounded faces of the Euclidean Voronoi diagram of n objects are in 1-1 correspondence with the faces of their convex hull, we get a lower bound on the size of the Voronoi diagram of n ellipsoids of $\mathbb{R}^d$. (Hint: consider n ellipsoids inscribed in a $(d-1)$-sphere S and intersecting S along great n $(d-2)$-spheres $\sigma_1, \ldots, \sigma_n$. The arrangement of the σ_i has $\Theta(n^{d-1})$ faces.)

2.3 Affine Voronoi Diagrams

We first introduce Euclidean Voronoi diagrams of points and establish a correspondence between those diagrams and convex polyhedra in one dimension higher. Polarity allows to associate to a Voronoi diagram its dual cell complex, called a Delaunay triangulation.

Almost identical results can be obtained for power (or Laguerre) diagrams where points are replaced by hyperspheres and the Euclidean distance by the power of a point to a hypersphere. Power diagrams constitute a natural extension of Euclidean Voronoi diagrams and are still affine diagrams. In fact, we will see that any affine diagram is the power diagram of a finite set of hyperspheres.

2.3.1 Euclidean Voronoi Diagrams of Points

Let $\mathcal{P} = \{p_1, \ldots, p_n\}$ be a set of points of $\mathbb{R}^d$. To each p_i, we associate its Voronoi region $V(p_i)$

$$V(p_i) = \{x \in \mathbb{R}^d : \|x - p_i\| \leq \|x - p_j\|, \forall j \leq n\}.$$

The region $V(p_i)$ is the intersection of $n-1$ half-spaces. Each such half-space contains p_i and is bounded by the bisector of p_i and some other point of $\mathcal{P}$. Since the bisectors are hyperplanes, $V(p_i)$ is a convex polyhedron, possibly unbounded.

The *Euclidean Voronoi diagram* of $\mathcal{P}$, noted Vor($\mathcal{P}$), is the cell complex whose cells are the Voronoi regions and their faces. Equivalently, the Euclidean Voronoi diagram of $\mathcal{P}$ can be defined as the minimization diagram of the distance functions $\delta_i, \ldots, \delta_n$, where

$$\delta_i(x) = \|x - p_i\|.$$

In other words, the Euclidean Voronoi diagram of $\mathcal{P}$ is the minimization diagram of a set of functions whose graphs are vertical[1] cones of revolution of

---

[1] By vertical, we mean that the axis of revolution is perpendicular to $\mathbb{R}^d$.

$\mathbb{R}^{d+1}$. Since minimizing $\|x - p_i\|$ over i is the same as minimizing $(x - p_i)^2$, the Euclidean Voronoi diagram of $\mathcal{P}$ can alternatively be defined as the mimization diagram of the smooth functions $(x - p_i)^2$ whose graphs are translated copies of a vertical paraboloid of revolution of $\mathbb{R}^{d+1}$.

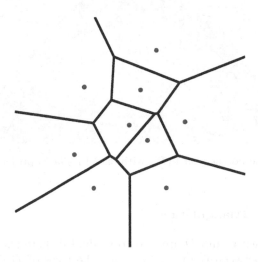

Fig. 2.2. The Voronoi diagram of a set of 9 points

Observing further that, for any x, $\arg\min_i (x - p_i)^2 = \arg\min_i (-2p_i \cdot x + p_i^2)$, we obtain that the Euclidean Voronoi diagram of $\mathcal{P}$ is the minimization diagram of a set of affine functions, namely the functions

$$d_i(x) = -2p_i \cdot x + p_i^2$$

whose graphs are hyperplanes of $\mathbb{R}^{d+1}$. Let us call h_{p_i}, $i = 1, \ldots, n$, those hyperplanes and let $h_{p_i}^-$ denote the half-space lying below h_{p_i}. The minimization diagram of the d_i is obtained by projecting the polyhedron

$$\mathcal{V}(\mathcal{P}) = h_{p_1}^- \cap \cdots \cap h_{p_n}^-.$$

vertically onto $\mathbb{R}^d$. See Fig. 2.3.

We have therefore proved the following theorem:

Theorem 2. *The faces of the Euclidean Voronoi diagram* $\mathrm{Vor}(\mathcal{P})$ *of a set of points* $\mathcal{P}$ *are the vertical projections of the faces of the convex polyhedron* $\mathcal{V}(\mathcal{P})$.

Exercise 3. Consider the maximization diagram obtained by projecting the faces of $h_{p_1}^+ \cap \cdots \cap h_{p_n}^+$ vertically. Characterize the points that belong to a face of this diagram in terms of the distance to the points of $\mathcal{P}$.

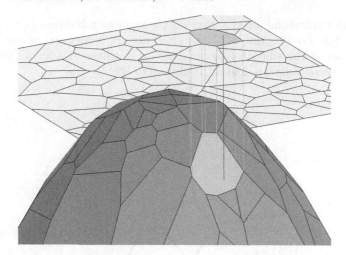

Fig. 2.3. The polyhedron $\mathcal{V}(\mathcal{P})$, with one of its faces projected onto $\mathbb{R}^d$

2.3.2 Delaunay Triangulation

Two cell complexes V and D are said to be dual if there exists an involutive correspondence between the faces of V and the faces of D that reverses the inclusions, i.e. for any two faces f and g of V, their dual faces f^* and g^* satisfy: $f \subset g \Rightarrow g^* \subset f^*$. We introduce now a cell complex that is *dual* to the Voronoi diagram of a finite set of points $\mathcal{P}$.

We assume for now that the set of points $\mathcal{P}$ is in *general position*, which means that no subset of $d+2$ points of $\mathcal{P}$ lie on a same hypersphere. Let f be a face of dimension k of the Voronoi diagram of $\mathcal{P}$. All points in the interior of f have the same subset $\mathcal{P}_f$ of closest points in $\mathcal{P}$. The face dual to f is the convex hull of $\mathcal{P}_f$. The *Delaunay triangulation of* $\mathcal{P}$, noted $\mathrm{Del}(\mathcal{P})$, is the cell complex consisting of all the dual faces. Because points of $\mathcal{P}$ are assumed to be in general position, $|\mathcal{P}_f| = d - k + 1$, all the faces of $\mathrm{Del}(\mathcal{P})$ are simplices and $\mathrm{Del}(\mathcal{P})$ is a simplicial complex. The fact that $\mathrm{Del}(\mathcal{P})$ is indeed a triangulation, i.e. a simplicial complex embedded in $\mathbb{R}^d$ and covering the convex hull of $\mathcal{P}$, will be proved now using a duality between points and hyperplanes in the so-called space of spheres.

Polarity

Let σ be the hypersphere of $\mathbb{R}^d$ of equation

$$\sigma(x) = (x - c)^2 - r^2 = x^2 - 2c \cdot x + s = 0,$$

where c is the center of σ, r its radius and $s = \sigma(0) = c^2 - r^2$.

We define the following bijective mapping

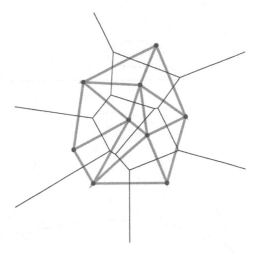

Fig. 2.4. The Delaunay triangulation of a point set (in bold) and its dual Voronoi diagram (thin lines)

$$\phi : \sigma \in \mathbb{R}^d \longrightarrow \phi(\sigma) = (c, -s) \in \mathbb{R}^{d+1}$$

that maps a hypersphere of $\mathbb{R}^d$ to a point of $\mathbb{R}^{d+1}$. We thus consider $\mathbb{R}^{d+1}$ as the images by ϕ of the hyperspheres of $\mathbb{R}^d$ and call $\mathbb{R}^{d+1}$ *the space of spheres*. We note $\phi(p) = (p, -p^2)$ the image by ϕ of a point, considered as a hypersphere of radius 0. Observe that $\phi(p)$ is a point of the paraboloid $\mathcal{Q}$ of $\mathbb{R}^{d+1}$ of equation $x^2 + x_{d+1} = 0$. The points of $\mathbb{R}^{d+1}$ that lie above $\mathcal{Q}$ are images of imaginary hyperspheres whose squared radii are negative. The points below $\mathcal{Q}$ are images of real hyperspheres.

We now introduced a mapping between points and hyperplanes of the space of spheres, known as *polarity*. Polarity associates to the point $\phi(\sigma)$ its *polar hyperplane* h_σ which is the hyperplane of $\mathbb{R}^{d+1}$ of equation $2c \cdot x + x_{d+1} - s = 0$. Observe that the intersection of h_σ with $\mathcal{Q}$ projects vertically onto σ, and that h_σ is the affine hull of the image by ϕ of the points of σ. If p is a point of $\mathbb{R}^d$, the polar hyperplane h_p of $\phi(p)$ is the hyperplane tangent to $\mathcal{Q}$ at $\phi(p)$.

We deduce the remarkable following property: $x \in \sigma$ if and only if $\phi(x) = (x, -x^2) \in h_\sigma$ and σ encloses x if and only if $\phi(x) \in h_\sigma^+$, where h_σ^+ (resp. h_σ^-) denotes the closed half-space above (resp. below) h_σ. Indeed

$$\sigma(x) = 0 \Longleftrightarrow x^2 - 2c \cdot x + s = 0 \Longleftrightarrow \phi(x) \in h_\sigma$$
$$\sigma(x) < 0 \Longleftrightarrow x^2 - 2c \cdot x + s < 0 \Longleftrightarrow \phi(x) \in \text{int } h_\sigma^+,$$

where int h_σ^+ denotes the open half-space above h_σ.

Polarity is an involution that preserves incidences and reverses inclusions. Indeed, if σ and σ' are two hyperspheres, we have

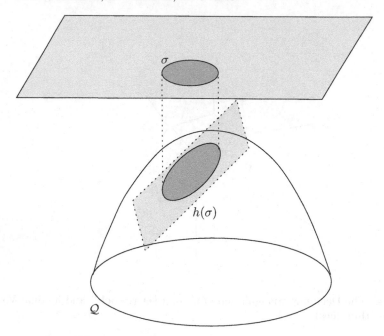

Fig. 2.5. The polar hyperplane of a sphere.

$$\phi(\sigma) \in h_{\sigma'} \iff 2c' \cdot c - s - s' = 0 \iff \phi(\sigma') \in h_\sigma$$
$$\phi(\sigma) \in h_{\sigma'}^+ \iff 2c' \cdot c - s - s' > 0 \iff \phi(\sigma') \in \mathrm{int}\, h_\sigma^+.$$

Consider now a set $\mathcal{P} = \{p_1, \ldots, p_n\}$ of n points and let $\mathcal{V}(\mathcal{P})$ denote, as in Sect. 2.3.1, the convex polyhedron defined as the intersection of the n halfspaces below the n polar hyperplanes $h_{p_1}, \ldots, h_{p_n}$. Let f be a face of $\mathcal{V}(\mathcal{P})$ and assume that f is contained in $k+1$ hyperplanes among the h_{p_i}. Without loss of generality, we denote those hyperplanes $h_{p_1}, \ldots, h_{p_{k+1}}$. Let σ denote a hypersphere of $\mathbb{R}^d$ such that $\phi(\sigma)$ belongs to the relative interior of f. From the above discussion, we have

$$\forall i, \ 1 \le i \le k+1, \ \phi(\sigma) \in h_{p_i} \iff \phi(p_i) \in h_\sigma \tag{2.1}$$
$$\forall i, \ k+1 < i \le n, \ \phi(\sigma) \in \mathrm{int}\, h_{p_i}^- \iff \phi(p_i) \in \mathrm{int}\, h_\sigma^- \tag{2.2}$$

Given a convex polyhedron $\mathcal{D}$, we say that a hyperplane h *supports* $\mathcal{D}$ if $\mathcal{D} \cap h$ is non-empty and $\mathcal{D}$ is included in one of the two halfspaces, h^+ or h^-, bounded by h. If h is a supporting hyperplane of $\mathcal{D}$, $g = \mathcal{D} \cap h$ is a face of $\mathcal{D}$. If $\mathcal{D} \subset h^-$, g is called an upper face of $\mathcal{D}$. The collection of all upper faces of $\mathcal{D}$ constitutes the *upper hull* of $\mathcal{D}$, which we denote by $\partial^+\mathcal{D}$.

Let $\mathcal{D}(\mathcal{P}) = \mathrm{conv}(\phi(\mathcal{P}))$ be the convex hull of the set $\phi(\mathcal{P})$ and consider again the face f of $\mathcal{V}(\mathcal{P})$ defined above. Write $\mathcal{P}_f = \{p_1, \ldots, p_{k+1}\}$. We deduce from (2.1) and (2.2) that, for any $\phi(\sigma)$ in the relative interior of f:

1. The hyperplane h_σ is a supporting hyperplane of $\mathcal{D}(\mathcal{P})$.
2. h_σ supports $\mathcal{D}(\mathcal{P})$ along the face $f^* = h_\sigma \cap \mathcal{D}(\mathcal{P}) = \text{conv}(\phi((\mathcal{P}_f)))$.
3. $\mathcal{D}(\mathcal{P}) \subset h_\sigma^-$ and f^* is a face of $\partial^+ \mathcal{D}(\mathcal{P})$.

To each face f of $\partial \mathcal{V}(\mathcal{P})$, we associate the face f^* of $\partial^+ \mathcal{D}$ obtained as described above. This correspondence between the faces of $\partial \mathcal{V}(\mathcal{P})$ and the faces of $\partial^+ \mathcal{D}(\mathcal{P})$ is bijective, preserves incidences and reverses inclusions, hence it is a duality.

The upper hull $\partial^+ \mathcal{D}(\mathcal{P})$ projects vertically onto a cell complex of $\mathbb{R}^d$ whose vertices are the points of $\mathcal{P}$. Because the projection is 1-1, this projected cell complex is properly embedded in $\mathbb{R}^d$ and, since the projection preserves convexity, it covers the convex hull of $\mathcal{P}$. Under the general position assumption, the convex polyhedron $\mathcal{D}(\mathcal{P})$ is simplicial and the projected complex is a triangulation of $\mathcal{P}$. The duality between the faces of $\partial \mathcal{V}(\mathcal{P})$ and the faces of $\partial^+ \mathcal{D}(\mathcal{P})$ implies that the projection of $\partial^+ \mathcal{D}(\mathcal{P})$ is the Delaunay triangulation $\text{Del}(\mathcal{P})$ of $\mathcal{P}$ introduced at the beginning of this section. This concludes the proof that, under the general position assumption, the Delaunay triangulation $\text{Del}(\mathcal{P})$ is a triangulation of $\mathcal{P}$. We have the following diagram:

$$\partial \mathcal{V}(\mathcal{P}) = \partial \left(h_{p_1}^- \cap \cdots \cap h_{p_n}^- \right) \quad \longleftrightarrow \quad \partial^+ \mathcal{D}(\mathcal{P}) = \partial^+ \left(\text{conv}(\phi(\mathcal{P})) \right)$$
$$\updownarrow \qquad\qquad\qquad\qquad\qquad\qquad\qquad \updownarrow$$
$$\text{Voronoi Diagram } \text{Vor}(\mathcal{P}) \quad \longleftrightarrow \quad \text{Delaunay Triangulation } \text{Del}(\mathcal{P})$$

It follows from the above correspondence that the combinatorial complexity of the Delaunay triangulation of n points is the same as the combinatorial complexity of its dual Voronoi diagram. Moreover, the Delaunay triangulation of n points of $\mathbb{R}^d$ can be deduced from the dual Voronoi diagram or vice versa in time proportional to its size. We also deduce from what precedes that computing the Delaunay triangulation of n points of $\mathbb{R}^d$ reduces to constructing the convex hull of n points of $\mathbb{R}^{d+1}$. The following theorem is then a direct consequence of known results on convex hulls [84].

Theorem 3. *The combinatorial complexity of the Voronoi diagram of n points of $\mathbb{R}^d$ and of their Delaunay triangulation is $\Theta\left(n^{\lfloor \frac{d+1}{2} \rfloor}\right)$. Both structures can be computed in optimal time $\Theta\left(n \log n + n^{\lfloor \frac{d+1}{2} \rfloor}\right)$.*

The bounds in this theorem are tight. In particular, the Voronoi diagram of n points of $\mathbb{R}^3$ may be quadratic (see Exercise 4). These bounds are worst-case bounds. Under some assumptions on the point distribution, better bounds can be obtained. For a set $\mathcal{P}$ of n points uniformly distributed in a ball of $\mathbb{R}^d$, the combinatorial complexity of the Voronoi diagram of $\mathcal{P}$ is $O(n)$ where the constant depends on the dimension d [129]. Other results are known for other point distributions [31, 33, 151].

In the discussion above, we have assumed that the points of $\mathcal{P}$ were in general position. If this is not the case, some faces of $\mathcal{D}(\mathcal{P})$ are not simplices, and the complex $\partial^+ \mathcal{D}(\mathcal{P})$ projects vertically onto a cell complex, dual to the

Voronoi diagram and called the *Delaunay complex*. The faces of the *Delaunay complex* are convex and any triangulation obtained by triangulating those faces is called a Delaunay triangulation. Since there are several ways of triangulating the faces of the Delaunay complex, the Delaunay triangulation of $\mathcal{P}$ is no longer unique.

Exercise 4. Show that if we take points on two non coplanar lines of $\mathbb{R}^3$, say $n_1 + 1$ on one of the lines and $n_2 + 1$ on the other, their Voronoi diagram has $n_1 n_2$ vertices.

Exercise 5. Let S be a hypersphere of $\mathbb{R}^d$ passing through $d + 1$ points $p_0, \ldots, p_d$. Show that a point p_{d+1} of $\mathbb{R}^d$ lies on S, in the interior of the ball B_S bounded by S or outside B_S, depending whether the determinant of the $(d + 2) \times (d + 2)$ matrix

$$\texttt{in\_sphere}(p_0, \ldots, p_{d+1}) = \begin{vmatrix} 1 & \cdots & 1 \\ p_0 & \cdots & p_{d+1} \\ p_0^2 & \cdots & p_{d+1}^2 \end{vmatrix}$$

is 0, negative or positive. This predicate is the only numerical operation that is required to check if a triangulation is a Delaunay triangulation.

Exercise 6. What are the preimages by ϕ of the points of $\mathbb{R}^{d+1}$ that lie on a line? (Distinguish the cases where the line intersects $\mathcal{Q}$ in 0, 1 or 2 points.)

Exercise 7. Project vertically the faces of the lower hull $\partial^-(\mathcal{D}(\mathcal{P})$. Show that we obtain a triangulation of the vertices of $\text{conv}(\mathcal{P})$ such that each ball circumscribing a simplex contains all the points of $\mathcal{P}$. Define a dual and make a link with Exercise 3.

Exercise 8 (Empty sphere property). Let s be any k-simplex with vertices in $\mathcal{P}$ that can be circumscribed by a a hypersphere that does not enclose any point of $\mathcal{P}$. Show that s is a face of a Delaunay triangulation of $\mathcal{P}$. Moreover, let $\mathcal{P}$ be a set of points and $\mathcal{T}$ a triangulation of $\mathcal{P}$ with the property that any hypersphere circumscribing a d-simplex of $\mathcal{T}$ does not enclose any point of $\mathcal{P}$. Show that $\mathcal{T}$ is a Delaunay triangulation of $\mathcal{P}$.

2.3.3 Power Diagrams

A construction similar to what we did for the Euclidean Voronoi diagrams of points and their dual Delaunay triangulations can be done for the so-called power or Laguerre diagrams. Here we take as our finite set of objects a set of hyperspheres (instead of points) and consider as distance function of a point x to a hypersphere σ the power of x to σ. As we will see, the class of power diagrams is identical to the class of affine diagrams, i.e. the diagrams whose bisectors are hyperplanes.

Definition of Power Diagrams

We call *power* of a point x to a hypersphere σ of center c and radius r the real number

$$\sigma(x) = (x - c)^2 - r^2.$$

Let $\mathcal{S} = \{\sigma_1, \ldots, \sigma_n\}$ be a set of hyperspheres of $\mathbb{R}^d$. We denote by c_i the center of σ_i, r_i its radius, $\sigma_i(x) = (x - c_i)^2 - r_i^2$ the power function to σ_i, and $s_i = c_i^2 - r_i^2$ the power of the origin. To each σ_i, we associate the region $L(\sigma_i)$ consisting of the points of $\mathbb{R}^d$ whose power to σ_i is not larger than their power to the other hyperspheres of $\mathcal{S}$:

$$L(\sigma_i) = \{x \in \mathbb{R}^d : \sigma_i(x) \le \sigma_j(x), 1 \le j \le n\}.$$

The set of points that have equal power to two hyperspheres σ_i and σ_j is a hyperplane, noted π_{ij}, called *the radical hyperplane* of σ_i and σ_j. Hyperplane π_{ij} is orthogonal to the line joining the centers of σ_i and σ_j. We denote by π_{ij}^i the half-space bounded by π_{ij} consisting of the points whose power to σ_i is smaller than their power to σ_j. The region $L(\sigma_i)$ is the intersection of all half-spaces π_{ij}^i, $j \ne i$. If this intersection is not empty, it is a convex polyhedron, possibly not bounded. We call *power regions* the non empty regions $L(\sigma_i)$.

We define the *power diagram* of $\mathcal{S}$, noted $\mathrm{Pow}(\mathcal{S})$, as the cell complex whose cells are the power regions and their faces. When all hyperspheres have the same radius, their power diagram is identical to the Voronoi diagram of their centers.

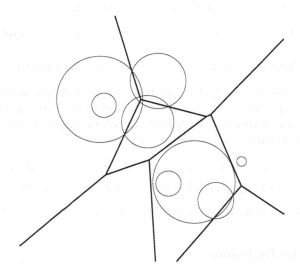

Fig. 2.6. A power diagram

Equivalently, the power diagram of $\mathcal{S}$ can be defined as the minimization diagram of the functions $\sigma_i, \ldots, \sigma_n$. Observing that for any x

$$\arg\min_i \sigma_i(x) = \arg\min_i(-2c_i \cdot x + s_i),$$

we obtain that the power diagram of $\mathcal{S}$ is the minimization diagram of the set of affine functions

$$d_i(x) = -2p_i \cdot x + s_i$$

whose graphs are hyperplanes of $\mathbb{R}^{d+1}$. Let us call h_{σ_i}, $i = 1, \ldots, n$, those hyperplanes and let $h_{\sigma_i}^-$ denote the half-space lying below h_{σ_i}. The minimization diagram of the δ_i is obtained by projecting vertically the convex polyhedron

$$\mathcal{L}(\mathcal{S}) = h_{p_1}^- \cap \cdots \cap h_{p_n}^-.$$

Theorem 4. *The faces of the power diagram* $\mathrm{Pow}(\mathcal{S})$ *of* $\mathcal{S}$ *are the vertical projections of the faces of the convex polyhedron* $\mathcal{L}(\mathcal{S})$.

Power diagrams are very similar to Voronoi diagrams: the only difference is that the hyperplanes supporting the faces of $\mathcal{L}(\mathcal{S})$ are not necessarily tangent to the paraboloid $\mathcal{Q}$ and that some hyperplane may not contribute a face. In other words, some hypersphere σ_i may have an empty power region (see the small circle in the upper left corner of Fig. 2.6).

By proceeding as in Sect. 2.3.2, we can define a convex polyhedron $\mathcal{R}(\mathcal{S})$ whose upper hull $\partial^+\mathcal{R}(\mathcal{S})$ is dual to $\partial\mathcal{L}(\mathcal{S})$. The vertical projection of the faces of $\partial^+\mathcal{R}(\mathcal{S})$ constitute the faces of a cell complex which, in general, is a simplicial complex. We call such a complex the *regular triangulation* of $\mathcal{S}$ and denote it by $\mathrm{Reg}(\mathcal{S})$. We have the following diagram:

$$
\begin{array}{ccc}
\partial\mathcal{L}(\mathcal{S}) = \partial\left(h_{\sigma_1}^- \cap \cdots \cap h_{\sigma_n}^-\right) & \longleftrightarrow & \partial^+\mathcal{R}(\mathcal{S}) = \partial^+\mathrm{conv}(\phi(\mathcal{S})) \\
\updownarrow & & \updownarrow \\
\text{Power diagram } \mathrm{Pow}(\mathcal{S}) & \longleftrightarrow & \text{Regular triangulation } \mathrm{Reg}(\mathcal{S})
\end{array}
$$

We deduce the following theorem that states that computing the power diagram of n hyperspheres of $\mathbb{R}^d$ (or equivalently its dual regular triangulation) has the same asymptotic complexity as computing the Euclidean Voronoi diagram or the Delaunay triangulation of n points of $\mathbb{R}^d$.

Theorem 5. *The combinatorial complexity of the power diagram of* n *hyperspheres of* $\mathbb{R}^d$ *and of its dual regular triangulation are* $\Theta\left(n^{\lfloor\frac{d+1}{2}\rfloor}\right)$. *Both structures can be computed in optimal time* $\Theta\left(n\log n + n^{\lfloor\frac{d+1}{2}\rfloor}\right)$.

Affine Voronoi Diagrams

Euclidean Voronoi diagrams of points and power diagrams of hyperspheres are two examples of minimization diagrams whose bisectors are hyperplanes. It is interesting to classify Voronoi diagrams with respect to their bisectors. A first important class of Voronoi diagrams is the class of *affine diagrams* which consists of all Voronoi diagrams whose bisectors are hyperplanes.

In Sect. 2.5, we will prove that any affine Voronoi diagram of $\mathbb{R}^d$ is identical to the power diagram of some set of hyperspheres of $\mathbb{R}^d$ (Theorem 13), therefore showing that the class of affine Voronoi diagrams is identical to the class of power diagrams.

Exercise 9. Show that the intersection of a power diagram with an affine subspace is still a power diagram and compute the corresponding spheres.

Exercise 10. Show that any power diagram of $\mathbb{R}^d$ is the intersection of a Voronoi diagram of $\mathbb{R}^{d+1}$ by a hyperplane.

Exercise 11. Show that the only numerical operation that is required to check if a triangulation is the regular triangulation of a set of hyperspheres σ_i is the evaluation of the sign of the determinant of the $(d+2) \times (d+2)$ matrix

$$
\text{power\_test}(\sigma_0, \ldots, \sigma_{d+1}) = \begin{vmatrix} 1 & \cdots & 1 \\ c_0 & \cdots & c_{d+1} \\ c_0^2 - r_0^2 & \cdots & c_{d+1}^2 - r_{d+1}^2 \end{vmatrix}
$$

where c_i and r_i are respectively the center and the radius of σ_i.

2.4 Voronoi Diagrams with Algebraic Bisectors

In this section, we introduce a first class of non-affine diagrams, namely the class of diagrams whose bisectors are algebraic hypersurfaces. We first consider the case of Möbius diagrams whose bisectors are hyperspheres and the case of *anisotropic* diagrams whose bisectors are quadratic hypersurfaces. These diagrams can be computed through linearization, a technique to be described in full generality in Sect. 2.5. Apollonius (or Johnson-Mehl) diagrams, although semi-algebraic and not algebraic, are also described in this section since they are closely related to Möbius diagrams and can also be linearized.

2.4.1 Möbius Diagrams

In this section, we introduce a class of non-affine Voronoi diagrams, the so-called Möbius diagrams, introduced by Boissonnat and Karavelas [63].

The class of Möbius diagrams includes affine diagrams. In fact, as we will see, the class of Möbius diagrams is identical to the class of diagrams whose bisectors are hyperspheres (or hyperplanes).

Definition of Möbius Diagrams

Let $\omega = \{\omega_1, \ldots, \omega_n\}$ be a set of so-called *Möbius sites* of $\mathbb{R}^d$, where ω_i is a triple (p_i, λ_i, μ_i) formed of a point p_i of $\mathbb{R}^d$, and two real numbers λ_i and μ_i.

For a point $x \in \mathbb{R}^d$, the distance $\delta_i(x)$ from x to the Möbius site ω_i is defined as

$$\delta_i(x) = \lambda_i(x - p_i)^2 - \mu_i.$$

Observe that the graph of δ_i is a paraboloid of revolution whose axis is vertical. The Möbius region of the Möbius site ω_i, $i = 1, \ldots, n$, is

$$M(\omega_i) = \{x \in \mathbb{R}^d : \delta_i(x) \leq \delta_j(x), 1 \leq j \leq n\}.$$

Observe that a Möbius region may be non-contractible and even disconnected.

The minimization diagram of the δ_i is called the *Möbius diagram* of ω and noted Möb (ω) (see Fig. 2.4.1).

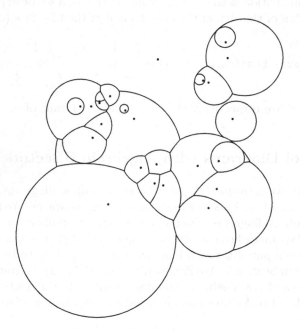

Fig. 2.7. A Möbius diagram

Möbius diagrams are generalizations of Euclidean Voronoi and power diagrams. In particular, if all λ_i are equal to some positive λ, the Möbius diagram coincides with the power diagram of a set of spheres $\{\sigma_i, i = 1, \ldots, n\}$, where σ_i is the sphere centered at p_i of squared radius μ_i/λ. If all μ_i are equal and all λ_i are positive, then the Möbius diagram coincides with the so-called *multiplicatively weighted Voronoi diagram* of the weighted points $(p_i, \sqrt{\lambda_i})$.

The following lemma states that the bisector of two Möbius sites is a hypersphere (possibly degenerated in a point or in a hyperplane). Its proof is straightforward.

Lemma 1. *Let $\omega_i = \{p_i, \lambda_i, \mu_i\}$ and $\omega_j = \{p_j, \lambda_j, \mu_j\}$, $\omega_i \neq \omega_j$ be two Möbius sites. The bisector σ_{ij} of ω_i and ω_j is the empty set, a single point, a hypersphere or a hyperplane.*

Möbius Diagrams and Power Diagrams

We now present an equivalence between Möbius diagrams in $\mathbb{R}^d$ and power diagrams in $\mathbb{R}^{d+1}$. This result is a direct generalization of a similar result for multiplicatively weighted diagrams [35]. Given a cell complex $\mathcal{C}$ covering a subspace X, we call *restriction of $\mathcal{C}$ to X* the subdivision of X whose faces are the intersections of the faces of $\mathcal{C}$ with X. The restriction of $\mathcal{C}$ to X is denoted by $\mathcal{C}_X$. Note that the restriction $\mathcal{C}_X$ is not, in general, a cell complex and that its faces may be non-contractible and even non-connected.

We associate to $\mathcal{W} = \{\omega_1, \ldots, \omega_n\}$ the set of hyperspheres $\Sigma = \{\Sigma_1, \ldots, \Sigma_n\}$ of $\mathbb{R}^{d+1}$ of equations

$$\Sigma_i(X) = X^2 - 2C_i \cdot X + s_i = 0,$$

where $C_i = (\lambda_i p_i, -\frac{\lambda_i}{2})$ and $s_i = \lambda_i\, p_i^2 - \mu_i$. We denote by $\mathcal{Q}$ the paraboloid of $\mathbb{R}^{d+1}$ of equation $x_{d+1} - x^2 = 0$.

Theorem 6 (Linearization). *The Möbius diagram Möb($\mathcal{W}$) of $\mathcal{W}$ is obtained by projecting vertically the faces of the restriction $\mathrm{Pow}_{\mathcal{Q}}(\Sigma)$ of the power diagram of Σ to $\mathcal{Q}$.*

Proof. If $x \in \mathbb{R}^d$ is closer to ω_i than to ω_j with respect to δ_M, we have for all $j = 1, \ldots, n$,

$$\lambda_i(x - p_i)^2 - \mu_i \leq \lambda_j(x - p_j)^2 - \mu_j$$
$$\Longleftrightarrow \lambda_i x^2 - 2\lambda_i p_i \cdot x + \lambda_i p_i^2 - \mu_i \leq \lambda_j x^2 - 2\lambda_j p_j \cdot x + \lambda_j p_j^2 - \mu_j$$
$$\Longleftrightarrow (x^2 + \tfrac{\lambda_i}{2})^2 + (x - \lambda_i p_i)^2 - \tfrac{\lambda_i^2}{4} - \lambda_i^2 p_i^2 + \lambda_i p_i^2 - \mu_i$$
$$\leq (x^2 + \tfrac{\lambda_j}{2})^2 + (x - \lambda_j p_j)^2 - \tfrac{\lambda_j^2}{4} - \lambda_j^2 p_j^2 + \lambda_j p_j^2 - \mu_j$$
$$\Longleftrightarrow (X - C_i)^2 - r_i^2 \leq (X - C_j)^2 - r_j^2$$
$$\Longleftrightarrow \Sigma_i(X) \leq \Sigma_j(X)$$

where $X = (x, x^2) \in \mathcal{Q} \subset \mathbb{R}^{d+1}$, $C_i = (\lambda_i p_i, -\frac{\lambda_i}{2}) \in \mathbb{R}^{d+1}$ and $r_i^2 = \lambda_i^2 p_i^2 + \frac{\lambda_i^2}{4} - \lambda_i p_i^2 + \mu_i$. The above inequality shows that x is closer to ω_i than to ω_j if and only if X belongs to the power region of Σ_i in the power diagram of the hyperspheres Σ_j, $j = 1, \ldots, n$. As X belongs to $\mathcal{Q}$ and projects vertically onto x, we have proved the result.

Corollary 1. *Let Σ be a finite set of hyperspheres of $\mathbb{R}^{d+1}$, $\mathrm{Pow}(\Sigma)$ its power diagram and $\mathrm{Pow}_{\mathcal{Q}}(\Sigma)$ the restriction of $\mathrm{Pow}(\Sigma)$ to $\mathcal{Q}$. The vertical projection of $\mathrm{Pow}_{\mathcal{Q}}(\Sigma)$ is the Möbius diagram Möb($\mathcal{W}$) of a set of Möbius sites of $\mathbb{R}^d$.*

Easy computations give $\mathcal{W}$.

Combinatorial and Algorithmic Properties

It follows from Theorem 6 that the combinatorial complexity of the Möbius diagram of n Möbius sites in $\mathbb{R}^d$ is $O(n^{\lfloor \frac{d}{2} \rfloor + 1})$. This bound is tight since Aurenhammer [35] has shown that it is tight for multiplicatively weighted Voronoi diagrams.

We easily deduce from the proof of the Linearization Theorem 6 an algorithm for constructing Möbius diagrams. First, we compute the power diagram of the hyperspheres Σ_i of $\mathbb{R}^{d+1}$, intersect each of the faces of this diagram with the paraboloid $\mathcal{Q}$ and then project the result on $\mathbb{R}^d$.

Theorem 7. *Let $\mathcal{W}$ be a set of n Möbius sites in $\mathbb{R}^d$, $d \geq 2$. The Möbius diagram $\mathrm{M\ddot{o}b}(\mathcal{W})$ of $\mathcal{W}$ can be constructed in worst-case optimal time $\Theta(n \log n + n^{\lfloor \frac{d}{2} \rfloor + 1})$.*

Another consequence of the linearization theorem is the fact that any Möbius diagram can be represented as a simplicial complex $T_{\mathcal{Q}}$ embedded in $\mathbb{R}^{d+1}$. $T_{\mathcal{Q}}$ is a sub-complex of the regular triangulation T dual to the power diagram $\mathrm{Pow}(\Sigma_i)$ of the hyperspheres Σ_i. Since T is embedded in $\mathbb{R}^{d+1}$, $T_{\mathcal{Q}}$ is a simplicial complex of $\mathbb{R}^{d+1}$. More precisely, $T_{\mathcal{Q}}$ consists of the faces of T that are dual to the faces of $\mathrm{Pow}_{\mathcal{Q}}(\Sigma)$, i.e. the faces of the power diagram that intersect $\mathcal{Q}$. We will call $T_{\mathcal{Q}}$ the *dual* of $\mathrm{Pow}_{\mathcal{Q}}(\Sigma)$. Observe that since, in general, no vertex of $\mathrm{Pow}(\Sigma)$ lies on $\mathcal{Q}$, $T_{\mathcal{Q}}$ is a d-dimensional simplicial complex (embedded in $\mathbb{R}^{d+1}$).

Moreover, if the faces of $\mathrm{Pow}(\Sigma)$ intersect $\mathcal{Q}$ transversally and along topological balls, then, by a result of Edelsbrunner and Shah [138], $T_{\mathcal{Q}}$ is homeomorphic to $\mathcal{Q}$ and therefore to $\mathbb{R}^d$. It should be noted that this result states that the simplicial complex $T_{\mathcal{Q}}$ has the topology of $\mathbb{R}^d$. This result, however, is mainly combinatorial, and does not imply that the embedding of $T_{\mathcal{Q}}$ into $\mathbb{R}^{d+1}$ as a sub-complex of the regular triangulation T may be projected in a 1-1 manner onto $\mathbb{R}^d$.

Spherical Voronoi Diagrams

Lemma 1 states that the bisectors of two Möbius sites is a hypersphere (possibly degenerated in a hyperplane). More generally, let us consider the Voronoi diagrams such that, for any two objects o_i and o_j of $\mathcal{O}$, the bisector $\sigma_{ij} = \{x \in \mathbb{R}^d, \delta_i(x) = \delta_j(x)\}$ is a hypersphere. Such a diagram is called a *spherical Voronoi diagram.*

In Sect. 2.5, we will prove that any spherical Voronoi diagram of $\mathbb{R}^d$ is a Möbius diagram (Theorem 15).

Möbius transformations are the transformations that preserve hyperspheres. An example of a Möbius transformation is the inversion with respect to a hypersphere. If the hypersphere is centered at c and has radius r, the inversion associates to a point $x \in \mathbb{R}^d$ its image

$$x' = c + \frac{r(x - c)}{(x - c)^2}.$$

Moreover, it is known that any Möbius transformation is the composition of up to four inversions [108]. An immediate consequence of Theorem 15 is that the set of Möbius diagrams in $\mathbb{R}^d$ is stable under Möbius transformations, hence their name.

Möbius Diagrams on Spheres

Given a set $\mathcal{W}$ of n Möbius sites of $\mathbb{R}^{d+1}$, the restriction of their Möbius diagram to a hypersphere $\mathbb{S}^d$ is called a Möbius diagram on $\mathbb{S}^d$. It can be shown that such a diagram is also the restriction of a power diagram of hyperspheres of $\mathbb{R}^{d+1}$ to $\mathbb{S}^d$ (Exercise 14) .

We define spherical diagrams on $\mathbb{S}^d$ as the diagrams on $\mathbb{S}^d$ whose bisectors are hyperspheres of $\mathbb{S}^d$ and that satisfy two properties detailed in Sect. 2.5.1 and 2.5.2. See Exercise 16 for more details on these conditions. This exercise proves that the restriction of a Möbius diagram, i.e. a Möbius diagram on $\mathbb{S}^d$, is a spherical diagram.

Let us now prove the converse: any spherical diagram on $\mathbb{S}^d$ is a Möbius diagram on $\mathbb{S}^d$. Let h be a hyperplane of $\mathbb{R}^{d+1}$. The stereographic projection that maps $\mathbb{S}^d$ to h maps any spherical diagram $\mathcal{D}$ on $\mathbb{S}^d$ to some spherical diagram $\mathcal{D}'$ on h. Theorem 15 implies that this $\mathcal{D}'$ is in fact a Möbius diagram. Exercise 12 shows that $\mathcal{D}$, which is the image of $\mathcal{D}'$ by the inverse of the stereographic projection, is the restriction of some power diagram of $\mathbb{R}^{d+1}$ to $\mathbb{S}^d$. Exercise 14 then proves that it is indeed a Möbius diagram.

Exercise 12. Show that the linearization theorem and its corollary still hold if one replaces the paraboloid $\mathcal{Q}$ by any hypersphere of $\mathbb{R}^{d+1}$ and the vertical projection by the corresponding stereographic projection.

Exercise 13. Show that the intersection of a Möbius diagram in $\mathbb{R}^d$ with a k-flat or a k-sphere σ is a Möbius diagram in σ.

Exercise 14. Show that the restriction of a Möbius diagram of n Möbius sites to a hypersphere $\Sigma \subset \mathbb{R}^d$ (i.e. a Möbius diagram on Σ) is identical to the restriction of a power diagram of n hyperspheres of $\mathbb{R}^d$ with Σ, and vice versa.

Exercise 15. The predicates needed for constructing a Möbius diagram are those needed to construct $\mathrm{Pow}(\Sigma)$ and those that decide whether a face of $\mathrm{Pow}(\Sigma)$ intersects $\mathcal{Q}$ or not. Write the corresponding algebraic expressions.

Exercise 16. Explain how the two conditions A.C. and L.C.C. presented in Sect. 2.5.1 and 2.5.2 are to be adapted to the case of spherical diagrams on a sphere (Hint: consider the L.C.C. condition as a pencil condition, and define a pencil of circles on a sphere as the intersection of a pencil of hyperplanes with this sphere).

Note that the restriction of a Voronoi diagram (affine or not) to $\mathbb{S}^d$ always satisfies this adapted version of A.C. and prove that the restriction of an affine Voronoi diagram to $\mathbb{S}^d$ satisfies L.C.C. so that Exercise 14 allows to conclude that the restriction of a Möbius diagram to $\mathbb{S}^d$, i.e. Möbius diagram on $\mathbb{S}^d$, is a spherical diagram.

2.4.2 Anisotropic Diagrams

The definition of anisotropic Voronoi diagrams presented in this section is a slight extension of the definition proposed by Labelle and Shewchuk [234]. The objects are points and the distance to a point is a quadratic form with an additive weight.

Anisotropic diagrams appear to be a natural generalization of Möbius diagrams and reduce to Möbius diagrams when the matrices are taken to be a scalar times the identity matrix. As will be shown, the class of anisotropic diagrams is identical to the class of diagrams whose bisectors are quadratic hypersurfaces.

Definition and linearization

Consider a finite set of anisotropic sites $S = \{s_1, \ldots, s_n\}$. Each site s_i, $i = 1, \ldots, n$, is a triple (p_i, M_i, π_i) formed by a point $p_i \in \mathbb{R}^d$, a $d \times d$ symmetric positive definite matrix M_i and a scalar weight π_i. The distance $\delta_i(x)$ of point $x \in \mathbb{R}^d$ to site s_i is defined by

$$\delta_i(x) = (x - p_i)^t M_i (x - p_i) - \pi_i.$$

The anisotropic Voronoi region of site s is then defined as

$$AV(s_i) = \{x \in \mathbb{R}^d, \delta_i(x) \leq \delta_j(x), \forall 1 \leq j \leq n\},$$

The *anisotropic Voronoi diagram* is the minimization diagram of the functions $\delta_i(x)$.

Let $D = \frac{d(d+3)}{2}$. To each point $x = (x_1, \ldots, x_d) \in \mathbb{R}^d$, we associate the two points

$$\tilde{\phi}(x) = (x_r x_i, 1 \leq r \leq s \leq d) \in \mathbb{R}^{\frac{d(d+1)}{2}}$$
$$\hat{\phi}(x) = (x, \tilde{\phi}(x)) \in \mathbb{R}^D,$$

and we denote by $\mathcal{Q}$ the d-manifold of $\mathbb{R}^D$ defined as

$$\mathcal{Q} = \left\{ \hat{\phi}(x), \ x \in \mathbb{R}^d \right\}.$$

To each site $s_i = (p_i, M_i, \pi_i) \in S$, we associate:

1. the point $\tilde{m}_i \in \mathbb{R}^{\frac{d(d+1)}{2}}$ defined as

$$\tilde{m}_i^{u,u} = -\frac{1}{2} M_i^{u,u}, \quad \text{for } 1 \le u \le d;$$
$$\tilde{m}_i^{u,v} = -M_i^{u,v}, \quad \text{for } 1 \le u < v \le d,$$

2. the point $\hat{p}_i = (M_i p_i, \tilde{m}_i)$,
3. the sphere σ_i of center $\hat{p}_i$ and radius $\sqrt{\|\hat{p}_i\|^2 - p_i^t M_i p_i - \pi_i}$.

Let Π be the projection $\hat{y} = (y, \tilde{y}) \in \mathbb{R}^D \mapsto y \in \mathbb{R}^d$ and let Σ be the set of spheres σ_i, $i = 1, \ldots, n$.

Theorem 8 (Linearization). *The anisotropic diagram of S is the image by Π of the restriction of the power diagram $\mathrm{Pow}(\Sigma)$ to the d-manifold $\mathcal{Q}$.*

Proof. We have

$$\begin{aligned} \delta_i(x) &= (x - p_i)^t M (x - p_i) - \pi_i \\ &= x^t M_i x - 2 p_i^t M_i x + p_i^t M_i p_i - \pi_i \\ &= -2 \hat{p}_i^t \hat{\phi}(x) + p_i^t M_i p_i - \pi_i \end{aligned}$$

This implies that $\delta_i(x) < \delta_j(x)$ if and only if

$$(\hat{\phi}(x) - \hat{p}_i)^2 - (\hat{p}_i^2 - p_i^t M_i p_i - \pi_i) < (\hat{\phi}(x) - \hat{p}_j)^2 - (\hat{p}_j^2 - p_j^t M_j p_j - \pi_j)$$

Hence, x is closer to s_i than to s_j if and only if the power of $\hat{\phi}(x)$ to σ_i is smaller than its power to σ_j. Equivalently, a point $\hat{\phi}(x) \in \mathcal{Q}$ belongs to the power cell of $\sigma(s_i)$ if and only if its projection $x = \Pi(\hat{\phi}(x))$ belongs to the anisotropic Voronoi region $AV(s_i)$.

We easily deduce the following theorem.

Theorem 9. *The Voronoi diagram of n anisotropic sites of $\mathbb{R}^d$ can be computed in time $O(n^{\lfloor \frac{D+1}{2} \rfloor})$ where $D = \frac{d(d+3)}{2}$.*

This result is to be compared to Theorem 1 which provides a better combinatorial bound. We let as an open question to fill the gap between those two bounds.

Quadratic Voronoi Diagrams

The bisectors of anisotropic diagrams, as defined in the previous section, are quadratic hypersurfaces. A minimization diagram whose bisectors are hyperquadrics is called a *quadratic Voronoi diagram*. In Sect. 2.5, we will prove that any quadratic Voronoi diagram is the anisotropic Voronoi diagram of a set of anisotropic sites (Theorem 16).

2.4.3 Apollonius Diagrams

In this section, we present diagrams that are closely related to Möbius diagrams: namely, the Euclidean Voronoi diagrams of hyperspheres, also called Apollonius or Johnson-Mehl diagrams. Contrary to Möbius and anisotropic diagrams, the bisectors of Apollonius diagrams are not algebraic hypersurfaces since the bisector between two hyperspheres is only one sheet of a hyperboloid. As a consequence, Apollonius diagrams cannot be linearized in the same way as Möbius and anisotropic diagrams. Nevertheless, another linearization scheme can be applied, leading to interesting combinatorial and algorithmic results.

Definition of Apollonius Diagrams

Let us consider a finite set of weighted points $S = \{\sigma_0, \sigma_1, \ldots, \sigma_n\}$ where $\sigma_i = (p_i, r_i)$, $p_i \in \mathbb{R}^d$ and $r_i \in \mathbb{R}$. We define the distance from x to σ_i as

$$\delta_i(x) = \|x - p_i\| - r_i.$$

This distance is also called the *additively weighted* distance from x to the weighted point σ_i. The minimization diagram of the distance functions δ_i, $i = 1, \ldots, n$, is called the additively weighted Voronoi diagram, or the Apollonius diagram of S. We denote it by $\text{Apo}(S)$ (see Fig. 2.8).

The Apollonius region $A(\sigma_i)$ of σ_i is defined as

$$A(\sigma_i) = \{x \in \mathbb{R}^d, \delta_i(x) \leq \delta_j(x)\}.$$

It is easy to see that $A(\sigma_i)$ is either empty or star-shaped from p_i. The boundary of $A(\sigma_i)$ may have a complicated structure. In fact, as we will see, the boundary of $A(\sigma_i)$ has the same combinatorial structure as a Möbius diagram in $\mathbb{R}^{d-1}$.

Since the diagram is not changed if we replace all r_i by $r_i + r$ for any $r \in \mathbb{R}$, we can assume, without loss of generality, that all r_i are non negative. The weighted points are then hyperspheres and the distance to a weighted point is the signed Euclidean distance to the corresponding hypersphere, counted positively outside the hypersphere and negatively inside the hypersphere.

Observe that, in the plane, a vertex of an Apollonius diagram is the center of a circle tangent to three circles of S (assuming all r_i non negative). Computing such a point is known as Apollonius' Tenth Problem, hence the name of the diagram.

Apollonius Diagrams and Power Diagrams

The graph of the distance function $\delta_i(x)$ is the half-cone of revolution C_i of equation

$$C_i : x_{d+1} = \|x - p_i\| - r_i, \quad x_{d+1} + r_i \geq 0$$

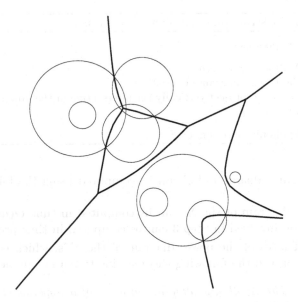

Fig. 2.8. The Apollonius diagram of a set of circles. Compare with the power diagram of the same set of circles in Fig. 2.6

The bisector of two hyperspheres of S is thus the projection of the intersection of two half-cones. This intersection is a quadratic hypersurface (in fact, a sheet of a two sheet hyperboloid) contained in a hyperplane. Indeed, we have

$$\mathcal{C}_1 \; : \; (x_{d+1} + r_1)^2 = (x - p_1)^2, \quad x_{d+1} + r_1 > 0,$$
$$\mathcal{C}_2 \; : \; (x_{d+1} + r_2)^2 = (x - p_2)^2, \quad x_{d+1} + r_2 > 0.$$

The intersection of the two half-cones is contained in the hyperplane h_{12} whose equation is obtained by subtracting the two sides of the above equations:

$$h_{12} \; : \quad -2(p_1 - p_2) \cdot x - 2(r_1 - r_2)x_{d+1} + p_1^2 - r_1^2 - p_2^2 + r_2^2 = 0.$$

This shows that there exists a correspondence between the diagram $\mathrm{Apo}(S)$ and the power diagram of the hyperspheres Σ_i in $\mathbb{R}^{d+1}$ ($i = 1, \ldots, n$), where Σ_i is centered at (p_i, r_i) and has radius $r_i\sqrt{2}$. More precisely, $A(\sigma_i)$ is the projection of the intersection of the half-cone $\mathcal{C}_i$ with the power region $L(\Sigma_i)$. Indeed, x is in $A(\sigma_i)$ if and only if the projection X_i of x onto $\mathcal{C}_i$ has a smaller x_{d+1}-coordinate than the projections of x onto the other half-cones $\mathcal{C}_j$, $j \neq i$. In other words, the coordinates (x, x_{d+1}) of X_i must obey

$$(x_{d+1} + r_i)^2 = (x - p_i)^2$$
$$(x_{d+1} + r_j)^2 \leq (x - p_j)^2 \quad \text{for any } j \neq i,$$

and by subtracting both sides, it follows that $\Sigma_i(X_i) \leq \Sigma_j(X_i)$ for all j.

Algorithm 2 Construction of Apollonius diagrams

INPUT: a set of hyperspheres S

1. Compute Σ_i, for $i = 1, \ldots, n$;
2. Compute the power diagram of the Σ_i's;
3. For all $i = 1, \ldots, n$, project vertically the intersection of the power region $L(\Sigma_i)$ with the half-cone C_i.

OUTPUT: the Apollonius diagram of S.

The Apollonius diagram of S can be computed using the following algorithm:

The power diagram of the Σ_i can be computed in time $O(n^{\lfloor \frac{d}{2} \rfloor + 1} \log n)$. The intersection involved in Step 3 can be computed in time proportional to the number of faces of the power diagram of the Σ_i's, which is $O(n^{\lfloor \frac{d}{2} \rfloor + 1})$. We have thus proved the following theorem due to Aurenhammer [35]:

Theorem 10. *The Apollonius diagram of a set of n hyperspheres in $\mathbb{R}^d$ has complexity $O(n^{\lfloor \frac{d}{2} \rfloor + 1})$ and can be computed in time $O(n^{\lfloor \frac{d}{2} \rfloor + 1} \log n)$.*

This result is optimal in odd dimensions, since the bounds above coincide with the corresponding bounds for the Voronoi diagram of points under the Euclidean distance. It is not optimal in dimension 2 (see Exercise 20). We also conjecture that it is not optimal in any even dimension.

Computing a Single Apollonius Region

We now establish a correspondence, due to Boissonnat and Karavelas [63], between a single Apollonius region and a Möbius diagram on a hypersphere.

To give the intuition behind the result, we consider first the case where one of the hyperspheres, say σ_0, is a hyperplane, i.e. a hypersphere of infinite radius. We take for σ_0 the hyperplane $x_d = 0$, and assume that all the other hyperspheres lie the half-space $x_d > 0$. The distance $\delta_0(x)$ from a point $x \in \mathbb{R}^d$ to σ_0 is defined as the Euclidean distance.

The points that are at equal distance from σ_0 and σ_i, $i > 0$, belong to a paraboloid of revolution with vertical axis. Consider such a paraboloid as the graph of a $(d-1)$-variate function ϑ_i defined over $\mathbb{R}^{d-1}$. If follows from Sect. 2.4.1 that the minimization diagram of the ϑ_i, $i = 1, \ldots, n$, is a Möbius diagram (see Fig. 2.9).

Easy computations give the associated weighted points. Write $p_i = (p_i', p_i'')$, $p_i' \in \mathbb{R}^{d-1}$, $p_i'' \in \mathbb{R}$, $i > 0$ and let $\mathcal{W} = \{\omega_1, \ldots, \omega_n\}$ be the set of Möbius sites of $\mathbb{R}^d$ where $\omega_i = \{p_i', \lambda_i, \mu_i\}$, and

$$\lambda_i = \frac{1}{r_i + p_i''}, \qquad \mu_i = r_i - p_i'', \qquad i > 0.$$

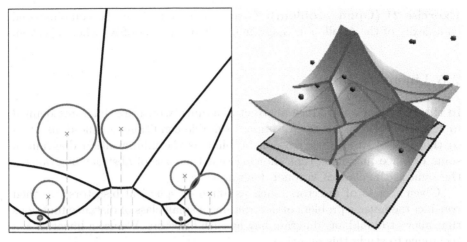

Fig. 2.9. A cell in an Apollonius diagram of hyperspheres projects vertically onto a Möbius diagram in σ_0

We let as an exercise to verify that the vertical projection of the boundary of the Apollonius region $A(\sigma_0)$ of σ_0 onto σ_0 is the Möbius diagram of $\mathcal{W}$.

We have assumed that one of the hyperspheres was a hyperplane. We now consider the case of hyperspheres of finite radii. The crucial observation is that the radial projection of $A(\sigma_0) \cap A(\sigma_i) \cap A(\sigma_j)$ onto σ_0, if not empty, is a hypersphere. It follows that the radial projection of the boundary of $A(\sigma_0)$ onto σ_0 is a Möbius diagram on σ_0.

Such a Möbius diagram on σ_0 can be computed by constructing the restriction of the power diagram of n hyperspheres of $\mathbb{R}^d$ with the hypersphere σ_0 (see Exercise 14).

Theorem 11. *Let $\mathcal{S}$ be a set of n hyperspheres in $\mathbb{R}^d$. The worst-case complexity of a single Apollonius region in the diagram of n hyperspheres of $\mathbb{R}^d$ is $\Theta(n^{\lfloor \frac{d+1}{2} \rfloor})$. Such a region can be computed in optimal time $\Theta(n \log n + n^{\lfloor \frac{d+1}{2} \rfloor})$.*

Exercise 17. Show that the cell of hypersphere σ_i in the Apollonius diagram of $\mathcal{S}$ is empty if and only if σ_i is inside another hypersphere σ_j.

Exercise 18. The predicates required to construct an Apollonius region are multivariate polynomials of degree at most 8 and 16 when $d = 2$ and 3 respectively. Detail these predicates [62].

Exercise 19. Show that the convex hull of a finite number of hyperspheres can be deduced from the restriction of a power diagram to a unit hypersphere [62].

Exercise 20. Prove that the combinatorial complexity of the Apollonius diagram of n circles in the plane has linear size.

Exercise 21 (Open problem). Give a tight bound on the combinatorial complexity of the Apollonius diagram of n hyperspheres of $\mathbb{R}^d$ when d is even.

2.5 Linearization

In this section, we introduce abstract diagrams, which are diagrams defined in terms of their bisectors. We impose suitable conditions on these bisectors so that any abstract diagram can be built as the minimization diagram of some distance functions, thus showing that the class of abstract diagrams is the same as the class of Voronoi diagrams.

Given a class of bisectors, such as affine or spherical bisectors, we then consider the inverse problem of determining a small class of distance functions that allows to build any diagram having such bisectors. We use a linearization technique to study this question.

2.5.1 Abstract Diagrams

Voronoi diagrams have been defined (see Sect. 2.2) as the minimization diagram of a finite set of continuous functions $\{\delta_1, \ldots, \delta_n\}$. It is convenient to interpret each δ_i as the distance function to an abstract object o_i, $i = 1, \ldots, n$. We define the bisector of two objects o_i and o_j of $\mathcal{O} = \{o_1, \ldots, o_n\}$ as

$$b_{ij} = \{x \in \mathbb{R}^d, \delta_i(x) = \delta_j(x)\}.$$

The bisector b_{ij} subdivides $\mathbb{R}^d$ into two open regions: one, b_{ij}^i, consisting of the points of $\mathbb{R}^d$ that are closer to o_i than to o_j, and the other one, b_{ij}^j, consisting of the points of $\mathbb{R}^d$ that are closer to o_j than to o_i. We can then define the Voronoi region of o_i as the intersection of the regions b_{ij}^i for all $j \neq i$. The union of the closures of these Voronoi regions covers $\mathbb{R}^d$. Furthermore, if we assume that the bisectors are $(d-1)$-manifolds, the Voronoi regions then have disjoint interiors and we can define the closed region associated to b_{ij}^i as $\bar{b}_{ij}^i = b_{ij}^i \cup b_{ij}$.

In a way similar to Klein [230], we now define diagrams in terms of bisectors instead of distance functions. Let $B = \{b_{ij}, i \neq j\}$ be a set of closed $(d-1)$-manifolds without boundary. We always assume in the following that $b_{ij} = b_{ji}$ for all $i \neq j$. We assume further that, for all distinct i, j, k, the following incidence condition (I.C.) holds:

$$b_{ij} \cap b_{jk} = b_{jk} \cap b_{ki} \quad (I.C.)$$

This incidence condition is obviously needed for B to be the set of bisectors of some distance functions.

By Jordan's theorem, each element of B subdivides $\mathbb{R}^d$ into at least two connected components and crossing a bisector b_{ij} implies moving into another connected component of $\mathbb{R}^d \setminus b_{ij}$. Hence, once a connected component of $\mathbb{R}^d \setminus b_{ij}$ is declared to belong to i, the assignments of all the other connected components of $\mathbb{R}^d \setminus b_{ij}$ to i or j are determined.

Given a set of bisectors $B = \{b_{ij}, i \neq j\}$, an *assignment* on B associates to each connected component of $\mathbb{R}^d \setminus b_{ij}$ a label i or j so that two adjacent connected components have different labels.

Once an assignment on B is defined, the elements of B are called *oriented bisectors*.

Given B, let us now consider such an assignment and study whether it may derive from some distance functions. In other words, we want to know whether there exists a set $\Delta = \{\delta_1, \ldots, \delta_n\}$ of distance functions such that

1. the set of bisectors of Δ is B;
2. for all $i \neq j$, a connected component C of $\mathbb{R}^d \setminus b_{ij}$ is labeled by i if and only if

$$\forall x \in C, \delta_i(x) \leq \delta_j(x).$$

We define the *region* of object o_i as $\cap_{j \neq i} \bar{b}_{ij}^i$.

A necessary condition for the considered assignment to derive from some distance functions is that the regions of any subdiagram cover $\mathbb{R}^d$. We call this condition the assignment condition (A.C.):

$$\forall I \subset \{1, \ldots, n\}, \cup_{i \in I} \cap_{j \in I \setminus \{i\}} \bar{b}_{ij}^i = \mathbb{R}^d \quad (A.C.)$$

Given a set of bisectors $B = \{b_{ij}, i \neq j\}$ and an assignment satisfying I.C. and A.C., the *abstract diagram* of $\mathcal{O}$ is the subdivision of $\mathbb{R}^d$ consisting of the regions of the objects of $\mathcal{O}$ and of their faces. The name *abstract Voronoi diagram* was coined by Klein [230], referring to similar objects in the plane.

For any set of distance functions δ_i, we can define the corresponding set of oriented bisectors. Obviously, I.C. and A.C. are satisfied and the abstract diagram defined by this set is exactly the minimization diagram for the distance functions δ_i. Hence any Voronoi diagram allows us to define a corresponding abstract diagram. Let us now prove the converse: any abstract diagram can be constructed as a Voronoi diagram.

Specifically, we prove that I.C. and A.C. are sufficient conditions for an abstract diagram to be the minimization diagram of some distance functions, thus proving the equivalence between abstract diagrams and Voronoi diagrams. We need the following technical lemmas.

Lemma 2. *The assignment condition implies that for any distinct i, j, k, we have*

$$b_{ij}^j \cap b_{jk}^k \cap b_{ki}^i = \emptyset.$$

Proof. A.C. implies that $\mathbb{R}^d = \cup_{1 \leq i \leq n} \cap_{j \neq i} \bar{b}_{ij}^i \subset \bar{b}_{ij}^i \cup \bar{b}_{jk}^j \cup \bar{b}_{ki}^k$. Hence, $\bar{b}_{ij}^i \cup \bar{b}_{jk}^j \cup \bar{b}_{ki}^k = \mathbb{R}^d$. Taking the complementary sets, we obtain $b_{ij}^j \cap b_{jk}^k \cap b_{ki}^i = \emptyset$.

Lemma 3. *For any distinct* i, j, k, *we have*

$$b_{ij} \cap b_{jk}^k \subset b_{ik}^k \quad and \quad b_{ij} \cap \bar{b}_{jk}^k \subset \bar{b}_{ik}^k \qquad (2.3)$$

$$b_{ij} \cap b_{jk}^j \subset b_{ik}^i \quad and \quad b_{ij} \cap \bar{b}_{jk}^j \subset \bar{b}_{ik}^i \qquad (2.4)$$

Proof. Let us first prove that $b_{ij} \cap b_{jk}^k \subset b_{ik}^k$:

Consider $x \in b_{ij} \cap b_{jk}^k$. Assume, for a contradiction, that $x \notin b_{ik}^k$. It follows that $x \in \bar{b}_{ik}^k$, but x cannot lie on b_{ik}, because this would imply that $x \in b_{ik} \cap b_{ij}$, which does not intersect b_{jk}^k. Hence, $x \in b_{ik}^i$ and therefore, $x \in b_{ij} \cap b_{jk}^k \cap b_{ik}^i$. We can then find x' in the neighborhood of x such that $x' \in b_{ij}^j \cap b_{jk}^k \cap b_{ki}^i$, contradicting Lemma 2.

Let us now prove that $b_{ij} \cap \bar{b}_{jk}^k \subset \bar{b}_{ik}^k$. We have proved the inclusion for $b_{ij} \cap b_{jk}^k$. It remains to prove that $b_{ij} \cap b_{jk} \subset \bar{b}_{ik}^k$ which is trivially true, by I.C. The two other inclusions are proved in a similar way.

We can now prove a lemma stating a transitivity relation:

Lemma 4. *For any distinct* i, j, k, *we have* $b_{ij}^i \cap b_{jk}^j \subset b_{ik}^i$.

Proof. Let $x \in b_{ij}^i \cap b_{jk}^j$. Assume, for a contradiction, that $x \notin b_{ik}^i$. If $x \in b_{ik}^k$, we have $b_{ij}^i \cap b_{jk}^j \cap b_{ik}^k \neq \emptyset$, contradicting Lemma 2. Therefore, x has to belong to b_{ik}, which implies that $x \in b_{ij}^i \cap b_{ik} \subset b_{kj}^k$ by Lemma 3. This contradicts $x \in b_{jk}^j$. We deduce that $x \in b_{ik}^i$, as needed.

The following lemma states that at most two assignments are likely to derive from some Voronoi diagram.

Lemma 5. *For a given set B satisfying I.C. and assuming that we never have $b_{ij} \subset b_{ik}$ for $j \neq k$, there are at most two ways of labeling the connected components of each $\mathbb{R}^d \setminus b_{ij}$ as b_{ij}^i and b_{ij}^j such that A.C. is verified.*

Proof. First assume that the sides b_{12}^1 and b_{12}^2 have been assigned. Consider now the labeling of the sides of b_{1i}, for some $i > 2$: let x be a point in the non empty set $b_{2i} \setminus b_{12}$. First assume that $x \in b_{12}^1$. Lemma 3 then implies that $x \in b_{1i}^1$. Conversely, if $x \in b_{12}^2$, $x \in b_{1i}^i$. In both cases, the assignment of the sides of b_{1i} is determined.

All other assignments are determined in a similar way. One can easily see that reversing the sides of b_{12} reverses all the assignments. Thus, we have at most two possible global assignments.

Theorem 12. *Given a set of bisectors $B = \{b_{ij}, 1 \leq i \neq j \leq n\}$ that satisfies the incidence condition (I.C.) and an assignment that satisfies the assignment condition (A.C.), there exists a set of distance functions $\{\delta_i, 1 \leq i \leq n\}$ defining the same bisectors and assignments.*

Proof. Let δ_1 be any real continuous function over $\mathbb{R}^d$. Let $j > 1$ and assume the following induction property: for all $i < j$, the functions δ_i have already been constructed so that

$$\forall i, i' < j, \quad \delta_i(x) \leq \delta_{i'}(x) \Leftrightarrow x \in \bar{b}^i_{ii'}.$$

Let us build δ_j. We consider the arrangement of all bisectors b_{ij}, for $i < j$: for each $I \subset J = \{1, \ldots, j-1\}$, we define $V_I = (\cap_{i \in I} \bar{b}^i_{ij}) \cap (\cap_{k \in J \setminus I} \bar{b}^j_{jk})$. The set V_I is a non necessarily connected region of the arrangement where we need $\delta_j > \delta_i$ if $i \in I$ and $\delta_j < \delta_i$ if $i \in J \setminus I$. This leads us to the following construction.

The interior of V_I is $\text{int } V_I = (\cap_{i \in I} b^i_{ij}) \cap (\cap_{k \in J \setminus I} b^j_{jk})$. Lemma 4 and the induction hypothesis imply that

$$\forall i \in I, \forall k \in J \setminus I, \forall x \in \text{int } V_I, \delta_i(x) < \delta_k(x).$$

In particular, if we define $\nu_I = \min_{k \in J \setminus I} \delta_k$ and $\mu_I = \max_{i \in I} \delta_i$ on V_I, we have $\mu_I < \nu_I$ on $\text{int } V_I$.

Let us now consider some point x on the boundary of V_I. We distinguish two cases. We can first assume that $x \in b_{ij}$ for some $i \in I$. Then, by Lemma 3, for any $i' \in I \setminus \{i\}$, $x \in b_{ij} \cap \bar{b}^{i'}_{i'j} \subset \bar{b}^{i'}_{i'i}$ so that $\delta_{i'}(x) \leq \delta_i(x)$. It follows that $\mu_I(x) = \delta_i(x)$.

Consider now the case when $x \in \partial V_I \cap b_{jk}$ with $k \in J \setminus I$, we have $\nu_I(x) = \delta_k(x)$. Finally, if $x \in \partial V_I \cap b_{ij} \cap b_{jk}$ with $i \in I$ and $k \in J \setminus I$, we have $\mu_I(x) = \delta_i(x)$ and $\nu_I(x) = \delta_k(x)$. By the induction hypothesis, $\delta_i(x) = \delta_k(x)$, which implies that $\mu_I(x) = \nu_I(x)$.

It follows that we can define a continuous function ρ on ∂V_I in the following way:

$$\rho_I(x) = \mu_I(x) \text{ if } \exists i \in I, x \in b_{ij}$$
$$= \nu_I(x) \text{ if } \exists k \in J \setminus I, x \in b_{jk}$$

Furthermore, on $\partial V_I \cap b_{ij} = \partial V_{I \setminus \{i\}} \cap b_{ij}$, if $i \in I$, we have

$$\rho_I(x) = \mu_I(x) = \nu_{I \setminus \{i\}}(x) = \rho_{I \setminus \{i\}}(x). \qquad (2.5)$$

The definitions of the ρ_I are therefore consistent, and we can now use these functions to prove that the following definition of δ_j satisfies the induction property.

Finally, we require δ_j to be any continuous function verifying

$$\mu_I < \delta_j < \nu_I$$

on each $\text{int } V_I$. By continuity of δ_j, we deduce from 2.5 that if $x \in \partial V_I \cap b_{jk} = \partial V_{I \setminus \{i\}} \cap b_{ij}$ with $k \in J \setminus I$, we have $\rho_I(x) = \mu_I(x) = \nu_{I \setminus \{i\}}(x) = \rho_{I \setminus \{i\}}(x) = \delta_j(x)$.

It follows that on each V_I, for all $i < j$, $\delta_i(x) < \delta_j(x)$ iff $x \in b^i_{ij}$ and $\delta_i(x) = \delta_j(x)$ iff $x \in b_{ij}$. The induction follows.

One can prove that, in the proof of Lemma 5, the assignment we build satisfies the consequences of A.C. stated in Lemmas 2, 3 and 4. The proof of Theorem 12 does not need A.C. but only the consequences of A.C. stated in those three lemmas. It follows that any of the two possible assignments determined in the proof of Lemma 5 allows the construction of distance functions, as in Theorem 12, which implies that A.C. is indeed verified. We thus obtain a stronger version of Lemma 5.

Lemma 6. *For a given set B satisfying I.C. and assuming that we never have $b_{ij} \subset b_{ik}$ for $j \neq k$, there are exactly two ways of labeling the connected components of each $\mathbb{R}^d \setminus b_{ij}$ as b_{ij}^i and b_{ij}^j such that A.C. is verified.*

Theorem 12 proves the equivalence between Voronoi diagrams and abstract diagrams by constructing a suitable set of distance functions. In the case of affine bisectors, the following result of Aurenhammer [35] allows us to choose the distance functions in a smaller class than the class of continuous functions.

Theorem 13. *Any abstract diagram of $\mathbb{R}^d$ with affine bisectors is identical to the power diagram of some set of spheres of $\mathbb{R}^d$.*

Proof. In this proof, we first assume that the affine bisectors are in general position, i.e. four of them cannot have a common subspace of co-dimension 2: the general result easily follows by passing to the limit.

Let $B = \{b_{ij}, 1 \leq i \neq j \leq n\}$ be such a set. We identify $\mathbb{R}^d$ with the hyperplane $x_{d+1} = 0$ of $\mathbb{R}^{d+1}$. Assume that we can find a set of hyperplanes $\{H_i, 1 \leq i \leq n\}$ of $\mathbb{R}^{d+1}$ such that the intersection $H_i \cap H_j$ projects onto b_{ij}. Sect. 2.3 then shows that the power diagram of the set of spheres $\{\sigma_i, 1 \leq i \leq n\}$ obtained by projecting the intersection of paraboloid $\mathcal{Q}$ with each H_i onto $\mathbb{R}^d$ admits B as its set of bisectors[2] (see Fig. 2.5).

Let us now construct such a set of hyperplanes, before considering the question of the assignment condition.

Let H_1 and H_2 be two non-vertical hyperplanes of $\mathbb{R}^d$ such that $H_1 \cap H_2$ projects vertically onto b_{12}. We now define the H_i for $i > 2$: let Δ_i^1 be the maximal subspace of H_1 that projects onto b_{1i} and let Δ_i^2 be the maximal subspace of H_2 that projects onto b_{2i}. Both Δ_i^1 and Δ_i^2 have dimension $d-1$. I.C. implies that $b_{12} \cap b_{2i} \cap b_{i1}$ has co-dimension 2 in $\mathbb{R}^d$. Thus $\Delta_i^1 \cap \Delta_i^2$, its preimage on H_1 (or H_2) by the vertical projection, has the same dimension $d-2$. This proves that Δ_i^1 and Δ_i^2 span a hyperplane H_i of $\mathbb{R}^{d+1}$. The fact that $H_i \neq H_1$ and $H_i \neq H_2$ easily follows from the general position assumption.

We still have to prove that $H_i \cap H_j$ projects onto b_{ij} for $i \neq j > 2$. I.C. ensures that the projection of $H_i \cap H_j$ contains the projection of $H_i \cap H_j \cap H_1$ and the projection of $H_i \cap H_j \cap H_2$, which are known to be $b_{ij} \cap b_{1i}$ and $b_{ij} \cap b_{2i}$, by construction. The general position assumption implies that there is only

---

[2]We may translate the hyperplanes vertically in order to have a non-empty intersection, or we may consider imaginary spheres with negative squared radii.

one hyperplane of $\mathbb{R}^d$, namely b_{ij}, containing both $b_{ij} \cap b_{1i}$ and $b_{ij} \cap b_{2i}$. This is the projection of $H_i \cap H_j$.

As we have seen, building this set of hyperplanes of $\mathbb{R}^{d+1}$ amounts to building a family of spheres whose power diagram admits B as its set of bisectors. At the beginning of the construction, while choosing H_1 and H_2, we may obtain any of the two possible labellings of the sides of b_{12}. Since there is no other degree of freedom, this choice determines all the assignments. Lemma 5 shows that there are at most two possible assignments satisfying A.C., which proves we can build a set of spheres satisfying any of the possible assignments. The result follows.

Exercise 22. Consider the diagram obtained from the Euclidean Voronoi diagram of n points by taking the other assignment. Characterize a region in this diagram in terms of distances to the points and make a link with Exercise 3.

2.5.2 Inverse Problem

We now assume that each bisector is defined as the zero-set of some real-valued function over $\mathbb{R}^d$, called a *bisector-function* in the following. Let us denote by B the set of bisector-functions. By convention, for any bisector-function β_{ij}, we assume that

$$b_{ij}^i = \{x \in \mathbb{R}^d : \beta_{ij}(x) < 0\} \text{ and } b_{ij}^j = \{x \in \mathbb{R}^d : \beta_{ij}(x) > 0\}.$$

We now define an algebraic equivalent of the incidence relation in terms of pencil of functions: we say that B satisfies the linear combination condition (L.C.C.) if, for any distinct i, j, k, β_{ki} belongs to the pencil defined by β_{ij} and β_{jk}, i.e.

$$\exists (\lambda, \mu) \in \mathbb{R}^2 \quad \beta_{ki} = \lambda \beta_{ij} + \mu \beta_{jk} \quad (L.C.C.)$$

Note that L.C.C. implies I.C. and that in the case of affine bisectors L.C.C. is equivalent to I.C. Furthermore, it should be noted that, in the case of Voronoi diagrams, the bisector-functions defined as $\beta_{ij} = \delta_i - \delta_j$ obviously satisfy L.C.C.

We now prove that we can view diagrams satisfying L.C.C. as diagrams that can be linearized.

Definition 1. *A diagram $\mathcal{D}$ of n objects in some space E is said to be a pullback of a diagram $\mathcal{D}'$ of m objects in space F by a function $\phi : E \to F$ if $m = n$ and if, for any distinct i, j, we have*

$$b_{ij}^i = \phi^{-1}(c_{ij}^i)$$

where b_{ij}^i denotes the set of points closer to i than to j in $\mathcal{D}$ and c_{ij}^i denotes the set of points closer to i than to j in $\mathcal{D}'$.

Theorem 14. *Let $B = \{\beta_{ij}\}$ be a set of real-valued bisector-functions over $\mathbb{R}^d$ satisfying L.C.C. and A.C. Let V be any vector space of real functions over $\mathbb{R}^d$ that contains B and constant functions.*

If N is the dimension of V, the diagram defined by B is the pullback by some continuous function of an affine diagram in dimension $N - 1$.

More explicitly, there exist a set $C = \{\psi_{ij} \cdot X + c_{ij}\}$ of oriented affine hyperplanes of $\mathbb{R}^{N-1}$ satisfying I.C. and A.C. and a continuous function ϕ: $\mathbb{R}^d \to \mathbb{R}^{N-1}$ such that for all $i \neq j$,

$$\bar{b}_{ij}^i = \{x \in \mathbb{R}^d, \beta_{ij}(x) \leq 0\} = \phi^{-1}\{y \in \mathbb{R}^{N-1}, \psi_{ij}(x) \leq c_{ij}\}.$$

Proof. Let $(\gamma_0, \ldots, \gamma_{N-1})$ be a basis of V such that γ_0 is the constant function equal to 1.

Consider the evaluation application,

$$\phi : x \in \mathbb{R}^d \mapsto (\gamma_1(x), \ldots, \gamma_{N-1}(x)) \in \mathbb{R}^{N-1}.$$

If point x belongs to some b_{ij}^i, we have $\beta_{ij}(x) < 0$. Furthermore, there exists real coefficients $\lambda_{ij}^0, \ldots, \lambda_{ij}^{N-1}$ such that $\beta_{ij} = \sum_{k=0}^{N-1} \lambda_{ij}^k \gamma_k$. The image $\phi(x)$ of x thus belongs to the affine half-space B_{ij}^i of $\mathbb{R}^{N-1}$ of equation

$$\sum_{k=1}^{N-1} \lambda_{ij}^k X_k < -\lambda_{ij}^0.$$

In this way, we can define all the affine half-spaces B_{ij}^i of $\mathbb{R}^{N-1}$ for $i \neq j$: B_{ij} is an oriented affine hyperplane with normal vector $(\lambda_{ij}^1, \ldots, \lambda_{ij}^{N-1})$ and constant term λ_{ij}^0. Plainly, L.C.C. on the β_{ij} translates into I.C. on the B_{ij}, and we have

$$\bar{b}_{ij}^i = \{x \in \mathbb{R}^d, \beta_{ij}(x) \leq 0\} = \phi^{-1}\{y \in \mathbb{R}^{N-1}, B_{ij}(x) \leq -\lambda_{ij}^n\} \qquad (2.6)$$

Finally, let us prove that A.C. is also satisfied. Lemma 6 states that the B_{ij} have exactly two inverse assignments satisfying A.C. Furthermore, Equation 2.6 implies that any of these two assignments defines an assignment for the b_{ij} that also satisfies A.C. It follows that if the current assignment did not satisfy A.C., there would be more than two assignments for the b_{ij} that satisfy A.C. This proves that A.C. is also satisfied by the B_{ij} and concludes the proof.

We can now use Theorem 13 and specialize Theorem 14 to the specific case of diagrams whose bisectors are hyperspheres or hyperquadrics, or, more generally, to the case of diagrams whose class of bisectors spans a finite dimensional vector space.

Theorem 15. *Any abstract diagram of* $\mathbb{R}^d$ *with spherical bisectors such that the corresponding degree 2 polynomials satisfy L.C.C. is a Möbius diagram.*

Proof. Since the spherical bisectors satisfy L.C.C., we can apply Theorem 14 and Theorem 13. Function ϕ of Theorem 14 is simply the lifting mapping $x \mapsto (x, x^2)$, and we know from Theorem 13 that our diagram can be obtained as a power diagram pulled-back by ϕ. That is to say $\delta_i(x) = \Sigma_i(\phi(x))$, where Σ_i is a hypersphere in $\mathbb{R}^{d+1}$.

Another way to state this transformation is to consider the diagram with spherical bisectors in $\mathbb{R}^d$ as the projection by ϕ^{-1} of the restriction of the power diagram of the hyperspheres Σ_i to the paraboloid $\phi(\mathbb{R}^d) \subset \mathbb{R}^{d+1}$ of equation $x_{d+1} = x^2$.

Assume that the center of Σ_j is $(u_1^j, \ldots, u_{d+1}^j)$, and that the squared radius of Σ_j is w^j. We denote by Σ_j the power to Σ_j. Distance δ_j can be expressed in terms of these parameters:

$$\delta_j(x) = \Sigma_j(\phi(x)) = \sum_{1 \leq i \leq d} (x_i - u_i^j)^2 + ((\sum_{1 \leq i \leq d} x_i^2) - u_{d+1}^j)^2 - w^j.$$

Subtracting from each δ_j the same term $(\sum_{1 \leq i \leq D} x_i^2)^2$ leads to a new set of distance functions that define the same minimization diagram as the δ_j. In this way, we obtain new distance functions which are exactly the ones defining Möbius diagrams.

This proves that any diagram whose bisectors are hyperspheres can be constructed as a Möbius diagram.

The proof of the following theorem is similar to the previous one:

Theorem 16. *Any abstract diagram of* $\mathbb{R}^d$ *with quadratic bisectors such that the corresponding degree 2 polynomials satisfy L.C.C. is an anisotropic Voronoi diagram.*

Exercise 23. Explain why, in Theorem 15, it is important to specify which bisector-functions satisfy L.C.C. instead of mentioning only the bisectors (Hint: Theorem 12 implies that there always exist some bisector-functions with the same zero-sets that satisfy L.C.C.)

2.6 Incremental Voronoi Algorithms

Incremental constructions consist in adding the objects one by one in the Voronoi diagram, updating the diagram at each insertion. Incremental algorithms are well known and highly popular for constructing Euclidean Voronoi diagrams of points and power diagrams of spheres in any dimension. Because the whole diagram can have to be modified at each insertion, incremental algorithms have a poor worst-case complexity. However most of the insertions result only in local modifications and the worst-case complexity does not reflect the actual complexity of the algorithm in most practical situations. To provide more realistic results, incremental constructions are analyzed in the randomized framework. This framework makes no assumption on the input object set but analyzes the expected complexity of the algorithm assuming that the objects are inserted in random order, each ordering sequence being equally likely. The following theorem, whose proof can be found in many textbooks (see e.g. [67]) recalls that state-of-the-art incremental constructions of Voronoi diagrams of points and power diagrams have an optimal randomized complexity.

Theorem 17. *The Euclidean Voronoi diagram of n points in $\mathbb{R}^d$ and the power diagram of n spheres in $\mathbb{R}^d$ can be constructed by an incremental algorithm in randomized time $O\left(n \log n + n^{\left\lceil \frac{d+1}{2} \right\rceil}\right)$.*

Owing to the linearization techniques of Sect. 2.5, this theorem yields complexity bounds for the construction of linearizable diagrams such as Möbius, anisotropic or Apollonius diagrams. Incremental constructions also apply to the construction of Voronoi diagrams for which no linearization scheme exists. This is for instance the case for the 2-dimensional Euclidean Voronoi diagrams of line segments. The efficiency of the incremental approach merely relies on the fact that the cells of the diagram are simply connected and that the 1-skeleton of the diagram, (i. e. the union of its edges and vertices) is a connected set. Unfortunately, these two conditions are seldom met except for planar Euclidean diagrams. Let us take Apollonius diagrams as an illustration. Each cell of an Apollonius diagram is star shaped with respect to the center of the associated sphere and is thus simply connected. In the planar case, Apollonius bisectors are unbounded hyperbolic arcs and the 1-skeleton can easily be made connected by adding a curve at infinity. The added curve can be seen as the bisector separating any input object from an added fictitious object. In 3-dimensional space, the skeleton of Apollonius diagrams is not connected: indeed, we know from Sect. 2.4.3 that the faces of a single cell are in 1-1 correspondence with the faces of a 2-dimensional Möbius diagram and therefore may include isolated loops.

As a consequence, the rest of this section focuses on planar Euclidean diagrams. After some definitions, the section recalls the incremental construction of Voronoi diagrams, outlines the topological conditions under which this approach is efficient and gives some examples. The efficiency of incremental algorithms also greatly relies on the availability of some point location data structure to answer nearest neighbor queries. A general data structure, the Voronoi hierarchy, is described at the end of the section. The last subsection lists the main predicates involved in the incremental construction of Voronoi diagrams.

2.6.1 Planar Euclidean diagrams

To be able to handle planar objects that possibly intersect, the distance functions that we consider in this section are signed Euclidean distance functions, i.e. the distance $\delta_i(x)$ from a point x to an object o_i is:

$$\delta_i(x) = \begin{cases} \min_{y \in \bar{o}_i} \|y - x\|, & \text{if } x \notin o \\ -\min_{y \in \bar{o}_i^c} \|y - x\|, & \text{if } x \in o \end{cases}$$

where $\bar{o}_i$ is the closure of o_i and $\bar{o}_i^c$ the closure of the complement of o_i. Note that the distance used to define Apollonius diagrams matches this definition. Then, given a finite set $\mathcal{O}$ of planar objects and $o_i \in \mathcal{O}$, we define the Voronoi region of o_i as the locus of points closer to o_i than to any other object in $\mathcal{O}$

$$V(o_i) = \{x \in \mathbb{R}^2 : \delta_i(x) \leq \delta_j(x), \forall o_j \in \mathcal{O}\}.$$

Voronoi edges are defined as the locus of points equidistant to two objects $\mathcal{O}$ and closer to these two objects than to any other object in $\mathcal{O}$, and Voronoi vertices are the locus of points equidistant to three or more objects and closer to these objects than to any other object in $\mathcal{O}$. The Voronoi diagram $\text{Vor}(\mathcal{O})$ is the planar subdivision induced by the Voronoi regions, edges and vertices.

The incremental construction described below relies on the three following topological properties of the diagram that are assumed to be met for any set of input objects:

1. The diagram is assumed to be a *nice diagram*, i. e. a diagram in which edges and vertices are respectively 0 and 1-dimensional sets.
2. The cells are assumed to be simply connected.
3. The 1-*skeleton* of the diagram is connected.

Owing to Euler formula, Properties 1 and 2 imply that the Voronoi diagram of n objects is a planar map of complexity $O(n)$. Property 3 is generally not granted for any input set. Think for example of a set of points on a line. However, in the planar case, this condition can be easily enforced as soon as Properties 1 and 2 are met. Indeed, if the cells are simply connected, there is no bounded bisector and the 1-*skeleton* can be connected by adding a curve at infinity. The added curve can be seen as the bisector separating any input object from an added fictitious object. The resulting diagram is called the *compactified* version of the diagram.

2.6.2 Incremental Construction

We assume that the Voronoi diagram of any input set we consider is a nice diagram with simply connected cells and a connected 1-skeleton. Each step of the incremental construction takes as input the Voronoi diagram $\text{Vor}(\mathcal{O}_{i-1})$ of a current set of objects $\mathcal{O}_{i-1}$ and an object $o_i \notin \mathcal{O}_{i-1}$, and aims to construct the Voronoi diagram $\text{Vor}(\mathcal{O}_i)$ of the set $\mathcal{O}_i = \mathcal{O}_{i-1} \cup \{o_i\}$. In the following, we note $V(o, \mathcal{O}_{i-1})$ the region of an object o in the diagram $\text{Vor}(\mathcal{O}_{i-1})$ and $V(o, \mathcal{O}_i)$ the region of o in $\text{Vor}(\mathcal{O}_i)$.

We note $\text{Skel}(\mathcal{O}_{i-1})$ the 1-skeleton of $\text{Vor}(\mathcal{O}_{i-1})$. Let x be a point of $\text{Skel}(\mathcal{O}_{i-1})$. We note $\mathcal{N}(x, \mathcal{O}_{i-1})$ the nearest neighbors of x in $\mathcal{O}_{i-1}$, i.e. the subset of objects of $\mathcal{O}_{i-1}$ that are closest to x. The point x is said to be in conflict with o_i if x is closer to o_i than to $\mathcal{N}(x, \mathcal{O}_{i-1})$. Hence, the part of the skeleton that conflicts with o_i, called the *conflict skeleton* for short, is exactly the intersection of the skeleton $\text{Skel}(\mathcal{O}_{i-1})$ with the region of o_i in $\text{Vor}(\mathcal{O}_i)$. See Fig. 2.11.

The conflict skeleton is a subgraph of $\text{Skel}(\mathcal{O}_{i-1})$ and the endpoints of this subgraph are the vertices of the region $V(o_i, \mathcal{O}_i)$. If $V(o_i, \mathcal{O}_i)$ is not empty, the conflict skeleton is not empty either. Indeed, an empty conflict skeleton would imply that $V(o_i, \mathcal{O}_i)$ is included in a single region $V(o, \mathcal{O}_{i-1})$ of the diagram $\text{Vor}(\mathcal{O}_{i-1})$ and the region $V(o, \mathcal{O}_i)$ would not be simply connected. Furthermore, the following lemma, due to Klein et al. [231], proves that the conflict skeleton is a connected subgraph of $\text{Skel}(\mathcal{O}_{i-1})$.

Lemma 7. *If, for any input set, the Voronoi diagram is a nice diagram with simply connected regions and a connected 1-skeleton, the conflict skeleton of an additional object is connected.*

Proof. We use the above notation and assume, for a contradiction, that the conflict skeleton of o_i, which is $\text{Skel}(\mathcal{O}_{i-1}) \cap V(o_i, \mathcal{O}_i)$, consists of several disjoint connected components $Sk_1, Sk_2, \ldots, Sk_\ell$. Each connected component Sk_j has to intersect the boundary of the new region $V(o_i, \mathcal{O}_i)$, otherwise $\text{Skel}(\mathcal{O}_{i-1})$ would not be connected, a contradiction. Then, if $\ell \geq 2$, there exists a path $\mathcal{C}$ in $V(o_i, \mathcal{O}_i)$ connecting two points x and y on the boundary of $V(o_i, \mathcal{O}_i)$ and separating Sk_1 and Sk_2, see Fig. 2.10. The path $\mathcal{C}$ does not intersect $\text{Skel}(\mathcal{O}_{i-1})$ and is therefore included in the region $V(o, \mathcal{O}_{i-1})$ of some object o of $\mathcal{O}_{i-1}$. Since, points arbitrarily close to x and y but outside $V(o_i, \mathcal{O}_i)$ belong to $V(o, \mathcal{O}_{i-1})$, x and y can be joined by a simple path $\mathcal{D}$ in $V(o, \mathcal{O}_i) \subset V(o, \mathcal{O}_{i-1})$. The simple closed curve $\mathcal{C} \cup \mathcal{D}$ is contained in $V(o, \mathcal{O}_{i-1})$ and encloses Sk_1 or Sk_2, which contradicts the fact that $\text{Skel}(\mathcal{O}_{i-1})$ is connected.

Once the conflict skeleton is known, the Voronoi diagram $\text{Vor}(\mathcal{O}_{i-1})$ can be updated, leading to $\text{Vor}(\mathcal{O}_i)$. This is done by Procedure 3.

Procedure 4 describes a step of the incremental construction.

In the sequel, the incremental construction is analyzed in the randomized setting. It is assumed that each object has constant complexity, which implies that each operation involving a constant number of objects is performed in constant time. Because the conflict skeleton is connected, Substep 2 can be performed by traversing the graph $\text{Skel}(\mathcal{O}_{i-1})$ in time proportional to the number of edges involved in the conflict skeleton. These edges will be either deleted or shortened in the new diagram. Substep 3 takes time proportional

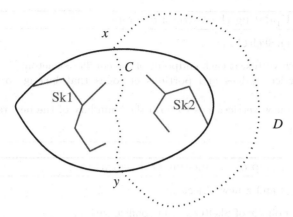

Fig. 2.10. For the proof of Lemma 7

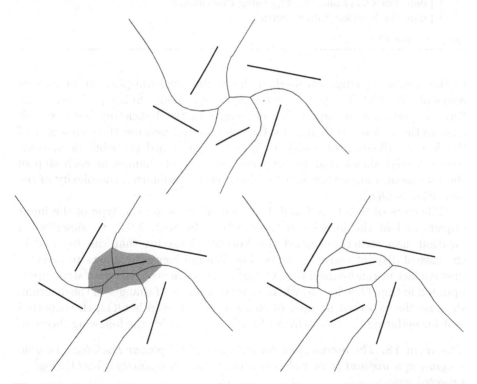

Fig. 2.11. Incremental construction of the Voronoi diagram of disjoint line segments

---

Procedure 3 Updating the Voronoi diagram

INPUT: $\mathrm{Vor}(\mathcal{O}_{i-1}), \mathrm{Skel}(\mathcal{O}_{i-1})$

1. Create a new vertex at each endpoint of the conflict skeleton;
2. Remove vertices, edges and portions of edges that belong to $\mathrm{Skel}(\mathcal{O}_{i-1}) \cap V(o_i, \mathcal{O}_i)$;
3. Connect the new vertices s as to form the boundary of the new region.

OUTPUT: $\mathrm{Vor}(\mathcal{O}_i)$

Procedure 4 A step of the incremental algorithm

INPUT: $\mathrm{Vor}(\mathcal{O}_{i-1})$ and a new object o_i

1. Find a first point x of $\mathrm{Skel}(\mathcal{O}_{i-1})$ in conflict with o_i;
2. Compute the whole conflict skeleton;
3. Update $\mathrm{Vor}(\mathcal{O}_{i-1})$ into $\mathrm{Vor}(\mathcal{O}_i)$ using Procedure 3;
4. Update the location data structure.

OUTPUT: $\mathrm{Vor}(\mathcal{O}_i)$

---

to the number of edges involved in the conflict skeleton plus the number of edges of $V(o_i, \mathcal{O}_i)$. The latter are the new edges. Hence Substeps 2 and 3 take time proportional to the number of changes in the 1-skeleton. Because each edge in the skeleton is defined by four objects and because the complexity of the Voronoi diagram of n objects is $O(n)$, a standard probabilistic analysis (see e.g. [67]) shows that the expected number of changes at each step of the incremental algorithm is $O(1)$. The overall randomized complexity of the algorithm is $O(n)$.

The costs of Substeps 1 and 4 depend of course on the type of the input objects and of the location data structure. In Sect. 2.6.3, we described a location data structure, called the Voronoi hierarchy, that can be used in the case of disjoint convex objects. The Voronoi hierarchy allows to detect a first conflict in randomized time $O(\log^2 n)$. At each step, the data structure is updated in time $O(m \log^2 n)$ where m is the number of changes in the diagram. Because the expected number of changes at each step is $O(1)$, the expected cost for updating the hierarchy is $O(\log^2 n)$. This yields the following theorem.

Theorem 18. *The incremental construction of the planar Euclidean Voronoi diagram of n disjoint convex objects with constant complexity takes $O(n \log^2 n)$ expected time.*

Note that the incremental construction described here is *on-line*, meaning that the algorithm does not need to know the whole set of objects right from the beginning. If the whole set of objects is known in advance, localization can be made easy and the maintenance of a location data structure is no longer required [20]. The idea consists in picking one witness point inside each object and in building first the Voronoi diagram of witness points. In a second phase, each witness point is replaced in turn by the corresponding object. When replacing a witness point by the corresponding object, any point on the boundary of the cell of the witness point belongs to the conflict skeleton. The algorithm is no longer on-line but its randomized complexity is reduced to $O(n \log n)$.

Voronoi Diagrams of Line Segments

The above incremental construction applies to the Voronoi diagram of disjoint line segments. Indeed, in the case of disjoint line segments, the bisector curves are unbounded simple curves, each composed of at most seven line segments and parabolic arcs (see e.g. [67, Chap. 19]). Hence, Voronoi vertices and edges are respectively 0 and 1-dimensional sets. Furthermore, each region in the diagram is weakly star shaped with respect to its generating segment, meaning that the segment joining any point in the region to its closest point on the associated segment is included in the region. It follows that Voronoi regions are simply connected.

If the segments are allowed to share endpoints, the Voronoi diagram exhibits 2-dimensional Voronoi edges, hence violating the definition of nice Voronoi diagrams. A way to circumvent this problem consists in considering that each segment is composed of three distinct objects: the two endpoints and the open segment. If the two endpoints of a segment are inserted in the diagram prior to the open segment, the incremental construction encounters no 2-dimensional bisecting region and the algorithm presented above can be used.

Voronoi Diagrams of Curved Segments

Voronoi diagrams of disjoint curved segments have been studied by Alt, Cheong and Vigneron [20]. Alt, Cheong and Vigneron introduce the notion of harmless curved segments defined as follows. A curved segment is said to be *convex* when the region bounded by the curved segment and the line segment joining its endpoints is convex. A *spiral* arc is a convex curved segment with monotonously increasing curvature. A *harmless* curved segment is either a line segment or a circular arc or a spiral arc. It can be shown that, if the input curved segments are split into harmless sub-segments and if each open curved sub-segment and its two endpoints are considered as three distinct sites, the Voronoi diagram is a nice diagram with simply connected regions. The incremental construction paradigm described above therefore applies.

Voronoi Diagrams of Convex Objects

The case of disjoint smooth convex objects is quite similar to the case of disjoint segments. The bisecting curves between two such objects is a 1-dimensional curve. Furthermore, each Voronoi region is weakly star shaped

with respect to the medial axis of its object, hence simply connected. Therefore, the Voronoi diagram can be built using the incremental algorithm. Note
that the Voronoi diagram of disjoint smooth convex objects could also be obtained by applying the incremental algorithm to the curved segments forming
the boundaries of the objects. However, this approach requires the subdivision of the boundary of each object into harmless parts and yields a Voronoi
diagram which is a refinement of the diagram of the input objects.

If we still assume the objects to be smooth and convex but allow them
to intersect, things become more difficult. Karavelas and Yvinec [217] have
shown that the Voronoi regions remain simply connected if and only if the
objects of $\mathcal{O}$ are pseudo-disks, meaning that the boundaries of any two objects
of $\mathcal{O}$ intersect in at most two points. The above incremental algorithm can be
adapted to work in this case. However, since the distance is a signed distance,
some sites may have an empty region, which makes the algorithm slightly
more complicated. Note that this may only happen when some of the objects
are included in the union of others. The algorithm has to check that the new
object has a non-empty region and must handle the case where the insertion
of a new object causes the region of some other object to vanish. Karavelas
and Yvinec [217] showed that there is no use to maintain a location data
structure in this case because each insertion takes linear time anyway.

The algorithm can be generalized to the case of convex objects with piecewise smooth pseudo-circular boundaries. As in the case of segments, the main
problem comes from the fact that sharp corners on the boundaries of objects
yield 2-dimensional bisectors. This problem can be handled as in the case of
line segments and planar curves by considering each corner as an object on
its own.

2.6.3 The Voronoi Hierarchy

The first step when inserting a new object o_i consists in finding one point
of the current skeleton $\mathrm{Skel}(\mathcal{O}_{i-1})$ in conflict with o_i. If the objects do not
intersect, this is done by searching the object o of $\mathcal{O}_{i-1}$ nearest to a point
x of o_i. Indeed, if the objects do not intersect, x belongs to the region of o
in $\mathrm{Vor}(\mathcal{O}_{i-1})$ and to the region of o_i in $\mathrm{Vor}(\mathcal{O}_i)$. Therefore o_i has to be a
neighbor of o in $\mathrm{Vor}(\mathcal{O}_i)$ and some point on the boundary of $V(o, \mathcal{O}_{i-1})$ is in
conflict with o_i. If the objects intersect, things are slightly more complicated
but nearest object queries can still be used to find out whether the new object
is hidden and, if not, to find a first conflict (see [217]).

Let us describe now how to find the object o of a set $\mathcal{O}$ closest to a query
point x. A simple strategy is to perform a walk in the Voronoi diagram $\mathrm{Vor}(\mathcal{O})$.
The walk starts at any region of the diagram. When the walk visits the region
$V(o)$ of an object o it considers in turn each of the neighboring regions. If
one of the neighbors of o, say o', is closer to x than o, the walk steps to
the region $V(o')$. If none of the neighbors of o in $\mathrm{Vor}(\mathcal{O})$ is closer to x than
o, then o is the object closest to x and the walk ends. Because the distance

between x and the objects of the visited regions is decreasing, the walk cannot loop and is bound to end. However, the walk may visit all the regions before ending. The Voronoi hierarchy [217] is a randomized data structure that makes this strategy more efficient. The Voronoi hierarchy can be considered as a 2-dimensional version of the skip lists introduced by Pugh [291] and generalizes the Delaunay hierarchy described in [115].

For a set of objects $\mathcal{O}$, the Voronoi hierarchy $\mathcal{H}_\mathcal{V}(\mathcal{O})$ is a sequence of Voronoi diagrams $\mathrm{Vor}(\Theta_\ell)$, $\ell = 0, \ldots, L$, built for subsets of $\mathcal{O}$ forming a hierarchy, i.e, $\mathcal{O} = \Theta_0 \supseteq \Theta_1 \supseteq \cdots \supseteq \Theta_L$.

The hierarchy $\mathcal{H}_\mathcal{V}(\mathcal{O})$ is built together with the Voronoi diagram $\mathrm{Vor}(\mathcal{O})$ according to the following rules:

1. Every object of $\mathcal{O}$ is inserted in $\mathrm{Vor}(\Theta_0) = \mathrm{Vor}(\mathcal{O})$;
2. An object o that has been inserted in $\mathrm{Vor}(\Theta_\ell)$, is inserted in $\mathrm{Vor}(\Theta_{\ell+1})$ with probability β.

To answer nearest object queries, the Voronoi hierarchy works as follows. Let us call θ_ℓ the object of Θ_ℓ closest to the query point x. First, a simple walk is performed in the top-most diagram to find θ_L. Then, at each level $\ell = L - 1, \ldots, 0$, a simple walk is performed in $\mathrm{Vor}(\Theta_\ell)$ from $\theta_{\ell+1}$ to θ_ℓ (see Fig. 2.12).

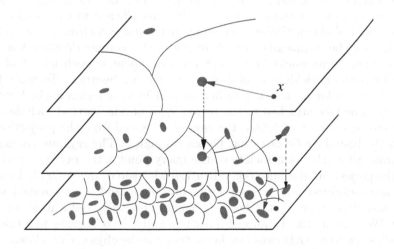

Fig. 2.12. Locating x using the Voronoi hierarchy

It is easy to show that the expected size of $\mathcal{H}_\mathcal{V}(\mathcal{O})$ is $O(\frac{n}{1-\beta})$, and that the expected number of levels in $\mathcal{H}_\mathcal{V}(\mathcal{O})$ is $O(\log_{1/\beta} n)$. Moreover, the following lemma proves that the expected number of steps performed by the walk at each level is constant.

Lemma 8. *Let x be a point in the plane. Let θ_ℓ (resp. $\theta_{\ell+1}$) be the object closest to x in Θ_ℓ (resp. $\Theta_{\ell+1}$). Then the expected number of Voronoi regions visited during the walk in $\mathrm{Vor}(\Theta_\ell)$ from $\theta_{\ell+1}$ to θ_ℓ is $O(1/\beta)$.*

Proof. The objects whose regions are visited at level ℓ are closer to x than $\theta_{\ell+1}$. Consequently, if, among the objects of Θ_ℓ, $\theta_{\ell+1}$ is the k-th closest object to x, the walk in $\mathrm{Vor}(\Theta_\ell)$ performs at most k steps.

Let us show that $\theta_{\ell+1}$ is the k-th closest object to x in Θ_ℓ with probability $\beta(1-\beta)^{k-1}$. Such a case occurs if and only if the two following conditions are satisfied:

1. Object $\theta_{\ell+1}$ has been inserted in $\Theta_{\ell+1}$
2. None of the $k-1$ objects of Θ_ℓ that are closer to x than $\theta_{\ell+1}$ has been inserted in $\Theta_{\ell+1}$.

The first condition occurs with probability β and the second with probability $(1-\beta)^{k-1}$.

Let n_l be the number of objects in Θ_ℓ. The expected number N_ℓ of objects that are visited at level ℓ is bounded as follows:

$$N_\ell \leq \sum_{k=1}^{n_\ell} k(1-\beta)^{k-1}\beta < \beta \sum_{k=1}^{\infty} k(1-\beta)^{k-1} = \frac{1}{\beta},$$

We still have to bound the time spent in each of the visited regions. Let o be the site of a visited region in $\mathrm{Vor}(\Theta_\ell)$. It is not efficient to consider in turn each neighbor o' of o in $\mathrm{Vor}(\Theta_\ell)$ and compare the distances from x to o and o'. Indeed, since the complexity of each region in the Voronoi diagram $\mathrm{Vor}(\Theta_\ell)$ may be $\Omega(n_\ell)$, this would imply that the time spent at each level ℓ of the hierarchy is $O(n)$, yielding a total of $O(n)$ time per insertion. To avoid this cost, a balanced binary tree is attached to each Voronoi region in the Voronoi hierarchy. The tree attached to the region $V_\ell(o)$ of o in $\mathrm{Vor}(\Theta_\ell)$ includes, for each Voronoi vertex v of $V_\ell(o_i)$, the ray $[v_o, v)$ issued from the projection of v onto the boundary ∂o of o that passes through v. The rays are sorted by directions. Similarly, we associate to the query point x the ray $[x_o, x)$ where x_o is the projection of x onto ∂o. When $V_\ell(o)$ is visited, ray $[x_o, x)$ is located in the associated tree and we get the two rays $[v_o, v)$ and $[w_o, w)$ immediately before and after $[x_o, x)$. Let o' be the neighbor of o whose cell is incident to v and w. We compare the distances from x to o and o'. If o' is closer to x than o, the walk steps to o'. Otherwise, we know that o is the object of Θ_ℓ closest to x and the walk halts (see [217] for details). Hence, visiting the Voronoi region of o_i in $\mathrm{Vor}(\Theta_\ell)$ reduces to querying the tree and comparing the distances from x to o and o' which takes $O(\log n_\ell)$ time.

Lemma 9. *Using a hierarchy of Voronoi diagrams, nearest neighbor queries can be answered in expected time $O(\log^2 n)$.*

It has been shown [217] that the expected cost of updating the Voronoi hierarchy when inserting an object is $O(\log^2 n)$.

Exercise 24. Show that the planar Euclidean Voronoi diagram of n points can be computed on line in $O(n \log n)$ time
(Hint: in the case of points, using a Delaunay hierarchy instead of the Voronoi hierarchy, nearest neighbor queries can be answered in $O(\log n)$. Upon insertion the structure is updated in randomized time $O(\log n)$. See [115])

Exercise 25. Show that the planar Euclidean Voronoi diagram of n line segments can be computed on line in $O(n \log n)$ time. See e.g.[67] for a solution.

Exercise 26. Show that the planar euclidean diagram of n disjoint convex objects can be computed using predicates that involve only four objects.

Exercise 27. Provide a detailed description of the predicates of the incremental Voronoi diagram construction and a way to implement them efficiently for various types of simple objects (e.g., line segments, circles). See [219, 218].

Exercise 28 (2D Abstract Voronoi diagrams). Klein et al. [231] have defined abstract Voronoi diagrams in dimension 2 using bisecting curves. Each bisecting curve b_{ij} is assumed to be an infinite curve separating the plane in two regions affected respectively to o_i and o_j and the Voronoi regions are defined as in Sect. 2.5.1. Klein et al. assume that the affectation fulfills the assignment condition but they do not assume the incidence condition. Instead they assume that each pair of bisecting curves intersect in only a finite number of connected components and that the interior of Voronoi regions are path-connected. Show that, under Klein et al. assumptions, the transitivity relation of Lemma 4 is satisfied and that Voronoi regions are simply connected.

Exercise 29. Let $\mathcal{O}$ be a set of planar convex objects that may intersect and may not form a pseudo-circle set. Show that the Voronoi diagram $\mathrm{Vor}(\mathcal{O})$ may exhibit disconnected Voronoi regions. Propose an extension of the incremental algorithm to build the restriction $\mathrm{Vor}(\mathcal{O}) \cap U^c$ of the Voronoi diagram $\mathrm{Vor}(\mathcal{O})$ to the region U^c which is the complement of the union of the objects. The solution can be found in [217].

Exercise 30. Describes the geometric predicates required to implement the incremental algorithm. Provide algebraic expressions for the case of circles and line segments [219, 146].

2.7 Medial Axis

In this section, we introduce the concept of Medial Axis of a bounded set Ω, which can be seen as an extension of the notion of Voronoi diagram to infinite sets. Interestingly, it is possible to construct certified approximations of the medial axis of quite general sets efficiently. One approach to be described consists in sampling the boundary of Ω and then computing an appropriate subset of the Voronoi diagram of the sample which approximates the medial axis. Hence the problem of approximating the medial axis of Ω boils down to sampling the boundary of Ω, a problem that is closely related to mesh generation (see Chap. 5). Other informations on the medial axis can be found in Chap. 6.

2.7.1 Medial Axis and Lower Envelope

The *medial axis* of an open set Ω, denoted by $\mathcal{M}(\Omega)$, is defined as the the set of points of Ω that have more than one nearest neighbor on the boundary of Ω. Nearest refers in this section to the Euclidean distance although the results may be extended to other distance functions. A *medial sphere* σ is a sphere centered at a point c of the medial axis and passing through the nearest neighbors of c on $\partial\Omega$. Those points where σ is tangent (in the sense that σ does not enclose any point of $\partial\Omega$) to $\partial\Omega$ are called the *contact points* of σ.

The concept of medial axis can be considered as an extension of the notion of Voronoi diagram to infinite sets. Let o be a point of the boundary of Ω and δ_o be the distance function to o defined over Ω

$$\forall x \in \Omega : \ \delta_o(x) = \|x - o\|.$$

The *lower envelope* of the infinite set of functions δ_o is defined as

$$\Delta^- = \inf_{o \in \partial\Omega} \delta_o.$$

Following what we did for Voronoi diagrams (Sect. 2.2), we define, for any point $x \in \Omega$, its index set $I(x)$ as the set of all o such that $\Delta^-(x) = \delta_o(x)$. The set of points x such that $|I(x)| > 1$ constitutes the medial axis of Ω.

Computing the medial axis is difficult in general. If Ω is defined as a semi-algebraic set, i.e. a finite collection of algebraic equations and inequalities, $\mathcal{M}(\Omega)$ is also a semi-algebraic set that can therefore be computed using techniques from real algebraic geometry [32, 44]. This general approach, however, leads to algorithms of very high complexity. Theorem 1 can also be used but, still, working out the algebraic issues is a formidable task. Effective implementations are currently limited to simple objects. If Ω is a planar domain bounded by line segments and circular arcs, one can apply the results of Sect. 2.6. Further results can be found in [159].

An alternative and more practical approach consists in departing from the requirement to compute the medial axis exactly. In Sect. 2.7.2, we describe a method that approximates the medial axis of an object by first sampling its boundary, and then computing and pruning the Voronoi diagram of the sample.

2.7.2 Approximation of the Medial Axis

Approximating the medial axis of a set is a non trivial issue since sets that are close for the Hausdorff distance may have very different medial axes. This is illustrated in Fig. 2.13. Let S be a closed curve, $\Omega = \mathbb{R}^2 \setminus S$ and $\mathcal{P}$ be a finite set of points approximating S. As can be seen on the figure, the skeleton of the Voronoi diagram of $\mathcal{P}$ is far from the medial axis of Ω since there are many long branches with no counterpart in $\mathcal{M}(\Omega)$. These branches are Voronoi edges whose dual Delaunay edges are small (their lengths tend to 0 when the sampling density increases). In other words, the medial axis is not continuous under the Hausdorff distance. Notice however that if we remove the long branches in Fig. 2.13, we obtain a good approximation of the medial axis of S. This observation will be made precise in Lemma 10 below. It leads to an approximate algorithm that first sample S and extract from the Euclidean Voronoi diagram of the sample a sub-complex that approximates the medial axis of S.

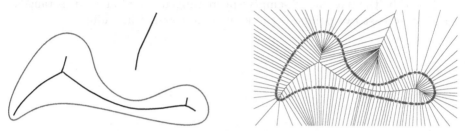

Fig. 2.13. On the left side, a closed curve S and the medial axis of $\Omega = \mathbb{R}^2 \setminus S$. On the right side, a dense sample E of points and its Voronoi diagram. The medial axis of $\mathbb{R}^2 \setminus E$ (which is the 1-skeleton of $\mathrm{Vor}(E)$) is very different from the medial axis of Ω

Given an open set Ω and a point x on the medial axis of Ω, we define $D(x)$ as the diameter of the smallest closed ball containing the contact points of the medial sphere centered at x. We define the λ-medial axis of Ω, denoted $\mathcal{M}_\lambda(\Omega)$ as the subset of the medial axis of Ω consisting of points x such that $D(x) \geq \lambda$.

Let Ω and Ω' be two open sets and let S and S' denote their boundaries. We assume that S and S' are compact and that their Hausdorff distance $d_H(S, S')$ is at most ε: any point of S is at distance at most ε from a point of S' and vice versa. Notice that we do not specify S nor S' to be finite or infinite point sets. For convenience, we rename medial axis of S (resp. S') and write $\mathcal{M}(S)$ (resp. $\mathcal{M}(S')$) the medial axis of $\mathbb{R}^3 \setminus S$. Similarly, we rename medial axis of S' and write $\mathcal{M}(S')$ the medial axis of $\mathbb{R}^3 \setminus S$.

The following lemma says that the λ-medial axis of S is close to the medial axis of S', provided that λ is sufficiently large. It should be emphasized that close here refers to the one-sided Hausdorff distance: the medial axis of S' is not necessarily close to the λ-medial axis of S although, by exchanging the roles of S and S', the lemma states that the λ'-medial axis of S' is close to the medial axis of S for a sufficiently large λ'. We will go back to this point later.

We say that a ball is S-empty if its interior does not intersect S. The sphere bounding a S-empty ball is called a S-empty sphere.

Lemma 10. *Let σ be a S-empty sphere centered at c, of radius r, intersecting S in two points x and y. If $\varepsilon < \frac{r}{2}$ and $l \stackrel{\text{def}}{=} \frac{\|x-y\|}{4} \geq \sqrt{\varepsilon r(1 - \frac{\varepsilon}{r})}$, there exists an S'-empty sphere tangent to S' in two points whose center c' and radius r' are such that $|1 - \frac{r'}{r}|$ and $\frac{\|c'-c\|}{r}$ are at most $\delta = \frac{\varepsilon r}{l^2 - \varepsilon r + \varepsilon^2}$.*

Proof. Let σ'' be the maximal S'-empty sphere centered at c and let r'' be its radius (see Fig. 2.14). The Hausdorff distance between S and S' being at most ε, we have $|r - r''| \leq \varepsilon$. Let y' be a point of $\sigma'' \cap S'$.

Let σ' be the maximal S'-empty sphere tangent to σ'' at y'. σ' is tangent to S' at at least two points. Let c' be its center and r' its radius.

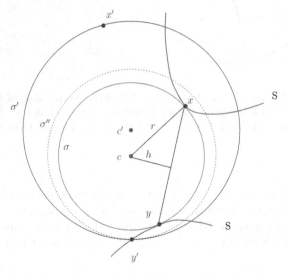

Fig. 2.14. S is the continuous curve. x' and y' belong to S'

Noting $h = \sqrt{r^2 - \frac{1}{4}\|x - y\|^2} = \sqrt{r^2 - 4l^2}$ the distance from c to line xy, we have

$$d(c', S) \leq \min(\|c' - x\|, \|c' - y\|)$$

$$\leq \sqrt{(\|c - c'\| + h)^2 + \frac{1}{4}\|x - y\|^2}$$

$$= \sqrt{r^2 + \|c - c'\|^2 + 2h\|c - c'\|}.$$

On the other hand,

$$d(c', S') = \|c' - y'\| = \|c - c'\| + r'' \geq \|c - c'\| + r - \varepsilon.$$

From these two inequalities, we deduce

$$\|c - c'\| + r - \varepsilon \leq d(c', S') \leq d(c', S) + \varepsilon \leq \sqrt{r^2 + \|c - c'\|^2 + 2h\|c - c'\|} + \varepsilon.$$

and, since, by assumption, $r > 2\varepsilon$, we get $\|c - c'\| (r - 2\varepsilon - h) \leq 2\varepsilon(r - \varepsilon)$. Moreover, by assumption, $l \geq \sqrt{\varepsilon r(1 - \frac{\varepsilon}{r})}$ which implies $r - 2\varepsilon - h \geq 0$. Indeed,

$$h = \sqrt{r^2 - 4l^2} \leq \sqrt{r^2 - 4\varepsilon r + 4\varepsilon^2} = r - 2\varepsilon.$$

We then deduce

$$\|c - c'\| \leq \frac{2\varepsilon(r - \varepsilon)}{r - h - 2\varepsilon} \leq \frac{2\varepsilon r}{r - \sqrt{r^2 - 4l^2} - 2\varepsilon}.$$

We then get

$$\frac{\|c - c'\|}{r} \leq \frac{2\varepsilon(r + \sqrt{r^2 - 4l^2} - 2\varepsilon)}{(r - 2\varepsilon)^2 - (r^2 - 4l^2)} \leq \frac{\varepsilon r}{l^2 - \varepsilon r + \varepsilon^2}.$$

The same bound plainly holds for $|1 - \frac{r'}{r}|$.

We consider the case where S is a surface of $\mathbb{R}^3$ and S' is a finite set of points on S at Hausdorff distance at most ε from S. To avoid confusion, we rename S' as $\mathcal{P}$. As already noticed, $\mathcal{M}(\mathcal{P})$ is the 1-skeleton of the Voronoi diagram $\mathrm{Vor}(\mathcal{P})$, called simply the Voronoi diagram of $\mathcal{P}$ in this section. The set of contact points of a medial sphere centered at a point $c \in \mathcal{M}(\mathcal{P})$ is the set of closest points of c in $\mathcal{P}$. Any point in the relative interior of a face of $\mathrm{Vor}(\mathcal{P})$ has the same closest points in $\mathcal{P}$. It follows that the λ-medial axis of $\mathcal{P}$ is the subset of the faces of $\mathrm{Vor}(\mathcal{P})$ whose contact points cannot be enclosed in a ball of diameter λ.

Lemma 10 then says that any Delaunay sphere of $\mathrm{Del}(\mathcal{P})$ passing through two sample points that are sufficiently far apart, is close to a medial sphere of S (for the Hausdorff distance). We have therefore bounded the *one-sided* Hausdorff distance from the λ-medial axis (Voronoi diagram) of an ε-sample $\mathcal{P}$ of S to the medial axis of S, when λ is sufficiently large with respect to ε. If we apply the lemma the other way around, we see that, for sufficiently large λ', $\mathcal{M}_{\lambda'}(S)$ is close to $\mathcal{M}(\mathcal{P})$. However, as observed above (Fig. 2.13), we cannot hope to bound the two-sided Hausdorff distance between $\mathcal{M}(S)$ and $\mathcal{M}(\mathcal{P})$.

The above lemma can be strengthened as recently shown by Chazal and Lieutier [83, 32]. They proved that the λ-medial axis of S is close to the λ'-medial axis of $\mathcal{P}$ for a sufficiently large λ and some positive λ' that depends on λ and ε. More precisely, let D be the diameter of S and k'' a positive constant. They showed that there exist three functions of ε, $k(\varepsilon) = 15\sqrt{2}\sqrt[4]{D^3\varepsilon}$, $k'(\varepsilon) = 10\sqrt{3}\sqrt[4]{D^3\varepsilon}$ and $k''(\varepsilon) = k''\sqrt[4]{D^3\varepsilon}$, such that

$$\mathcal{M}_{k(\varepsilon)}(S) \subset \mathcal{M}_{k'(\varepsilon)}(\mathcal{P}) \oplus B_{2\sqrt{D\varepsilon}} \subset \mathcal{M}_{k''(\varepsilon)}(S) \oplus B_{4\sqrt{D\varepsilon}}.$$

Here, B_r denotes the ball centered at the origin of radius r, and $\oplus$ the Minkowski sum.

Consider now a family of point sets $\mathcal{P}_\varepsilon$ parametrized by ε such that $d_H(S, \mathcal{P}_\varepsilon) \leq \varepsilon$ and let ε tends to 0. Because $\mathcal{M}_\eta(S)$ tends to $\mathcal{M}(S)$ when η tends to 0, we deduce from the above inequalities, that

$$\lim_{\varepsilon \to 0} d_H(\mathcal{M}(S), \mathcal{M}_{10\sqrt[4]{9D^3\varepsilon}}(\mathcal{P}_\varepsilon)) = 0.$$

The above discussion provides an algorithm to approximate the medial axis of S within any specified error (see Algorithm 5).

Algorithm 5 Approximation of the Medial Axis

INPUT: A surface S and a positive real ε

1. Sample S so as to obtain a sample $\mathcal{P}$ such that $d_H(S, \mathcal{P}) \leq \varepsilon$;
2. Construct the Voronoi diagram of $\mathcal{P}$;
3. Remove from the diagram the faces for which the diameter of the set of contact points is smaller than $10\sqrt[4]{9D^3\varepsilon}$.

OUTPUT: A PL approximation of $\mathcal{M}(S)$

The main issue is therefore to compute a sample of points on S (step 1). If S is a surface of $\mathbb{R}^3$, one can use a surface mesh generator to mesh S and take for $\mathcal{P}$ the vertices of the mesh. Various algorithms can be found in Chap. 5 and we refer to that chapter for a thorough description and analysis of these algorithms. Especially attractive in the context of medial axis approximation, are the algorithms that are based on the 3-dimensional Delaunay triangulation since we get the Voronoi diagram of the sample points (step 2) at no additional cost. An example obtained with the surface mesh generator of Boissonnat and Oudot [65] is shown in Fig. 2.15.

Exercise 31. Let O be a bounded open set. Show that $\mathcal{M}(O)$ is a retract of O (and therefore has the same homotopy type as O) [243].

2.8 Voronoi Diagrams in CGAL

The Computational Geometry Algorithms Library CGAL [2] offers severals packages to compute Voronoi diagrams. Euclidean Voronoi diagrams of points and power diagrams are represented through their dual Delaunay and regular triangulations. CGAL provides Delaunay and regular triangulations in $\mathbb{R}^2$ and $\mathbb{R}^3$. The implementation is based on a randomized incremental algorithm

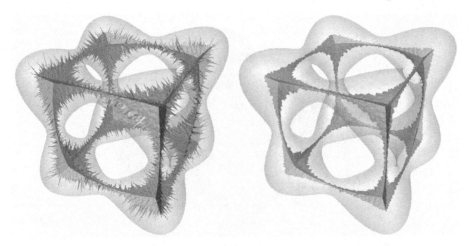

Fig. 2.15. Two λ-medial axes of the same shape, with λ increasing from left to right, computed as a subset of the Voronoi diagram of a sample of the boundary (courtesy of Steve Oudot)

using a variant of the Voronoi hierarchy described in Sect. 2.6. Delaunay triangulations are also provided in higher dimensions.

The library also contains packages to compute Voronoi diagrams of line segments [215] and Apollonius diagrams in $\mathbb{R}^2$ [216]. Those packages implement the incremental algorithm described in Sect. 2.6. A prototype implementation of Möbius diagrams in $\mathbb{R}^2$ also exists. This prototype computes the Möbius diagram as the projection of the intersection of a 3-dimensional power diagram with a paraboloid, as described in Sect. 2.4.1. This prototype also serves as the basis for the developement of a CGAL package for 3-dimensional Apollonius diagrams, where the boundary of each cell is computed as a 2-dimensional Möbius diagram, following the results of Sect. 2.4.3 [62]. See Fig. 2.8.

2.9 Applications

Euclidean and affine Voronoi diagrams have numerous applications we do not discuss here. The interested reader can consult other chapters of the book, most notably Chap. 5 on surface meshing and Chap. 6 on reconstruction. Other applications can be found in the surveys and the textbooks mentionned in the introduction.

Additively and multiplicatively weighted distances arise when modeling growing processes and have important applications in biology, ecology and other fields. Consider a number of crystals, all growing at the same rate, and all starting at the same time : one gets a number of growing circles. As these circles meet, they draw a Euclidean Voronoi diagram. In reality, crystals start

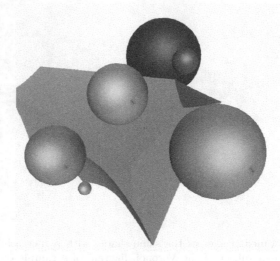

Fig. 2.16. A cell in an Apollonius diagram of spheres

growing at different times. If they still grow at the same rate, they will meet along an Apollonius diagram. This growth model is known as the Johnson-Mehl model in cell biology. In other contexts, all the crystals start at the same time, but grow at different rates. Now we get what is called the multiplicatively weighted Voronoi diagram, a special case of Möbius diagrams.

Spheres are common models for a variety of objects such as particles, atoms or beads. Hence, Apollonius diagrams have been used in physics, material sciences, molecular biology and chemistry [245, 339, 227, 228]. They have also been used for sphere packing [246] and shortest paths computations [256].

Euclidean Voronoi diagrams of non punctual objects find applications in robot motion planning [237, 197]. Medial axes are used for shape analysis [160], for computing offsets in Computer-Aided Design [118], and for mesh generation [290, 289, 316]. Medial axes are also used in character recognition, road network detection in geographic information systems, and other applications.

Acknowledgments

We thank D. Attali, C. Delage and M. Karavelas with whom part of the research reported in this chapter has been conducted. We also thank F. Chazal and A. Lieutier for fruitful discussions on the approximation of the medial axis.

3

Algebraic Issues in Computational Geometry

Bernard Mourrain*, Sylvain Pion, Susanne Schmitt, Jean-Pierre Técourt,
Elias Tsigaridas, and Nicola Wolpert

3.1 Introduction

Geometric modeling plays an increasing role in fields at the frontier between
computer science and mathematics. This is the case for example in CAGD
(Computer-aided Geometric design, where the objects of a scene or a piece to
be built are represented by parameterized curves or surfaces such as NURBS),
robotics or molecular biology (rebuilding of a molecule starting from the matrix of the distances between its atoms obtained by NMR).

The representation of shapes by piecewise-algebraic functions (such as B-
spline functions) provides models which are able to encode the geometry of
an object in a compact way. For instance, B-spline representations are heavily
used in Computed Aided Geometric Design, being now a standard for this
area. Recently, we also observe a new trend involving the use of patches of
implicit surfaces. This includes in particular the representation by quadrics,
which are more natural objects than meshes for the representation of curved
shapes.

From a practical point of view, critical operations such as computing intersection curves of parameterized surfaces are performed on these geometric
models. This intersection problem, as a typical example linking together geometry, algebra and numeric computation, received a lot of attention in the past
literature. See for instance [158, 280, 233]. It requires robust methods, for solving (semi)-algebraic problems. Different techniques (subdivision, lattice evaluation, marching methods) have been developed [278, 176, 14, 191, 190, 280].
A critical question is to certify or to control the topology of the result.

From a theoretical point of view, the study of algebraic surfaces is also a
fascinating area where important developments of mathematics such as singularity theory interact with visualization problems and the rendering of mathematical objects. The classification of singularities [29] provides simple algebraic formulas for complicated shapes, which geometrically may be difficult to

---

* Chapter coordinator

handle. Such models can be visualized through techniques such as ray-tracing[1] in order to produce beautiful pictures of these singularities. Many open questions, related for instance to the topological types of real algebraic curves or surfaces, remain to be solved in this area. Computation tools, which allow to treat such algebraic models are thus important to understand their geometric properties.

In this chapter, we will describe methods for the treatment of algebraic models. We focus on the problem of computing the topology of implicit curves or surfaces. Our objective is to devise certified and output-sensitive methods, in order to combine control and efficiency. We distinguish two types of subproblems:

- the construction of new geometric objects such as points of intersection,
- predicates such as the comparison of coordinates of intersection points.

In the first case, a good approximation of the exact algebraic object, which usually cannot be described explicitly by an analytic formula, may be enough. On the contrary for the second subproblem, the result has to be exact in order to avoid incoherence problems, which might be dangerous from an implementation point of view, leading to well known non-robustness issues.

These two types of geometric problems, which appear for instance in arrangement computations (see Chapter 1) lead to the solution of algebraic questions. In particular, the construction or the comparison of coordinates of points of intersections of two curves or three surfaces involve computations with algebraic numbers. In the next section, we will describe exact methods for their treatment. Then we show how to apply these tools to compute the topology of implicit curves. This presentation includes effective aspects and pointers to software. It does not include proofs, which can be found in the cited literature.

3.2 Computers and Numbers

Geometric computation is closely tied to arithmetic, as the Ancient Greeks (in particular Pythagoras of Samos and Hippasus of Metapontum) observed a long time ago. This has been formalized more recently by Hilbert [205], who showed how geometric hypotheses are correlated with the arithmetic properties of the underlying field. For instance, it is well-known that Pappus' theorem is equivalent to the commutativity property of the underlying arithmetic field. When we want to do geometric computations on a computer, the situation becomes even more intricate. First, we cannot represent all real numbers on a computer.

Integers (even integers of unbounded size) are the basis of arithmetic on a computer. These integers are (usually) represented in the binary system as an

---

[1]see e.g. http://www.algebraicsurface.net/

array of bits; an integer n has (bit) size $O(\log |n|)$. Under this notion, integers are no longer constant size objects thus arithmetic operations on them are performed in non-constant time: for two integer of bit size $O(\log |n|)$ addition or subtraction can be done in linear time with respect to their size, i.e $O(\log |n|)$ and multiplication or division can be done in $O(\log |n| \log \log |n| \log \log \log |n|)$. Therefore, depending on the context, manipulating multi precision integers can be expensive. Dedicated libraries such as GMP[6] however have been tuned to treat such large integers.

Similarly, rational numbers can be manipulated as pairs of integer numbers. As in Pythagoras' philosophy, these numbers can be considered as the foundations of computer arithmetic. That is why, hereafter, we will consider that our input (which as we will see in the next sections corresponds to the coefficients of a polynomial equation) will be represented with rational numbers $\in \mathbb{Q}$. In other words, we will consider that the input data of our algorithms are exact. From the complexity point of view, the cost of the operations on rationals is a simple consequence of the one on integers, however we can also point out that adding rationals roughly doubles their sizes, contrary to integers, so additional care has to be taken to get good performance with rationals.

When performing geometric computations, such as for instance computing intersections, the values that we need to manipulate are no longer rationals. We are facing *Pythagoras' dilemma*: how to deal with *non-commensurable* values, when only rational arithmetic is effectively available on a computer. In our context, these non-commensurable values are defined implicitly by equations whose coefficients are rationals. As we will see, they involve algebraic numbers. A classical way to deal with numbers which are not representable in the initial arithmetic model, is to approximate them. This is usually performed by floating point numbers. For instance, numerical approximations can be sufficient, for evaluation purposes, if one controls the error of approximation. And usually, computations with approximate values is much cheaper than with the exact representation. The important problem which has to be handled is then how to control the error.

Hereafter, we describe shortly this machine floating point arithmetic and interval arithmetic, for their use in geometric computation.

3.2.1 Machine Floating Point Numbers: the IEEE 754 norm

Besides multiple-precision arithmetic provided by various software libraries, modern processors directly provide in hardware some floating point arithmetic in a way which has been standardized as the IEEE 754 norm [212]. We briefly describe the parts of this norm which are interesting in the sequel.

The IEEE 754 norm offers several possible precisions. We are going to describe the details of the so-called double precision numbers, which correspond to the `double` built-in types of the C and C++ languages. These numbers are encoded in 64 bits: 1 bit for the sign, 11 bits for the exponent, and 52 bits for the mantissa.

For non-extreme values of the exponent, the real value corresponding to the encoding is simply: $(-1)^{sign} \times 1.mantissa \times 2^{exponent-1023}$. That is, there is an implicit 1 which is not represented in front of the mantissa, and the exponent value is shifted in order to be centered at zero.

Extreme values of the exponent are special: when it is zero, then the numbers are called denormalized values and the implicit 1 disappears, which leads to a nice property called gradual underflow. This property implies that there cannot be any underflow with the subtraction or the addition: $a - b = 0 \iff a = b$. The maximal value 2047 for the exponent is used to represent 4 different special values: $+\infty$, $-\infty$, qNAN, sNAN, depending on the sign bit and the value of the mantissa. Infinite values are generated by overflow situations, or when dividing by zero. A NaN (not a number) exists in two variants, quiet or signaling, and is used to represent the result of operations like $\infty - \infty$, $0 \times \infty$, $0/0$ and any operation taking a NaN as argument.

The following arithmetic operations are specified by the IEEE 754 standard: $+$, $-$, $\times$, $\div$, $\sqrt{}$. Their precise meaning depends on a rounding mode, which can have 4 values: to the nearest (with the *round-to-even* rule in case of a tie), towards zero, towards $+\infty$ and towards $-\infty$. This way, an arithmetic operation is decomposed into its exact real counterpart, and a rounding operation, which is going to choose the representable value in cases where the real exact value is not representable in the standard format. In the sequel, the arithmetic operations with directed rounding modes are going to be written as $\underline{+}$ and $\overline{\times}$, standing for addition rounded towards $+\infty$ and multiplication rounded towards $-\infty$ for example.

Finally, let us mention that the IEEE 754 norm is currently under revision, and we can expect that in the future more operations will be available in a standardized way.

3.2.2 Interval Arithmetic

Interval arithmetic is a well known technique to control accumulated rounding errors of floating point computations at run time. It is especially used in the field of interval analysis [257]. We use interval arithmetic here in the following way: we represent at run time the roundoff error associated with a variable x by two floating point numbers $\underline{x}$ and $\overline{x}$, such that the exact value of x lies in the interval $[\underline{x}, \overline{x}]$. This is denoted as the *inclusion property*.

All arithmetic operations on these intervals preserve this property. For example, the addition of x and y is performed by computing the interval $[\underline{x} \underline{+} \underline{y}, \overline{x} \overline{+} \overline{y}]$. The multiplication is slightly more complicated and is specified as

$$x \times y = [\min(\underline{x} \underline{\times} \underline{y}, \underline{x} \underline{\times} \overline{y}, \overline{x} \underline{\times} \underline{y}, \overline{x} \underline{\times} \overline{y}), \max(\underline{x} \overline{\times} \underline{y}, \underline{x} \overline{\times} \overline{y}, \overline{x} \overline{\times} \underline{y}, \overline{x} \overline{\times} \overline{y})].$$

The other basic arithmetic operations $(-, \div, \sqrt{})$ are defined on intervals in a similar way. More complex functions, like the trigonometric functions, can also be defined over intervals on mathematical grounds. However, the IEEE 754 standard does not specify their exact behavior for floating point computations, so it is harder to implement such interval functions in practice, although some libraries can help here.

Comparison functions on intervals are special, and several different semantics can be defined for them. What we are interested in here is to detect when a comparison of the exact value can be guaranteed by the intervals. Therefore looking at the intervals allows to conclude the order of the exact values in the following cases:

$$\overline{x} < \underline{y} \Rightarrow x < y \text{ is true}$$

$$\underline{x} >= \overline{y} \Rightarrow x < y \text{ is false}$$

$$\texttt{otherwise} \Rightarrow x < y \text{ is unknown}$$

The other comparison operators $(>, \leq, \geq, =, \neq)$ can be defined similarly.

From the implementation point of view, the difficulty lies in portability, since the IEEE 754 functions for changing the rounding modes tend to vary from system to system, and the behavior of some processors does not always match perfectly the standard. In practice, operations on intervals can be roughly 5–10 times slower than the corresponding operations on floating point numbers, this is what we observe on low degree geometric algorithms.

Interval arithmetic is very precise compared to other methods which consist in storing a central and an error values, as the IEEE 754 norm guarantees that, at each operation, the smallest interval is computed. It is possible to get more precision from it by using multiple-precision bounds, or by rewriting the expressions to improve numerical stability for some expressions [69] which improves the sharpness of the intervals.

3.2.3 Filters

Most algebraic computations are based on evaluating numerical quantities. Sometimes, like in geometric predicates, only signs of quantities are needed in the end.

Computing with multiple-precision arithmetic in order to achieve exactness is by nature costly, since arithmetic operations do not have unit cost, in contrast to floating-point computations. It is also common to observe that floating point computation almost always leads to correct results, because the error propagation is usually small enough that sign detection is exact. Wrong signs tend to happen when the polynomial value of which the sign is sought is zero, or small compared to the roundoff error propagation. Geometrically, this usually means a degenerate or nearly degenerate instance.

Arithmetic filtering techniques have been introduced in the last ten years [168] in order to take advantage of the efficiency of floating point computations, *but* by also providing a certificate allowing to determine whether the sign of the approximately computed value is the same as the exact sign.

In the case of *filter failure*, i.e., when the certificate cannot guarantee that the sign of the approximation is exact, then another method must be used to obtain the exact result: it can be a more precise filter, or it can be multiple-precision arithmetic directly.

From the complexity point of view, if the filter step succeeds often—which is expected—then the cost of the exact method will be amortized over many calls to the predicates. The probability that the filter succeeds is linked to two factors. The first is the shape of the predicate: how many arithmetic operations does it contain and how do they influence the roundoff-error (the degree of the predicate does not really matter in itself). The second factor is the distribution of the input data of the predicates, since filter failures are more common on degenerate or nearly degenerate cases.

There are various techniques which can be used to implement these filters. They vary by the cost of the computation of the certificate, and by their precision, i.e. their typical failure rate. Finding the optimal filter for a problem may not be easy, and in general, the best solution is to use a cascade of filters [74, 117]: first try the less precise and fastest one, and in case of failure, continue with a more precise and more costly one, etc. Detailed experiments illustrating this have been performed in the case of the 3D Delaunay triangulation used in surface reconstruction in [117].

We are now going to detail two important categories of filters: dynamic filters using interval arithmetic, and static filters based on static analysis of the shape of predicates.

Dynamic Filters

Interval arithmetic, as we previously described it in 3.2.2, can be used to write filters for the evaluation of signs of polynomial expressions, and even a bit more since division and square root are also defined.

Interval arithmetic is easy to use because no analysis of a particular polynomial expression is required, and it is enough to instantiate the polynomials with a given arithmetic without changing their evaluation order. It is also the most precise approach within the hardware precision since the IEEE 754 standard guarantees the smallest interval for each individual operation. We are next going to present a less precise but faster approach known as static filters.

Static Filters

Interval arithmetic computes the roundoff error at run time. Another idea which has been initially promoted by Fortune [168] is to pull more of the error computation off run time.

The basic idea is the following: if you know a bound b on the input variables $x_1, \ldots, x_n$ of the polynomial expression $P(x_1, \ldots, x_n)$, then it is possible to deduce a bound on the roundoff error ϵ_P that will occur during the evaluation of P. This can be shown inductively, by considering the roundoff error propagation bound of each operation, for example for the addition: suppose x and y are variables you want to add, b_x and b_y are bounds on $|x|$ and $|y|$ respectively, and ϵ_x and ϵ_y bounds on the roundoff errors done so far on x and y. Then it is easy to see that $|x+y|$ is bounded by $b_{x+y} = b_x + b_y$, and that the roundoff error is bounded by $\epsilon_x + \epsilon_y + b_{x+y} 2^{-53}$, considering IEEE 754 double precision floating point computations. Similar bounds can be computed for subtraction and multiplication. Division does not play nicely here because the result is not bounded.

This scheme can also be refined in several directions by:

- considering independent initial bounds on the input variables,
- computing the bounds on the input and the epsilons at run time, which is usually still fast since the polynomial expressions we are dealing with tend to be homogeneous due to their geometric nature [252],
- doing some caching on this last computation [117],

Such filters are very efficient when a bound on the input is known, because the only change compared to a simple floating point evaluation is the sign comparison which is made with a constant ϵ whereas it would be with 0 otherwise. Drawbacks of these methods are that they are less precise, and so they need to be complemented by dynamic filters to be efficient in general. They are also harder to program since they are more difficult to automatize (the shape of the predicates needs to be analyzed). This is why some automatic tools have been developed to generate them from the algebraic formulas of the predicates [169, 273, 74].

3.3 Effective Real Numbers

In this section, we will consider a special type of real numbers, which we call *effective real numbers*. We will be able to manipulate them effectively in geometric computations, because the following methods are available:

- an algorithm which computes a numerical approximation of them to any precision.
- an algorithm which compares them in an exact way.

We will see that working in this sub-class of real numbers, is enough to tackle the geometric problems that we want to solve. Namely, we are interested by computing intersection points of curves, arrangements of pieces of algebraic curves and surfaces, ... This leads to the resolution of polynomial equations.

Here are some notations. A polynomial over a ring $\mathbb{L}$ of coefficients is an expression of the form

$$f(x) = a_n x^n + \cdots + a_1 x + a_0$$

where the *coefficients* $a_n \neq 0, a_{n-1}, \ldots, a_1, a_0$ are elements of $\mathbb{L}$ and the *variable* x may be regarded as a formal symbol with an indeterminate meaning. The greatest power of x appeared in f (with an non zero coefficient) is called the *degree* of f, (n in our case since $a_n \neq 0$). It is denoted $\deg(f)$. The degree of the zero polynomial is equal to $-\infty$. The coefficient a_n is called the *leading coefficient*, and denoted $\mathrm{ldcf}(f)$. The ring of polynomials with coefficient in $\mathbb{L}$, is denoted $\mathbb{L}[x]$.

We call a polynomial $g \in \mathbb{L}[x]$ a *factor* of f if there exists another polynomial $g \in \mathbb{L}[x]$ with $f = g \cdot h$. In particular, if $f = 0$, then every $g \in \mathbb{L}[x]$ is a factor of f.

In the following, we will consider polynomials with coefficient in a unitary ring $\mathbb{L}$. For instance, we can image $\mathbb{L} = \mathbb{Z}$. We denote by $\mathbb{K}$ a field containing $\mathbb{L}$. In the following, we work most of the time with $\mathbb{K}$ the field of rational numbers or its algebraic closure (that is the smallest field containing all the roots of polynomials with rational coefficients). In some cases, the problem may depend on parameters $u_1, \ldots, u_n$ and so in theses cases the field $\mathbb{K}$ will be the fraction field $\mathbb{K} = \mathbb{Q}(u_1, \ldots, u_n)$. The algebraic closure of the field $\mathbb{K}$ is denoted $\overline{\mathbb{K}}$. (so image $\overline{\mathbb{K}} = \mathbb{C}$).

3.3.1 Algebraic Numbers

We recall here the basic definitions on algebraic numbers. An *algebraic number* over the field $\mathbb{K}$ is a root of a polynomial $p(x)$ with coefficients in $\mathbb{K}$ ($p(x) \in \mathbb{K}[x]$). An *algebraic integer* over the ring $\mathbb{L}$ is a root of a polynomial with coefficients in $\mathbb{L}$, where the leading coefficient is 1.

Let α be an algebraic number over $\mathbb{K}$ and $p(x) \in \mathbb{K}[x]$ be a polynomial of degree d with $p(\alpha) = 0$. If $p(x)$ is irreducible over $\mathbb{K}$ (it cannot be written in $\mathbb{K}[x]$ as the product of two polynomials which are both different from 1), it is called the *minimal polynomial* of α. The other roots $\alpha_2, \ldots, \alpha_d$ of the minimal polynomial in $\overline{\mathbb{K}}$ are the *conjugates* of α. The *degree* of the algebraic number α is the degree of the minimal polynomial defining α. Let $\alpha_1 = \alpha$, then the *norm* of α is

$$N(\alpha) = \prod_{i=1}^{d} |\alpha_i|$$

If α, β are algebraic numbers over $\mathbb{K}$, then $\alpha \pm \beta, \alpha \cdot \beta, \alpha/\beta$ (if $\beta \neq 0$) and $\sqrt[k]{\alpha}$ are algebraic numbers over $\mathbb{K}$. If α, β are algebraic integers over $\mathbb{L}$, then $\alpha \pm \beta, \alpha \cdot \beta$ and $\sqrt[k]{\alpha}$ are algebraic integers over $\mathbb{L}$.

For instance, $\gamma = 7$ is an algebraic integer over $\mathbb{Q}$ since it is the root of $x - 7 = 0$. Moreover, $\alpha = \sqrt{2}$ (resp. $\beta = \sqrt{3}$) is an algebraic integer over $\mathbb{Q}$, since it is the positive root of the (minimal) polynomial $x^2 - 2$, (resp. $x^2 - 3$) and $\alpha + \beta$ is a root of $(x^2 - 5)^2 - 24 = x^4 - 10x^2 + 1 = 0$. We observe in the last example, that the degree of the minimal polynomial of $\alpha + \beta$ is bounded by the product of the degrees of the minimal polynomials of α and β. This is a general result, which we deduce from the resultant properties (see

Section 3.4.1 and [236]). The same result is valid for the operations $-, \times, /$ on these algebraic numbers.

Let $p(x)$ be a polynomial with algebraic number as coefficients. Then the roots of $p(x)$ are algebraic numbers. If the coefficients of $p(x)$ are algebraic integers and the leading coefficient of $p(x)$ is 1, then the roots of $p(x)$ are algebraic integers.

We describe now two important methods to represent real algebraic numbers.

3.3.2 Isolating Interval Representation of Real Algebraic Numbers

A natural way to encode a *real* algebraic number α over $\mathbb{Q}$ is by using

- a polynomial $p(x)$ of $\mathbb{Q}[x]$, which vanishes at α, and
- an *isolating interval* $[a, b]$ containing α such that $a, b \in \mathbb{Q}$ and $p(x)$ has exactly one real root in $[a, b]$.

This representation is not unique, since the size of the interval $[a, b]$ can reduce to any $\epsilon > 0$ close to 0. If we assume moreover that p is a square-free polynomial (that is $\gcd(p, p') = 1$ or in other words that the roots of p are distinct), as we assume in what follows, then α is a simple root of p and p obtains different sign when evaluated over the endpoints of the isolating interval, i.e. $p(a)p(b) < 0$.

Besides isolating interval representation there are other representations of real algebraic numbers. The most important alternative is *Thom's encoding* [44]. The basic idea behind this representation is that the signs of all the derivatives of p obtained by evaluation over the real roots of p uniquely characterize (and order) the real roots. This representation besides the uniqueness property is also more general than the isolating interval representation. However we are not going into the details.

For higher (arbitrary) degree, isolating intervals will be computed by univariate polynomial solvers, that we will describe in Section 3.4.2. For polynomials of degree up to 4, real root isolation is more effective since it can be performed in constant time (see Section 3.4.3).

3.3.3 Symbolic Representation of Real Algebraic Numbers

The sum $\alpha + \beta$ of two algebraic numbers α, β of degree $\leq d$ is an algebraic number, whose minimal polynomial is of degree $\leq d^2$ (see Section 3.4.1 and [236]). Instead of computing this minimal polynomial (which might be costly to perform), we may wonder if we can use its symbolic representation (as an arithmetic tree) and develop methods which allow us to consider it as an effective algebraic numbers. In other words,

- how can we approximate such a number within an arbitrary precision?
- how can we compare two such numbers?

In this section, we describe the symbolic representation of such algebraic numbers and in the next section, we will show how to perform these operations.

A *real algebraic expression* is an arithmetic expression built from the integers using the operations $+, -, \cdot, /, \sqrt[k]{\ }$ and $\diamond$, representing a root of a univariate polynomial, and which is defined as follows: The syntax for the $\diamond$-operator is $\diamond(j, E_d, \ldots, E_0)$, where E_i are real algebraic expressions and $1 \leq j \leq d$ is an integer. It is representing the j-th real root (if it exists) of a polynomial with coefficients $(E_i)_{i=0}^d$.

The value $\mathrm{val}(E)$ of a real algebraic expression E is the real value given by the expression (if this is defined). For $E = \diamond(j, E_d, \ldots, E_0)$, the value $\mathrm{val}(E)$ is the j-th smallest real root of the polynomial

$$p(x) = \sum_{i=0}^{d} \mathrm{val}(E_i) x^i,$$

if it exists and if the values of all coefficients are defined.

We are representing real algebraic expressions as directed acyclic graphs. The inner nodes are the operations and the leaves are integers. Every node knows an interval containing the exact value represented by its subgraph. If further accuracy is needed, the values are approximated recursively with higher precision.

Operations are done by creating a new root node and building the graph structure. Then a first approximating interval is computed. Comparisons are done exactly. The algorithms involved in these comparisons will be described in the next section.

3.4 Computing with Algebraic Numbers

In the previous section, we have described how we encode real algebraic numbers. We are going to describe now the main tools and algorithms, which allow us to compute this representation, that is

- to isolate the real roots of a polynomial.

We are also going to see how to perform the main operations we are interested in for geometric computations, namely:

- the comparison of two algebraic numbers,
- and the sign evaluation of a polynomial expression.

3.4.1 Resultant

A fundamental tool for the manipulation and construction of algebraic numbers is the resultant. It allows us to answer the following question; When do two univariate polynomials f and g of positive degree have a non-constant common factor? Since every non-contant polynomial has a root in $\overline{\mathbb{K}}$, this is equivalent to the following question: When do two univariate polynomials f and g of positive degree have a common root in $\overline{\mathbb{K}}$? Here is a first answer:

Theorem 1. *Let $f, g \in \mathbb{K}[x]$ be two polynomials of degrees $\deg(f) = n > 0$ and $\deg(g) = m > 0$. Then f and g have a non-constant common factor if and only if there exist polynomials $A \in \mathbb{K}[x]$ and $B \in \mathbb{K}[x]$ with $\deg(A) < m$ and $\deg(B) < n$ which are not both zero and such that $Af + Bg = 0$.*

Now, in order to determine the existence of a common factor of

$$f(x) = f_n x^n + f_{n-1} x^{n-1} + \cdots + f_0, \quad f_n \neq 0, \ n > 0 \quad \text{and}$$
$$g(x) = g_m x^m + g_{m-1} x^{m-1} + \cdots + g_0, \quad g_m \neq 0, \ m > 0$$

we have to decide whether two polynomials A and B with the required properties can be found. This question can be answered with the help of linear algebra: A and B are polynomials of degree at most $m-1$ and $n-1$, and therefore there are all in all $m + n$ unknown coefficients $a_{m-1}, \ldots, a_0, b_{n-1}, \ldots, b_0$ of A and B:

$$A(x) = a_{m-1} x^{m-1} + \cdots + a_1 x + a_0$$
$$B(x) = b_{n-1} x^{n-1} + \cdots + b_1 x + b_0.$$

The polynomial $A(x)f(x) + B(x)g(x)$ has degree at most $n+m-1$ in x. Each of its coefficients has to be zero in order to achieve $Af + Bg = 0$:

$$
\begin{array}{lll}
f_n a_{m-1} & + \quad g_m b_{n-1} & = 0 \ (\text{coefficient of } x^{m+n-1}) \\
f_{n-1} a_{m-1} + f_n a_{m-2} & + \ g_{m-1} b_{n-1} + g_m b_{n-2} & = 0 \ (\text{coefficient of } x^{m+n-2}) \\
\ddots & \ddots & \vdots \\
f_0 a_0 & + \quad g_0 b_0 & = 0 \ (\text{coefficient of } x^0).
\end{array}
$$

We get $n + m$ linear equations in the unknowns a_i, b_i with coefficients f_i, g_j. Written in matrix style this system of linear equations has the form:

$$
(a_{m-1}, \ldots, a_0, b_{n-1}, \ldots, b_0) \cdot
\begin{pmatrix}
f_n & f_{n-1} & \cdots & f_0 & & & \\
 & f_n & f_{n-1} & \cdots & f_0 & & \\
 & & \ddots & \ddots & & \ddots & \\
 & & & f_n & f_{n-1} & \cdots & f_0 \\
g_m & g_{m-1} & \cdots & g_0 & & & \\
 & g_m & g_{m-1} & \cdots & g_0 & & \\
 & & \ddots & \ddots & & \ddots & \\
 & & & g_m & g_{m-1} & \cdots & g_0
\end{pmatrix}
= (0, \ldots, 0)
$$

where the empty positions are filled with zeroes. We know from linear algebra that this system of linear equations has a non-zero solution if and only if the determinant of the coefficient matrix is equal to zero. This leads to the following definition:

Definition 1. *The $(m+n) \times (m+n)$ coefficient matrix with m rows of f-entries and n rows of g-entries*

$$\mathrm{Syl}(f,g) := \begin{pmatrix} f_n & f_{n-1} & \cdots & f_0 & & & \\ & f_n & f_{n-1} & \cdots & f_0 & & \\ & & \ddots & \ddots & & \ddots & \\ & & & f_n & f_{n-1} & \cdots & f_0 \\ g_m & g_{m-1} & \cdots & g_0 & & & \\ & g_m & g_{m-1} & \cdots & g_0 & & \\ & & \ddots & \ddots & & \ddots & \\ & & & g_m & g_{m-1} & \cdots & g_0 \end{pmatrix}$$

where the empty positions are filled by zeroes is called the Sylvester *matrix of f and g. The determinant of the matrix is called the* resultant *of f and g:* $\mathrm{Res}(f,g) := \det(\mathrm{Syl}(f,g))$.

This resultant will also be denoted $\mathrm{Res}_x(f,g)$ if the coefficients a_i, b_j of f and g are themselves polynomials of other variables and we want to denote that we consider them as univariate polynomials with respect to x.

From the above observations we immediately obtain a criterion for testing whether two polynomials f and g have a non-constant common factor.

Proposition 1. *Given $f, g \in \mathbb{K}[x]$ of positive degree, the resultant $\mathrm{Res}(f,g) \in k$ is equal to zero if and only if f and g have a non-constant common factor. For $\mathbb{K} = \mathbb{C}$ the equality $\mathrm{Res}(f,g) = 0$ holds if and only if f and g have a common complex root.*

As a direct application, we see that if α (resp. β) is a root of the polynomial f (resp. g), then $\alpha + \beta$ is a root of the polynomial $\mathrm{Res}_x(f(x), g(u-x)) = R(u) = 0$ where we consider $g(u-x)$ as a polynomial in x (with coefficients which are polynomials in u). Indeed, the two polynomials $f(x)$ and $g(\alpha+\beta-x)$ have a common root $x = \alpha$, so that their resultant $R(\alpha + \beta)$ vanishes. Similarly, for $\beta \neq 0$, a defining polynomial of $\frac{\alpha}{\beta}$ is $\mathrm{Res}_x(f(x), g(x\,u)) = 0$. Though the resultant yields a direct way to compute a defining polynomial of sums, differences, products, divisions of algebraic numbers, alternative approaches, which are more interesting from a complexity point of view, have been considered (see for instance [17]). They are based on Newton sums and series expansions.

Properties of resultants.

We present some useful properties of the resultants.

Lemma 1. *Let $f, g \in \mathbb{K}[x]$ and $\alpha \in \bar{\mathbb{K}}$.*

  1. For $\deg(f) > 0$ and $\deg(g) = m > 0$ we have $\mathrm{Res}(\alpha \cdot f, g) = \alpha^m \cdot \mathrm{Res}(f,g)$.
  2. If $\deg(g) > 0$, then $\mathrm{Res}((x - \alpha) \cdot f, g) = g(\alpha) \cdot \mathrm{Res}(f,g)$.

The lemma leads to the following important characterization of resultants:

Theorem 2. *Let* $f, g \in \mathbb{K}[x]$, $f_n = \mathrm{ldcf}(f)$, $g_m = \mathrm{ldcf}(g)$, $\deg(f) = n > 0$, $\deg(g) = m > 0$, *with (complex) roots*

$$\alpha_1, \ldots, \alpha_n, \beta_1 \ldots, \beta_m \in \overline{\mathbb{K}}.$$

For the resultant of f *and* g *the following holds:*

$$\mathrm{Res}(f, g) \;=\; f_n^m g_m^n \prod_{i=1}^{n} \prod_{j=1}^{m} (\alpha_i - \beta_j) = f_n^m \prod_{i=1}^{m} g(\alpha_i) = (-1)^{mn} g_m^n \prod_{i=1}^{m} f(\beta_i).$$

Subresultants of Univariate Polynomials

Theorem 1 of the previous section can be generalized to

Theorem 3. *Let* $f, g \in \mathbb{K}[x]$ *be two polynomials of degrees* $\deg(f) = n > 0$ *and* $\deg(g) = m > 0$. *Then* f *and* g *have a common factor of degree greater than* $l \geq 0$ *if and only if there are polynomials* A *and* B *in* $\mathbb{K}[x]$, *with* $\deg(A) < m - l$ *and* $\deg(B) < n - l$ *which are not both zero, and such that* $Af + Bg = 0$.

As an immediate consequence we obtain a statement about the degree of the greatest common divisor of f and g:

Corollary 1. *The degree of the gcd of two polynomials* $f, g \in \mathbb{K}[x]$ *is equal to the smallest index* h *such that for all polynomials* A *and* $B \in \mathbb{K}[x]$, *with* $\deg(A) < m - h$ *and* $\deg(B) < n - h$: $Af + Bg \neq 0$.

This corollary can be reformulated in the following way:

Corollary 2. *The degree of the gcd of two polynomials* $f, g \in \mathbb{K}[x]$ *is equal to the smallest index* h *such that for all rational polynomials* A *and* B *with* $\deg(A) < m - h$ *and* $\deg(B) < n - h$: $\deg(Af + Bg) \geq h$.

We are interested in determining the degree of the greatest common divisor of two polynomials f and g. According to Corollary 2, we have to test in succession whether for $l = 1, 2, 3, \ldots$ there exist polynomials A and B, with the claimed restriction of the degrees such that the degree of $Af + Bg$ is strictly smaller than l. The first index h, for which this test gives a negative answer, is equal to the degree of the gcd. How can we perform such a test? We have seen in the previous section that the test for $l = 0$ can be made by testing whether the resultant of f and g is equal to zero. For $l = 1, 2, 3, \ldots$ we proceed in a similar way. Let l be a fixed index and let

$$f(x) = f_n x^n + f_{n-1} x^{n-1} + \cdots + f_0, \quad f_n \neq 0, \quad \text{and}$$
$$g(x) = g_m x^m + g_{m-1} x^{m-1} + \cdots + g_0, \quad g_m \neq 0.$$

We are looking for two polynomials

$$A(x) = a_{m-l-1}x^{m-l-1} + \cdots + a_1 x + a_0,$$
$$B(x) = b_{n-l-1}x^{n-l-1} + \cdots + b_1 x + b_0,$$

such that $\deg(Af + Bg) < l$. There are $m + n - 2l$ unknown coefficients $a_{m-l-1}, \ldots, a_0, b_{n-l-1}, \ldots, b_0$. The polynomial $A(x)f(x) + B(x)g(x)$ has degree at most $n+m-l-1$. The $m+n-2l$ coefficients of $x^l, x^{l+1}, \ldots, x^{m+n-l-1}$ have to be zero in order to achieve $\deg(Af + Bg) < l$. This leads to a linear system

$$(a_{m-l-1}, \ldots, a_0, b_{n-l-1}, \ldots, b_0) \cdot S_l = (0, \ldots, 0)$$

where S_l is the submatrix of the Sylvester matrix of f and g obtained by deleting the last $2l$ columns, the last l rows of f-entries, and the last l rows of g-entries. We call $\mathrm{sr}_l(f, g) = \det S_l$ the lth *subresultant* of f and g. For $l = 0$, the equality $\mathrm{Res}(f, g) = \mathrm{sr}_0(f, g)$ holds. In fact, S_l is a submatrix of S_i for $l > i \geq 0$. The $2l \times 2l$ minors of the submatrix of the Sylvester matrix of f and g obtained by deleting the last l rows of f-entries, can be collected in order to construct a polynomial, which has interesting properties. To be more specific, we need the following definition:

Definition 2 (Determinant polynomial).
Let $\mathcal{M}$ be a $s \times t$ matrix, $s \leq t$, over an integral domain $\mathbb{L}$. The determinant polynomial of $\mathcal{M}$ is:

$$\mathrm{detpol}(\mathcal{M}) = |\mathcal{M}_s| x^{t-s} + \cdots + |\mathcal{M}_t|$$

where $\mathcal{M}_j$ denotes the submatrix of $\mathcal{M}$ consisting of the first $s - 1$ columns followed by the jth column, for $s \leq j \leq t$.

Definition 3 (Subresultant).
Let $f, g \in \mathbb{K}[x]$, two polynomials with $\deg(f) = n > 0$, $\deg(g) = m > 0$. For $0 \leq l \leq \min(f, g)$, we define:

$$M_l = \mathrm{mat}(x^{n-l-1}f(x), x^{n-l-2}f(x), \ldots, f(x), x^{m-l-1}g(x), \ldots, g(x))$$

Then the lth subresultant polynomial of f and g is $\mathrm{Sr}_l(f, g)(x) = \mathrm{detpol}(M_l)$.

Notice that the coefficient of x^l in $\mathrm{Sr}_l(f, g)$ is the lth subresultant coefficient, denoted $\mathrm{sr}_l(f, g)$. Here is the main proposition [44, 330]:

Proposition 2. *Two polynomials f and g of positive degree have a gcd of degree h if and only if h is the least index l for which $\mathrm{sr}_l(f, g) \neq 0$. In this case, their gcd is $\mathrm{Sr}_l(f, g)(x)$.*

This yields Algorithm 6 for computing the square-free part of a polynomial. By Hadamard's identity (see [44]), the size of the coefficients of the subresultants is bounded linearly (up to a logarithmic factor) in terms of the size of the minors.

Algorithm 6 Square free part of a univariate polynomial

INPUT: a polynomial $f \in \mathbb{K}[x]$.

- Compute the last non-zero subresultant $\mathrm{Sr}(x)$ polynomial of $f(x)$ and $f'(x)$.
- Compute $f^r = f/\mathrm{Sr}(x)$.

OUTPUT: the square-free part f^r of f.

3.4.2 Isolation

We are now going to describe algorithms that compute isolating intervals for the real roots of polynomials of arbitrary degree with rational coefficients. In the next section we will present how isolating intervals can be computed directly for polynomials of degree up to 4. Two families of methods will be considered, one based on Descartes's rule and the other on Sturm sequences. Both produce certified isolation intervals for all the roots of a polynomial and have interesting complexity bounds.

Descartes' rule.

Consider a polynomial $p(x) \in \mathbb{R}[x]$, represented in the monomial basis by

$$p(x) = a_d \, x^d + \cdots + a_0,$$

with $a_d \neq 0$. A simple rule to estimate the number of real roots of p is obtained by counting the number of sign changes of its coefficients:

Definition 4 (Sign changes). *The number of sign changes $V(\mathbf{a})$ in a sequence $\mathbf{a} = \{a_0, \ldots, a_d\}$ in $\mathbb{R} - \{0\}$ is defined, by induction on d, by:*

- $V(a_0) = 0$,
- $V(a_0, \ldots, a_d) = V(a_1, \ldots, a_d) + 1$ *if* $a_0 a_1 < 0$,
- $V(a_0, \ldots, a_d) = V(a_1, \ldots, a_d)$ *if* $a_0 a_1 > 0$.

This definition extends to any finite sequence a of elements in $\mathbb{R}$ by considering the finite sequence b obtained by dropping the zeroes in a and defining $V(a) = V(b)$, with the convention $V(\emptyset) = 0$.

Here is a simple but interesting result, known as Descartes' lemma [114]:

Lemma 2 (Descartes). *Let $p(x) = a_d \, x^d + \cdots + a_0$, let $V(\mathbf{a})$ be the number of sign changes in the sequence $\mathbf{a} = \{a_0, a_1, \ldots, a_d\}$ and let $N_+(p)$ be the number of strictly positive roots of p (counted with multiplicity). Then*

$$N_+(p) = V(\mathbf{a}) - 2\,k,$$

with $k \in \mathbb{N}$.

This rule can be used to isolate the roots of a polynomial p, as shown hereafter.

Let us recall that a univariate polynomial $p(x)$ of degree d can also be represented as:

$$p(x) = \sum_{i=0}^{d} b_i\, B_d^i(x),$$

where $B_d^i(x) = \binom{d}{i}x^i(1-x)^{d-i}$. The polynomials $(B_d^i)_{i=0,\dots,d}$ form the so-called Bernstein basis in degree d on $[0,1]$. The sequence $\mathbf{b} = [b_i]_{i=0,\dots,d}$ is called the set of control coefficients on $[0,1]$. Similarly, we will say that a sequence $\mathbf{b}$ represents the polynomial function p on the interval $[a,b]$ if:

$$p(x) = \sum_{i=0}^{d} b_i\, \binom{d}{i}\frac{1}{(b-a)^d}\,(x-a)^i(b-x)^{d-i}.$$

The polynomials

$$B_d^i(x;a,b) := \binom{d}{i}\frac{1}{(b-a)^d}\,(x-a)^i(b-x)^{d-i}$$

form the Bernstein basis in degree d on $[a,b]$. We are going to consider the sequence of values $\mathbf{b}$ together with the corresponding interval $[a,b]$. A first property of this representation is that the derivative f' of f, is deduced easily from the control coefficients:

$$d\Delta\mathbf{b} := d(b_{i+1} - b_i)_{0\leqslant i\leqslant d-1},$$

where $\Delta\mathbf{b} = (\Delta b_0, \dots, \Delta b_{d-1})$ with $\Delta b_i = b_{i+1} - b_i$. Another fundamental algorithm that we will use is the de Casteljau algorithm [157]:

$$b_i^0 = b_i \quad i = 0, \dots, d$$
$$b_i^r = (1-t)\, b_i^{r-1} + t\, b_{i+1}^{r-1}(t) \quad i = 0, \dots, d-r$$

It allows us to compute the representation of p on the subdivided intervals $[a, (1-t)a + tb]$ and $[(1-t)a + tb, b]$. For a complete list of methods on this representation, we refer to [157].

By substituting x by $x = \frac{t}{1-t}$, the Descartes rule yields the following proposition [157], [295]:

Proposition 3. *The number of sign changes $V(\mathbf{b})$ of the control coefficients $\mathbf{b} = [b_i]_{i=0,\dots,d}$ of a univariate polynomial on $[0,1]$ bounds its number of real roots in $[0,1]$ and is equal to it modulo 2. Thus,*

- *if $V(\mathbf{b}) = 0$, there is no real root in $[0,1]$, and*
- *if $V(\mathbf{b}) = 1$, there is exactly one real root in $[0,1]$.*

Sturm sequences.

Another tool which can also be used for the isolation of real roots is described below.

Definition 5 (Sturm sequence). *Let p, q be two univariate polynomials. A polynomial sequence $f_0 = p, f_1 = q, \ldots, f_s$ is a Sturm sequence if:*

- *f_s divides all the f_i, $0 \le i \le s$. Let $\delta_i = f_i/f_s$, $0 \le i \le s$.*
- *If c is a real number such that $\delta_j(c) = 0$, $0 < j < s$, then $\delta_{j-1}(c)\delta_{j+1}(c) < 0$.*
- *If c is a real number such that $\delta_0(c) = 0$ then $\delta_0(x)\delta_1(x)$ has the sign of $x - c$ in a neighborhood of c.*

We will denote the Sturm sequence by $Sturm(p, q) = (p, q, f_2, \ldots, f_s)$. Such a Sturm sequence can be computed for instance, as follows:

$$f_0 = p, f_1 = q, f_2 = -\mathrm{rem}(f_0, f_1), \ldots f_{i+1} = -\mathrm{rem}(f_{i-1}, f_i), \ldots$$

where $\mathrm{rem}(f, g)$ is the reminder in the Euclidean division of the polynomial f by the polynomial g.

For any sequence S of real polynomials and $a \in \mathbb{R}$, we denote by $V(S, a)$ the number of sign variations of the values of the polynomials in S at a. Then, we have the well-known theorem of Sturm (see for instance [44]):

Proposition 4 (Sturm theorem). *Assume $S = Sturm(p, p'q)$ and $]a, b[$ is an interval such that $p(a)p(b) \neq 0$. The difference $V(S, a) - V(S, b)$ is equal to the difference between the number $Z_p(q > 0; a, b)$ of roots α of p in $]a, b[$ (without multiplicity) such that $q(\alpha) > 0$ and the number $Z_p(q < 0; a, b)$ of roots α of p in $]a, b[$ such that $q(\alpha) < 0$:*

$$V(S, a) - V(S, b) = Z_p(p; a, b) - Z_p(q < 0; a, b).$$

If we take $q = 1$, then we have that $V(S, a) - V(S, b)$ is the number of real roots of p in $]a, b[$.

An efficient way to compute a Sturm sequence is to compute a Sturm-Habicht (or Sylvester-Habicht) sequence. For that purpose, we recall here the definition of pseudo-remainder (see for instance [44]).

Definition 6 (Pseudo-remainder). *Let*

$$f(x) = a_p x^p + \cdots + a_0$$

$$g(x) = b_q x^q + \cdots + b_0$$

be two polynomials in $\mathbb{L}[x]$ where $\mathbb{L}$ is a ring. Note that the only denominators occurring in the euclidean division of f by g are b_q^i, $i \le p + q - 1$. The signed pseudo-remainder denoted $\mathrm{prem}(f, g)$, is the remainder in the Euclidean division of $b_q^d f$ by g, where d is the smallest even integer greater than or equal $p - q + 1$. Note that the euclidean division of $b_q^d f$ by g can be performed in $\mathbb{L}$ and that $\mathrm{prem}(f, g) \in \mathbb{L}[x]$.

Definition 7 (Sturm-Habicht sequence). *Let p and q be univariate polynomials, $d = \max(\deg(p), \deg(q) + 1)$, $\mathrm{coef}_k(p)$ the coefficient of x^k in p, and $\delta_k = (-1)^{k(k-1)/2}$.*

The Sturm-Habicht sequence of p and q is defined inductively as follows:

- $H_d = p$, $h_d = 1$.
- $H_{d-1} = q$.

Assume that we have computed $H_d, \ldots, H_{j-1}$, $h_d, \ldots, h_j$ with $h_j \neq 0$ and $H_{j-1} \neq 0$. Let $k = deg(H_{j-1})$. Then:

- *If $k < j - 1$, let $H_k = \delta_{j-k} \dfrac{\mathrm{coef}_k(H_{j-1})^{j-1-k}}{h_j^{j-1-k}} H_{j-1}$, $h_{j-1} = 1$.*

  For $l \in \mathbb{N}$ with $k < l < j - 1$, let $H_l = 0$, $h_l = 0$.
- *Let $h_k = \mathrm{coef}_k(H_k)$, $H_{k-1} = \delta_{j-k+2} \dfrac{\mathrm{prem}(H_j, H_{j-1})}{h_j^{j-k+1}}$ (where prem is the pseudo-remainder)*

The main interest of this construction is that the polynomials in the sequence are related to the subresultants of p and q (see Section 3.4.1). Therefore, if we are working with coefficients in a ring $\mathbb{L}$, the coefficients of these subresultant polynomials are in the ring $\mathbb{L}$. If for instance $\mathbb{L} = \mathbb{K}[u]$, the subresultant polynomials are polynomials in x, which coefficients are polynomials in the variable u. By substituting u by a value $u_0 \in \mathbb{K}$, we get by construction, the subresultant sequence, evaluated at $u = u_0$. In other words, subresultant sequences behave well under specialization.

This Sturm-(Habicht) sequence can also be useful for gcd computations, since the gcd corresponds to the last non-zero term of the sequence. In particular, it yields anotehr way to compute the square-free part $p/\gcd(p, p')$ of a polynomial $p \in \mathbb{Q}[x]$ (see Algorithm 6).

Isolation algorithm.

The idea behind both approaches that we are goping to present now, is to consider an interval that initialy contains all the real roots, and then repeatedly subdivide it until we obtain intervals that is guaranteed that they contain zero or one real root.

Both Descartes's and Sturm's approach gives a bound on the number of roots of a polynomial f on an interval I, that we denote hereafter by $V(f; I)$. For Descartes' approach, $V(f; I)$ is the number of sign changes in the coefficients of f in the Bernstein basis on I. This bound is equal to the number of roots modulo 2. For the Sturm approach, $V(f; I)$ is the difference of the number of sign changes of the Sturm sequence of f and f', evaluated over the end points of the interval. Notice that in this approach we count exactly the number of real roots of f in the interval I.

This leads to the algorithm 7.

In the Sturm approach, we compute first the Sturm sequence of f and f'. Then, in order to compute $V(f; I)$, we evaluate the sign of the polynomials of this Sturm sequence at the end points of the interval I. If we use

---

Algorithm 7 Isolation of real roots

INPUT: A polynomial $f \in \mathbb{Z}[x]$.

- Compute the square-free part f^r of f (Algorithm 6).
- Compute an interval I_0 containing the roots of f and initialize a queue Q with I_0.
- While Q is not empty,
  - Pop an interval I from Q and compute $V(f, I)$.
  - If $V(f; I) > 1$, split the interval in two subintervals I_- and I_+ and add them to Q.
  - If $V(f; I) = 0$, then remove the interval I.
  - If $V(f; I) = 1$, then output I.

OUTPUT: a list of intervals, with as many real roots of f than intervals and such that each interval contains exactly one real root of f.

---

the Sturm-Habicht construction, as $f \in \mathbb{Z}[x]$ the polynomials of the Sturm-Habicht sequence will also be in $\mathbb{Z}[x]$. Moreover, if the bit size of the coefficients of the polynomial f of degree d is bounded by τ, then the bit size of the co-efficients of the polynomials in the Sturm sequence are bounded by $\mathcal{O}(d\tau)$ (using Hadamard inequality). This algorithm can be refined by restricting the computation to the roots in a given interval.

For the Descartes' approach, we represent the polynomial f by an interval I and its coefficients in the Bernstein basis on I. Splitting the interval I consists in applying the de Casteljau algorithm in order to get the representation of f on the two subintervals I_-, I_+. Another variant consists in avoiding the square-free computation and stopping the subdivision if an interval is of size smaller than a given ϵ. This does not guarantees the result in the presence of a cluster of roots, but only bounds the multiplicity of this cluster.

A first version of the Bernstein approach was described in [235]. Its complexity was first analyzed in [268], and then improved in [262]. The proof of termination of the algorithm is based on a partial inverse of Descartes' rule, namely the two circle theorem [102, 262], which, roughly speaking, guarantees that the algorithm, after a number of steps, will produce polynomials with zero or one sign variation. Using a refined version of this theorem, and Davenport-Mahler bounds on the product of distances between pairs of (complex) roots of f (see [232]), one can even show that the number of subdivisions (for both approaches) is bounded by $\mathcal{O}(d\tau + d\log d)$, where d is the degree of f and τ a bound on the size of its coefficients (see also [313, 147]).

The correspondence with another algorithm based on Descartes' rule in the monomial basis, known as *modified Uspensky* algorithm [299], has been detailed in [262]. It is shown in particular, that de Casteljau algorithm can be reinterpreted as a shift of a univariate polynomial. This yields a bound $\mathcal{O}(d^4\tau^2)$ on the bit complexity for the isolation algorithm, which actually holds for both approaches. In practice, Descartes' approach is usually more efficient than the Sturm-Sequence approach (see [147] for experimentations).

Notice that Descartes' approach (with polynomials represented either in monomial or Bernstein basis) can be applied with interval arithmetic, by adapting the sign variation count [299, 262]. It also extends naturally to B-splines which are piecewise polynomial functions [157].

3.4.3 Algebraic Numbers of Small Degree

It is known that solving a polynomial equation by radicals is always possible for small degree ($d \leq 4$). However when we want to compute the real roots then we must do computations with complex numbers (for degrees 3 and 4). Thus even when explicit formulas are available it is not always possible or effective to do computations with them. Thus, we are not interested by describing the roots as explicit formulas, involving radicals for instance (since tis is not always possible).

Instead, our goal is to describe the roots of a polynomial by using only rational numbers. We will see that for small degree polynomials, of degree up to 4, we can compute the isolating interval representation of the real roots with minimal computations, while for arbitrary degree polynomials this task requires a more involved machinery (as presented in the previous section).

The interest of such approach has already been illustrated by intensive computations with algebraic numbers of degree 2. In this case, classical explicit formulas (involving simple square roots) allow us to represent the real roots and compute with them directly. Paradoxically, as it is shown in [116], it can be more interesting to use instead isolating interval representations, from an efficiency point of view. Such an approach has been recently extended for polynomials up to degree 4, in [149]. It yields a way to precompute isolating intervals offline, which proceeds as follows (see [149, 148] for more details; see also [219]).

Since algebraic numbers of degree up to 4 appear in many problems of computational geometry, for example arrangement of conic arcs in the plane, Voronoi diagrams of circles, etc, we present now how to compute the isolating interval representation of these numbers.

Given a polynomial $p \in \mathbb{Q}[x]$ of degree 4

$$p(x) = a\,x^4 - 4\,b\,x^3 + 6\,c\,x^2 - 4\,d\,x + e$$

with $a > 0$, we define the invariants

$$
\begin{aligned}
A &= ae - 4bd + 3c^2, & B &= ace + 2bcd - ad^2 - eb^2 - c^3, \\
\Delta_1 &= A^3 - 27B^2, & \Delta_2 &= b^2 - ac, \\
\Delta_3 &= c^2 - bd, & \Delta_4 &= d^2 - ce, \\
W_1 &= ad - bc, & W_2 &= be - cd, \\
W_3 &= ae - bd, & R &= -2b^3 - a^2d + 3abc, \\
T &= -9W_1^2 + 27\Delta_2\Delta_3 - 3W_3\Delta_2
\end{aligned}
$$

The following proposition gives the number of real roots of a quartic polynomial.

Proposition 5. *Let $p(x)$ be the quartic as above. The following table gives the numbers of real roots and their multiplicities in all cases. For example, $\{1,1,1,1\}$ means four simple real roots and $\{2,2\}$ means two double real roots.*

condition	real roots
$\Delta_1 > 0 \wedge T > 0 \wedge \Delta_2 > 0$	$\{1,1,1,1\}$
$\Delta_1 > 0 \wedge (T \leq 0 \vee \Delta_2 \leq 0)$	$\{\}$
$\Delta_1 < 0$	$\{1,1\}$
$\Delta_1 = 0 \wedge T > 0$	$\{2,1,1\}$
$\Delta_1 = 0 \wedge T < 0$	$\{2\}$
$\Delta_1 = 0 \wedge T = 0 \wedge \Delta_2 > 0 \wedge R = 0$	$\{2,2\}$
$\Delta_1 = 0 \wedge T = 0 \wedge \Delta_2 > 0 \wedge R \neq 0$	$\{3,1\}$
$\Delta_1 = 0 \wedge T = 0 \wedge \Delta_2 < 0$	$\{\}$
$\Delta_1 = 0 \wedge T = 0 \wedge \Delta_2 = 0$	$\{4\}$

The next theorem provides a tool to isolate these real roots.

Theorem 4. *Given a polynomial $p(x)$ with two successive real roots γ_1 and γ_2, and given two other polynomials $B(x)$ and $C(x)$, define*

$$A(x) := B(x)p'(x) + C(x)p(x).$$

Then $A(x)$ or $B(x)$ has at least one real root in the closed interval $[\gamma_1, \gamma_2]$.

The idea developed in [148] is to determine the number of real roots of the polynomial $p(x)$ using the invariants and then to find isolating numbers by providing clever choices for the polynomials $A(x)$ and $B(x)$, i.e the real roots of $A(x)$ and $B(x)$ are computed easily and ideally are rationals. of Theorem 4. In the case of two simple or four simple real roots, the isolating point that we get from the initial polynomial may not be rational. In [148] the authors show how to get a rational isolating point in these cases. The following polynomials are suggested:

$$A(x) = 3\Delta_2 x^2 + 3W_1 x - W_3$$
$$B(x) = ax - b$$
$$C(x) = -4a.$$

The isolating points are then in the set of real roots of $A(x)$ and $B(x)$. As a second possibility, they suggest to take the following three polynomials

$$A(x) = W_3 x^3 - 3W_2 x^2 - 3\Delta_4 x$$
$$B(x) = dx - e$$
$$C(x) = -4d$$

Note that the real roots of $A(x)$ might be real algebraic numbers of degree 2 in both cases, thus not rational numbers in general. However it was proven

[148] that using the ceiling of the real roots of $A(x)$ suffices in all the cases to compute isolating intervals for real roots of $p(x)$. The case of the cubic is even more straightforward, since only one choice of polynomials $A(x)$ and $B(x)$ is needed (see [148] for more details).

3.4.4 Comparison

An important operation which needs to be performed exactly on real algebraic numbers in geometric algorithms is the comparison. We describe here methods to achieve this goal.

Comparison of isolated interval algebraic numbers.

Let us describe briefly how we use Sturm's theorem to compare two algebraic numbers $\alpha = (p,]a, b[)$ and $\beta = (q,]c, d[)$, assuming for simplicity that α and β are *simple roots* of p and q. If $b < c$ (resp. $d < a$) we have $\alpha < \beta$ (resp. $\beta < \alpha$). Let us assume now that $a < c < b < d$ (the other cases being treated similarly). First we compute the sign s of $p(a)p(c)$. If $s < 0$, then we have $\alpha \in]a, c[$ and $\alpha < \beta$. If $s = 0$, we have $\alpha = c$ (since $\alpha \neq a$), which implies that $\alpha < \beta$. Otherwise $s > 0$, p has no root in the interval $[a, c]$. We compute $S = Sturm(p, p'q)$ and $v := V(S, c) - V(S, b)$. Let us assume first that $q(c) > 0$, $q(b) < 0$. Then if $v = 1$, by Sturm's theorem $q(\alpha) > 0$ and $\alpha < \beta$. If $v = -1$, $q(\alpha) < 0$ and $\alpha > \beta$. If $v = 0$, then $q(\alpha) = 0$ and $\alpha = \beta$. If now $q(c) < 0$, $q(b) > 0$, we negate the previous output. Finally, if $q(c)$ and $q(b)$ are of the same sign, then $\alpha < \beta$.

Regarding the complexity of this method, the effective computation of these sequences can be done using Sturm-Habicht sequences. For two polynomials p and q of degree bounded by d and coefficient bit size bounded by τ the bit size of the coefficients in the Sturm-Habicht sequence is $\mathcal{O}(d\tau)$ and the computation of the sequence is made with $\mathcal{O}(d^2)$ arithmetic operations [300, 44]. The reader may refer to [147], for a complete complexity analysis.

For algebraic numbers of small degree, these operations can even be precomputed, as described in [148], in order to accelerate the computation of the Sturm sequences of specific polynomials, taking advantage of the good specialization properties of the Sturm-Habicht sequences. Moreover, since the isolating interval representation of algebraic numbers of degree up to 4 can be computed in constant time we conclude that comparison of such numbers can also be performed in constant time.

For example [148], the Sturm sequence of a quartic $p(x)$ and its derivative $p'(x)$ is

$$S_0(x) = p(x)$$
$$S_1(x) = p'(x)$$
$$S_2(x) = 3\Delta_2 x^2 + 3W_1 x - W_3$$
$$S_3(x) = T_1 x + T_2$$
$$S_4(x) = -\Delta_1$$

where $T_1 = -W_3\Delta_2 - 3W_1^2 + 9\Delta_2\Delta_3$ and $T_2 = AW_1 - 9bB$. The Sturm sequences of two general quartic polynomials can be described similarly [148]. The precomputation of these Sturm sequences for polynomials of degree ≤ 4 reduces many of the algorithmic problems on small degree algebraic numbers to the computation of signs of algebraic expressions. The degree of these algebraic expressions is an indicator of the complexity of the approach [148]:

Theorem 5. *There is an algorithm that compares any two roots of two quartics using Sturm sequences and isolating intervals from above where the algebraic degree of the quantities involved is in the worst case at most* 14.

Notice that the comparison of two algebraic numbers can also be performed, using the Bernstein-basis approach. Indeed, suppose that the two algebraic numbers α and β are defined by the polynomials p and q. Then by computing their gcd and dividing by it, we may assume that p and q are co-prime. In this case, we can subdivide the isolation intervals of α and β using Algorithm 7, until say, q has a constant sign on the interval I_α defining α. Then I_α can be set-subtracted from the interval I_β defining β and the comparison of α and β reduces to the easy comparison of the non-overlapping intervals I_α and I_β.

Comparison of symbolic expressions.

Algorithms for the comparison of symbolic expressions representing algebraic numbers can also used. Let E be a real algebraic expression given as directed acyclic graph (dag) (source nodes are operands and internal nodes are operators). A *separation bound* $\text{sep}(E)$ is a real number $\text{sep}(E) > 0$, that can be computed easily from E, such that

$$\text{val}(E) = 0 \quad \text{or} \quad |\text{val}(E)| \geq \text{sep}(E).$$

Separation bounds allow us to determine the sign of an expression by means of numerical computations. An error bound Δ is initialized to some positive value, say $\Delta = 1$, and an approximation $\tilde{\xi}$ of $\xi = \text{val}(E)$ with $|\xi - \tilde{\xi}| \leq \Delta$ is computed using approximate arithmetic, say floating point arithmetic with arbitrary-length mantissa. If $|\tilde{\xi}| > \Delta$, the sign of ξ is equal to the sign of $\tilde{\xi}$. Otherwise, $|\tilde{\xi}| \leq \Delta$ and hence $|\xi| < 2\Delta$. If $2\Delta \leq \text{sep}(E)$, we have $\xi = 0$. If $2\Delta > \text{sep}(E)$, we halve Δ and repeat the process. The worst case complexity of the procedure just outlined is determined by the separation bound.

The following separation bound are taken from the article [73]. We need to define the weight $D(E)$ of an expression, which is an upper bound on the algebraic degree of the algebraic number defined by the expression. The *weight* $D(E)$ of an expression dag E is the product of the weights of the nodes and leaves of the dag. Leaves and $+$, $-$, $\cdot$ and $/$ operations have weight 1, a $\sqrt[k]{\,}$-node has weight k, and a $\diamond(j, E_d, \ldots)$-operation has weight d.

Let $p(x) = x^d + a_{d-1}x^{d-1} + \ldots + a_0$ be a polynomial with real coefficients a_i. A *root bound* $\Phi(a_{d-1}, \ldots, a_0)$ is an upper bound for the absolute value of all roots of $p(x)$. One example for a root bound is the Lagrange-Zassenhaus bound (see [345])

$$\Phi(\ldots, a_i, \ldots) = 2 \max_{i=0,\ldots,d-1} \left\{ \sqrt[d-i]{|a_i|} \right\}.$$

Theorem 6. *Let E be a real algebraic expression with* $\mathrm{val}(E) = \xi$. *Let* $u(E)$ *and* $l(E)$ *be defined inductively on the structure of E according to the following rules:*

	$u(E)$	$l(E)$
integer N	$|N|$	1
$E_1 \pm E_2$	$u(E_1) \cdot l(E_2) + l(E_1) \cdot u(E_2)$	$l(E_1) \cdot l(E_2)$
$E_1 \cdot E_2$	$u(E_1) \cdot u(E_2)$	$l(E_1) \cdot l(E_2)$
E_1/E_2	$u(E_1) \cdot l(E_2)$	$l(E_1) \cdot u(E_2)$
$\sqrt[k]{E_1}$ *and* $u(E_1) = 0$	0	1
$\sqrt[k]{E_1}$ *and* $u(E_1) \geq l(E_1)$	$\sqrt[k]{u(E_1) l(E_1)^{k-1}}$	$l(E_1)$
$\sqrt[k]{E_1}$ *and* $u(E_1) < l(E_1)$	$u(E_1)$	$\sqrt[k]{(u(E_1)^{k-1} l(E_1))}$
$\diamond(j, E_d, \ldots, E_0)$	$\Phi(\ldots, \left(l(E)^{d-i-1} \right.$ $\left. u(E_i) \prod_{k \neq i} l(E_k) \right), \ldots)$	$u(E_d) \prod_{k \neq d} l(E_k)$

Let $D(E)$ be the weight of E. Then either $\xi = 0$ or

$$\left(l(E) u(E)^{D(E)-1} \right)^{-1} \leq |\xi| \leq u(E) l(E)^{D(E)-1}$$

Separation bounds have been studied extensively in computer algebra [77, 254, 345, 304] as well as in computational geometry [75, 343, 73, 242].

The separation bounds given above can be improved using the fact that powers of integers can be factored out from the number *before* computing the separation bound [285, 306].

3.5 Multivariate Problems

Geometric problems in 3D often involve the solution of polynomial equations in several variables. This problem can be reduced to a univariate problem and thus to the manipulation of real algebraic numbers, as follows. We are interested in the case of zero-dimensional systems, i.e. systems that have only finitely many complex solutions.

We denote by $f_1 = 0, \ldots, f_m = 0$ the polynomial equations in $\mathbb{K}[\mathbf{x}] = \mathbb{K}[x_1, \ldots, x_n]$ that we want to solve. The quotient ring $\mathbb{K}[\mathbf{x}]/(f_1, \ldots, f_m)$ of polynomials modulo $f_1, \ldots, f_m \in \mathbb{K}[x_1, \ldots, x_n]$ is denoted by $\mathcal{A}$. In the case that we consider here, where the number of complex roots is finite, the quotient algebra $\mathcal{A}$ is a finite-dimensional vector space over $\mathbb{K}$.

We consider the operators of multiplication M_g by an element g in the ring $\mathcal{A}$:

$$M_g : \mathcal{A} \to \mathcal{A}$$
$$a \mapsto g\,a$$

Then, the algebraic solution of the system is performed by analyzing the eigenvalues and eigenvectors of these operators. It is based on the next theorem which involves the transposed of the operators M_g^t. By definition, M_g^t is acting on the set of linear forms $\widehat{\mathcal{A}}$ on $\mathcal{A}$. If ζ is a root of the polynomial system $f_1 = 0, \ldots, f_m = 0$, then the map $\mathbf{1}_\zeta : g \mapsto g(\zeta)$ is an element $\widehat{\mathcal{A}}$.

Here is the main theorem from which we deduce the root computation:

Theorem 7. *Assume that the set of complex solutions of $f_1 = 0, \ldots, f_s = 0$ is the finite set $\{\zeta_1, \ldots, \zeta_d\}$.*

1. *Let $g \in \mathcal{A}$. The eigenvalues of the operator M_g (and its transpose M_g^t) are $g(\zeta_1), \ldots, g(\zeta_d)$.*
2. *The common eigenvectors of (M_g^t) for $g \in \mathcal{A}$ are (up to a scalar) the evaluation $\mathbf{1}_{\zeta_1}, \ldots, \mathbf{1}_{\zeta_d}$.*

The first point of this theorem can be found in [39] and the second in [261]. This theorems implies in particular that the common eigenvectors of the transpose of the operators $M_{x_1}, \ldots, M_{x_n}$ of multiplication by $x_1, \ldots, x_n$, correspond to the linear forms which evaluate a polynomial at a root. A numerical approximation of these roots of the system can thus be obtained by computing the common eigenvectors of these operators, using standard linear algebra tools. See [39, 105, 261, 266, 328, 107, 143, 267] for more details on this approach.

These operators can also be used to describe the solution points as the image, by a rational map, of the roots of a univariate polynomial. In other words, the (real) coordinates of the solutions are rational functions evaluated at real algebraic numbers whose defining equations can be deduced explicitly from the matrices M_{x_i}. It leads to Algorithm 8, which yields the so-called *Rational Univariate Representation* (RUR) of the roots. For details on this

---

Algorithm 8 Rational Univariate Representation of the roots

INPUT: The tables $M_{x_1}, \ldots, M_{x_n}$ of multiplication by $x_1, \ldots, x_n$ in $\mathcal{A}$.

1. Compute the determinant $\Delta(\mathbf{u}) := \det(u_0 I + u_1 M_{x_1} + \cdots + u_n M_{x_n})$ and its square-free part $d(\mathbf{u})$.
2. Choose a generic $t \in \mathbb{K}^{n+1}$ and compute the $n+1$ first coefficients of

$$d(t + \mathbf{u}) = d_0(u_0) + u_1 d_1(u_0) + \cdots + u_n d_n(u_0) + \cdots$$

considered as a polynomial in $u_1, \ldots, u_n$.

OUTPUT: the roots of the system $f_1 = 0, \ldots, f_m = 0$ are

$$\zeta_1 = \frac{d_1(\alpha)}{d_0'(\alpha)}, \ldots, \zeta_n = \frac{d_n(\alpha)}{d_0'(\alpha)}$$

for all roots α of the univariate polynomial equation $d_0(\alpha) = 0$.

---

construction, see [247, 186, 298, 44, 142].

The generic condition required in Algorithm 8 on $t \in \mathbb{K}^{n+1}$ is that it *separates* the roots: $\sum_{i=0}^{n} \zeta_i t_i \neq \sum_{i=0}^{n} \zeta_i' t_i$ if ζ and ζ' are two distinct solutions of the system. Methods to find a generic t are described for instance in [298].

In order to get a minimal rational univariate representation, one can factorize $d_0(u_0)$ and keep the irreducible factors, which divide the numerator of the fraction obtained by substituting x_i with $\frac{d_i(u_0)}{d_0'(u_0)}$.

This Rational Univariate Representation (RUR) allows us to replace the treatment of solutions of a multivariate system by the manipulation of algebraic numbers of degree bounded above by the number of complex solutions of the system.

Another important aspect is that we can compute a RUR of a polynomial system with coefficients in an algebraic extension $\mathbb{Q}[\theta]$, for θ an algebraic number, i.e with coefficients in $\mathbb{Q}[x]/(p(x))$, where p is the minimal polynomial of θ. Computing the tables of multiplication and the RUR of the roots over $\mathbb{Q}[\theta]$ require field arithmetic operations and equality test in $\mathbb{Q}[\theta]$, which are performed easily by reduction modulo p.

3.6 Topology of Planar Implicit Curves

As an application of these algebraic techniques, we detail the computation of the topology of an implicit curve. It is a key ingredient of many geometric problems including arrangement computation on arcs of curves, intersection of surfaces, ... We consider first a curve $\mathcal{C}$ defined as the zero locus $\mathcal{V}(f)$ of a polynomial in two variables $f(x, y) \in \mathbb{Q}[x, y]$. We can assume that f is square-free (otherwise, we perform Algorithm 6). In Section 3.6.1, we are going to present from a geometric point of view the way the topology is computed. In this computation, we need to manipulate algebraic numbers. In Section 3.6.2, we describe different algebraic tools allowing to certify the computation. Finally, in Section 3.6.3, we present an alternative to the first algorithm. As the condition of genericity is costly, we propose an algorithm partially based on subdivision which deals with algebraic curves even not in generic position.

Before explaining how the algorithm works, we will give some definitions:

Definition 8 (Critical point). *A point (α, β) of $\mathcal{C} = \mathcal{V}(f)$ is x-critical if* $f(\alpha, \beta) = \frac{\partial f}{\partial y}(\alpha, \beta) = 0$.

Definition 9 (Singular point). *A point (α, β) of $\mathcal{C} = \mathcal{V}(f)$ is singular if* $f(\alpha, \beta) = \frac{\partial f}{\partial y}(\alpha, \beta) = \frac{\partial f}{\partial x}(\alpha, \beta) = 0$.

Definition 10 (Regular point). *A point (α, β) of $\mathcal{C} = \mathcal{V}(f)$ is regular if it is not singular.*

Definition 11 (Generic position). *The curve $\mathcal{C} = \mathcal{V}(f)$ is said to be in generic position if:*

- *The leading coefficient of f with respect to y (polynomial in x) has no real roots.*
- *For every α in $\mathbb{R}$, the number of critical points with x-coordinate α is at most 1.*

So in generic position, the curve has no vertical asymptote and its x-critical points have different x-coordinates.

3.6.1 The Algorithm from a Geometric Point of View

In this section, we are going to present the geometric idea permitting to recover the topology of the curve from the computation of some particular points.

Algorithm 9 Topology of an implicit planar curve

INPUT: a polynomial $f(x,y) \in \mathbb{Q}(x,y)$ defining a curve $\mathcal{C} \subset \mathbb{R}^2$ (up to a gcd-computation and a change of variables, we can assume f is square-free and is monic in y).

1. Compute the subresultant sequence of $f(x,y)$ and $\frac{\partial f}{\partial y}(x,y)$ viewed as polynomials in y.
2. Compute the x-critical points $\{P_i = (\alpha_i, \beta_i)\}$.
3. Check that the curve is in generic position (see Section 3.6.2) and if it is not we perform a random change of variables and restart from step 1.
4. For each critical point $P_i = (\alpha_i, \beta_i)$, compute the number of regular points with x-coordinate α_i which are above and below P_i using Sturm sequences.
5. Compute the number of arcs above a value between two successive abscissas α_i, α_{i+1}, which is constant. It can be done for example choosing a rational x-coordinate a between α_i and α_{i+1} and computing the number of real solutions of $f(a,y) = 0$ using Sturm sequences. Then we compute numerical approximations of those different points.
6. Construct the segments connecting the points computed before. For that purpose, consider a section $x = \alpha_i$, i.e. all the points of the curve with abscissas α_i and the next section $x = \alpha_{i+1}$ (see Figure 3.1).
   - Choose a rational point $a \in]\alpha_i, \alpha_{i+1}[$ and compute the section corresponding to $x = a$.
   - In the section $x = \alpha_i$, compute the number λ_i of points of $\mathcal{C}$ above (α_i, β_i) and μ_i the number of points below.
   - Connect the λ_i points above (α_i, β_i) with the λ_i points of largest y-coordinate of the section $x = a$, respecting the order on the y-coordinate.
   - Connect the μ_i points under (α_i, β_i) with the μ_i points of smaller y-coordinate of the section $x = a$, respecting the order on the y-coordinate. After that, connect the remaining points of the section $x = a$ to the critical point (α_i, β_i).

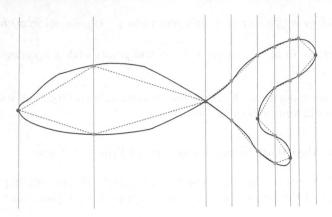

Fig. 3.1. Connections

3.6.2 Algebraic Ingredients

Computing the topology will lead us to the treatment of algebraic numbers. In this section, we come back to some delicate points of the algorithm. We show how to use the algebraic tools presented in the previous section to certify these steps.

Computing the critical points:

Proposition 6. *Let f be a square-free polynomial in $\mathbb{R}[x, y]$ of degree d in y, such that $\mathcal{C} = \mathcal{V}(f)$ is in generic position.*
Let Sr_j and sr_j denote the jth subresultant and subresultant coefficient of f and $\frac{\partial f}{\partial y}$ (considered as polynomials in y). If (α, β) is a critical point of $\mathcal{C}$, then there exists $k \in \mathbb{N}^*$ such that:

$$\mathrm{sr}_0(\alpha) = 0, \dots, \mathrm{sr}_{k-1}(\alpha) = 0, \mathrm{sr}_k(\alpha) \neq 0.$$

Moreover, we have: $\beta = \frac{-1}{k} \frac{\mathrm{sr}_{k,k-1}(\alpha)}{\mathrm{sr}_k(\alpha)}$, where $\mathrm{sr}_{k,k-1}(x)$ denotes the coefficient of y^{k-1} in $\mathrm{Sr}_{k-1}(x, y)$.

So, we observe that in generic position, the y-coordinate β of a critical point (α, β) is a rational expression of α.
We define inductively a family of polynomials $\Gamma_k(x)$:

$$\Phi_0(x) = \frac{\mathrm{sr}_0(x)}{\gcd(\mathrm{sr}_0(x), \mathrm{sr}_0'(x))}$$

$$\Phi_1(x) = \gcd(\Phi_0(x), \mathrm{sr}_1(x)) \qquad \Gamma_1 = \frac{\Phi_0(x)}{\Phi_1(x)}$$

$$\Phi_2(x) = \gcd(\Phi_1(x), \mathrm{sr}_2(x)) \qquad \Gamma_2 = \frac{\Phi_1(x)}{\Phi_2(x)}$$

$$\Phi_3(x) = \gcd(\Phi_2(x), \mathrm{sr}_3(x)) \qquad \Gamma_3 = \frac{\Phi_2(x)}{\Phi_3(x)}$$

$$\vdots \qquad\qquad\qquad\qquad \vdots$$

$$\Phi_0(x) = \gcd(\Phi_{d-2}(x), \mathrm{sr}_{d-1}(x)) \;\; \Gamma_{d-1} = \frac{\Phi_{d-2}(x)}{\Phi_{d-1}(x)}$$

We deduce that the square-free part $\Phi_0(x)$ of $\mathrm{sr}_0(x)$ can be written as $\Phi_0 = \Gamma_1(x).\Gamma_2(x) \cdots \Gamma_{d-1}(x)$. If (α, β) is a critical point of $\mathcal{C}$ in generic position, then α is a root of Φ_0. It is in fact the root of one and only one of the Γ_i. And if α is a root of Γ_k, we have that $\mathrm{sr}_1(\alpha) = 0, \ldots, \mathrm{sr}_{k-1}(\alpha) = 0, \mathrm{sr}_k(\alpha) \neq 0$. Thus $\gcd(f(\alpha, y), \frac{\partial f}{\partial y}(\alpha, y)) = \mathrm{Sr}_k(\alpha, y)$ and β is the only root of $\mathrm{Sr}_k(\alpha, y)$ (with multiplicity k).

Generic position: An important operation that we have to perform, is to check that $\mathcal{C}$ is in generic position. If α_i is the x-coordinate of a critical point, α_i is the root of a certain Γ_k and we note this root α_i^k. Then:

Proposition 7. *The curve $\mathcal{C}$ is in generic position if and only if for every α_i^k, the polynomial $\mathrm{Sr}_k(\alpha_i^k, y)$ has only one distinct root which is $\beta_i^k = \frac{-1}{k} \frac{\mathrm{sr}_{k,k-1}(\alpha)}{\mathrm{sr}_k(\alpha)}$.*

We have to check that $\mathrm{Sr}_k(\alpha_i^k, y)$ has only one distinct root where α_i^k is defined by a pair $(\Gamma_k,]a, b[)$, so that α_i^k is the only root of Γ_k in $]a, b[$. We compute a Sturm sequence for $\mathrm{Sr}_k(\alpha_i^k, y)$. It is a family of polynomials in $\mathbb{Q}[\alpha_i][y]$. So to apply Sturm's theorem, we need to compute signs of polynomials expressions in α_i^k. Such signs can be computed, using another Sturm sequence.

Number of points above and under a critical one: Assume $\mathcal{C}$ is in generic position. If $P_\alpha = (\alpha, \beta)$ is a critical point of $\mathcal{C}$, then we need to compute the number of regular points with x-coordinate α which are above and under P_α. We can assume that α is a root of Γ_k. Then y-coordinates of the regular points with abscissas α are the roots of the polynomial $F_k(\alpha, \beta, y) = \frac{f(\alpha,y)}{(y-\beta)^k}$. The coefficients of $F_k(\alpha, \beta, y)$ can be computed in an inductive way [185]. So as β is a rational expression of α we obtain the coefficients of $F_k(\alpha, \beta, y)$ as rational expressions in α. We determine the number of roots of this polynomial such that $y - \beta > 0$ (resp. $y - \beta < 0$). This can be computed again using Sturm sequences [106].

3.6.3 How to Avoid Genericity Conditions

We have seen that to check the genericity position of a curve, we had to compute several Sturm sequences, which can be costly. An important improvement would be able to deal with curves even not in generic position.

1. We compute the two resultants $\mathrm{Res}_y(f, \partial_y f)$ and $\mathrm{Res}_x(f, \partial_y f)$. This allows us to compute isolating boxes, containing at most one x-critical point of the curve. Let $B =]a, b[\times]c, d[$ be a box which contains an x-critical point (α, β). First, we refine (or delete) the box so that there is one and only one point with abscissa α in B. For that, we compute the Sturm sequence of $f(\alpha, y)$ and compute the number of changes of sign on $]c, d[$. We refine the box until the number of changes is at most one (if it is 0, the box is empty and we delete it).

2. Assume $P_{\alpha,1}, \ldots, P_{\alpha,k}$ are the x-critical points with abscissas α sorted according to their y-coordinate. After the first step $P_{\alpha,i}$ is isolated in a box $]a_i, b_i[\times]c_i, d_i[$. We compute the Sturm sequence of $f(\alpha, y)$ and $\partial_y f(\alpha, y)$ and compute the number n_k of points with y-coordinate bigger than d_k. Then n_k is the number of points above $P_{\alpha,k}$, $n_{k-1} - n_k - 1$ is the number of points between $P_{\alpha,k-1}$ and $P_{\alpha,k}$, $\ldots, n_1 - n_2 - 1$ is the number of points between $P_{\alpha,1}$ and $P_{\alpha,2}$. At last we compute the number of points with y-coordinate smaller than c_1. We have a family of boxes corresponding to x-critical points with abscissas α. Up to refinement, we assume all the boxes have the same x-coordinates a, b (the boxes are of the type $]a, b[\times]c, d[$). What we want, is to refine the boxes so that if the box is $B =]a, b[\times]c, d[$, then the curve $\mathcal{C}$ does not intersect the sides $[(a, c), (b, c)]$ and $[(a, d), (b, d)]$. For that, we compute the Sturm sequences of $f(x, c)$ and $f(x, d)$ and we refine the size of the box until the number of sign changes (for the interval $]a, b[$) is 0.
3. The next step consists of computing the number of points of intersection of $\mathcal{C}$ with the side $[(a, c), (a, d)]$ (resp. $[(b, c), (b, d)]$). This is done by computing the Sturm sequences of $f(a, y)$ and $\partial_y f(a, y)$ (resp. $f(b, y)$ and $\partial_y f(b, y)$) on $]c, d[$. We compute the number of points in an intermediate section $x = a \in \mathbb{Q}$ (using the Sturm Sequence of $f(x, y), \partial_y f(x, y)$ at $x = a$.
4. The connections of the different points is made similarly as in the generic case.

3.7 Topology of 3d Implicit Curves

We consider here an implicit curve in an affine space of dimension 3. By definition, it is an algebraic variety $\mathcal{C}_\mathbb{C} = \mathcal{V}(f_1, \ldots, f_m)$ $(f_i \in \mathbb{R}[x, y, z])$ of dimension 1 in $\mathbb{C}^3$. We denote by $I(\mathcal{C}_\mathbb{C}) \subset \mathbb{R}[x, y, z]$, the ideal of the curve $\mathcal{C}_\mathbb{C}$ (that is the set of polynomials which vanish on $\mathcal{C}_\mathbb{C}$) and by $g_1, \ldots, g_s \in \mathbb{R}[x, y, z]$ a set of generators: $I(\mathcal{C}_\mathbb{C}) = (g_1, \ldots, g_s)$. By Hilbert's Nullstellensatz [107, 202], we have $I(\mathcal{V}(f_1, \ldots, f_k)) = \sqrt{I} \subset \mathbb{R}[x, y, z]$. It can be proved [141, 202], that 3 polynomials g_1, g_2, $g_3 \in \mathbb{R}[x, y, z]$ are enough to generate $I(\mathcal{C}_\mathbb{C})$.

For simplicity, we will consider here that the curve is described as the intersection of two surfaces $P_1(x, y, z) = 0$, $P_2(x, y, z) = 0$, with $P_1, P_2 \in \mathbb{R}[x, y, z]$. We assume that the gcd of P_1 and P_2 in $\mathbb{R}[x, y, z]$ is 1, so that $\mathcal{V}(P_1, P_2) = \mathcal{C}_\mathbb{C}$ is of dimension 1, and all its irreducible components are of dimension 1. We are interested in describing the topology of the real part

$$\mathcal{C}_\mathbb{R} = \{(x, y, z) \in \mathbb{R}^3, \; P_1(x, y, z) = 0, \; P_2(x, y, z) = 0\},$$

that we will denote hereafter by $\mathcal{C}$.

In this section, we assume moreover that $I(\mathcal{C}) = (P_1, P_2)$ or equivalently that (P_1, P_2) is a *reduced ideal*, that is equal to its radical: $(P_1, P_2) = \sqrt{(P_1, P_2)}$.

We will not consider examples such as $P_1 = x^2+y^2-1, P_2 = x^2+y^2+z^2-1$, where $(P_1, P_2) = (x^2+y^2-1, z^2)$ and $I(\mathcal{C}) = (x^2+y^2-1, z)$, so that the curve $\mathcal{C}$ is defined "twice" by the equations $P_1 = 0, P_2 = 0$ (the two surfaces intersect tangentially along $\mathcal{C}$). Such a property can be tested by projecting into a generic direction and testing if the equation computed from the resultant of P_1, P_2 is square-free, or by more general methods such as computing the radical of (P_1, P_2) [192].

The general idea behind the algorithm that we are going to describe is as follows: we use a sweeping plane in a given direction (say parallel to the (y, z) plane) to detect the critical positions where *something* happens. We also compute the positions where something happens in projection on the (x, y) and (x, z) plane. Then, we connect the points of the curve of $\mathcal{C}$ on these critical planes. This yields a graph of points, connected by segments, with the same topology as the curve $\mathcal{C}$.

3.7.1 Critical Points and Generic Position

In this section, we make more precise what we mean by the points where *something* happens. These points will be called hereafter critical points.

Definition 12. Let $I(\mathcal{C}) = (g_1, g_2, \ldots, g_s)$ and let M be the $s \times 3$ Jacobian matrix with rows $\partial_x g_i, \partial_y g_i, \partial_z g_i$.

- A point $p \in \mathcal{C}$ is regular (or smooth) if the rank of M evaluated at p is 2.
- A point $p \in \mathcal{C}$ which is not regular is called singular.
- A point $p = (\alpha, \beta, \gamma) \in \mathcal{C}$ is x-critical (or critical for the projection on the x-axis) if the curve $\mathcal{C}$ is tangent at this point to a plane parallel to the (y, z)-plane i.e., the multiplicity of intersection of the plane with $I(\mathcal{C})$ at p is greater or equal to 2. The corresponding α is called an x-critical value.

A similar definition applies to the orthogonal projection onto the y and z axis or onto any line of the space. Notice that a singular point is critical for any direction of projection.

If $I(\mathcal{C}) = (P_1, P_2)$, then the x-critical points are the solutions of the system

$$P_1(x, y, z) = 0, \ \ P_2(x, y, z) = 0, \ \ (\partial_y P_1 \partial_z P_2 - \partial_y P_2 \partial_z P_1)(x, y, z) = 0. \quad (3.1)$$

In the case of a planar curve defined by $P(x, y) = z = 0$, with $P(x, y)$ square-free so that $I(\mathcal{C}) = (P(x, y), z)$, this yields the following definitions: a point (α, β)

- is *singular* if $P(\alpha, \beta) = \partial_x P(\alpha, \beta) = \partial_y P(\alpha, \beta) = 0$.
- is *x-critical* if $P(\alpha, \beta) = \partial_y P(\alpha, \beta) = 0$.

This allows us to describe the genericity condition that we require for the curve $\mathcal{C}$, in order to be able to apply the algorithm:

Definition 13 (Generic position). *Let*

$$N_x(\alpha) = \#\{ (\beta, \gamma) \in \mathbb{R}^2 \mid (\alpha, \beta, \gamma) \text{ is an } x\text{-critical point of } \mathcal{C} \}.$$

We say that $\mathcal{C}$ is in generic position for the x-direction, if

- $\forall \alpha \in \mathbb{R}$, $N_x(\alpha) \leqslant 1$, *and*
- *there is no asymptotic direction of $\mathcal{C}$ parallel to the (y, z)-plane.*

By a random change of variables, the curve can be put in a generic position. In practice, instead of changing the variables, we may choose a random direction for the sweeping plane.

3.7.2 The Projected Curves

The algorithm that we are going to describe, uses the singular points of the projection of $\mathcal{C}$ onto the (x, y) and (x, z)-planes. We denote by $\mathcal{C}'$ (resp. $\mathcal{C}''$) the projection of the curve $\mathcal{C}$ onto the (x, y) (resp. (x, z))-plane. The equation of the curve $\mathcal{C}'$ is obtained as follows. We decompose the polynomials P_1, P_2 in terms of the variable z:

$$P_1(x, y, z) = a_{d_1}(x, y)z^{d_1} + \ldots + a_0(x, y)$$
$$P_2(x, y, z) = b_{d_2}(x, y)z^{d_2} + \ldots + b_0(x, y)$$

with $\mathrm{a}_{d_1}(x, y) \neq 0$ and $\mathrm{b}_{d_2}(x, y) \neq 0$. Then, the resultant polynomial

$$G(x, y) = \mathrm{Res}_z(P_1, P_2)$$

vanishes on the projection of the curve $\mathcal{C}$ on the plane (x, y). Conversely, by the resultant theorem [236, 142] (see also Section 3.4.1), $a_{d_1}(x, y)$ and if the gcd $c(x, y)$ of $a_{d_1}(x, y)$ and $b_{d_2}(x, y)$ in $\mathbb{R}[x, y]$ is 1 then $a_{d_1}(x, y)$ and $b_{d_2}(x, y)$ do not vanish simultaneously on a component of dimension 1 of the projection $\mathcal{C}'$ of the curve $\mathcal{C}$. So $G(x, y) = 0$ defines $\mathcal{C}'$ and a finite number of additional points. If it's not the case, G is a non-trivial multiple of the implicit equation of $\mathcal{C}'$. Such a situation can be avoided, by a linear change of variables. Nevertheless, since the critical points of the curve defined by $G(x, y) = 0$ contains the critical points of $\mathcal{C}'$, we will see hereafter that this change of variables is not necessary.

Notice, that $G(x, y)$ is not necessarily a square-free polynomial. Consider for instance the case $P_1 = x^2 + y^2 - 1, P_2 = x^2 + y^2 + z^2 - 2$, where $G(x, y) = (x^2 + y^2 - 1)^2$. In this case, there are generically two (complex) points of $\mathcal{C}$ above a point of $\mathcal{C}'$.

We can easily compute the gcd of $G(x, y)$ and $\partial_y G(x, y)$ in order to get the square-free part $g(x, y) = G(x, y)/\gcd(G(x, y), \partial_y G(x, y))$ of $G(x, y)$.

Similarly, for the projection $\mathcal{C}''$ of $\mathcal{C}$ on the (x, z)-plane, we compute

$$H(x, z) = \mathrm{Res}_y(P_1, P_2),$$

and its square-free part $h(x, z)$ from the gcd of $H(x, z)$ and $\partial_z H(x, z)$. The equation $h(x, z) = 0$ defines a curve which is exactly $\mathcal{C}''$, if the gcd of the leading components of P_1, P_2 in y is 1. Its set of singular points contains those of $\mathcal{C}''$.

In order to analyze locally the projection of the curve $\mathcal{C}$, we recall the following definition:

Definition 14. *[338] Let X be an algebraic subset of $\mathbb{R}^n$ and let p be a point of X. The tangent cone at p to X is the set of points u in $\mathbb{R}^n$ such that there exists a sequence of points x_k of X converging to p and a sequence of real numbers t_k such that $\lim_{k \to +\infty} t_k(x_k - p) = u$.*

Notice, that at a smooth point of $\mathcal{C}$, the tangent cone is a line.

Proposition 8. *Let $p' = (\alpha, \beta)$ be an x-critical point of $\mathcal{C}'$, which is not singular. Then α is the x-coordinate of an x-critical point of $\mathcal{C}$.*

3.7.3 Lifting a Point of the Projected Curve

The problem we want to tackle here is the following: Assume we are given two surfaces defined by two implicit equations $P_1 = 0$ and $P_2 = 0$. Let us consider the projection of the curve of intersection of the two surfaces on the (x, y)-plane.

> *Starting from a point (x_0, y_0) of the projected curve, how can we find (a numerical approximation of) the z-coordinate of the point(s) above (x_0, y_0)?*

We note $p(z) = P_1(x_0, y_0, z)$, $q(z) = P_2(x_0, y_0, z)$ and $d = \deg(p)$, $d' = \deg(q)$. Consider the Sylvester submatrix $\mathrm{Syl}_1(x_0, y_0)$ of the mapping

$$\mathbb{R}[z]_{d'-2} \oplus \mathbb{R}[z]_{d-2} \longrightarrow \mathbb{R}[z]_{d+d'-2}$$
$$(u, v) \qquad \mapsto \qquad p\,u + q\,v$$

If ξ is a common root of p and q then $(1, \xi, \ldots, \xi^{d+d'-2})$ is in the kernel of the transpose of $\mathrm{Syl}_1(x_0, y_0)$. If we assume that $\mathrm{Syl}_1(x_0, y_0)$ is of maximal rank, and if Δ_i denotes the minor of $\mathrm{Syl}_1(x_0, y_0)$ obtained by removing the row i, then the (non-zero) vector $[\Delta_1, -\Delta_2, \ldots, (-1)^{d+d'-1}\Delta_{d+d'-1}]$ is in the kernel of the transpose of $\mathrm{Syl}_1(x_0, y_0)$. Thus $(1, \xi, \ldots, \xi^{d+d'-2})$ and $[\Delta_1, -\Delta_2, \ldots, (-1)^{d+d'-1}\Delta_{d+d'-2}]$ are linearly dependent. We deduce that $\xi = -\frac{\Delta_{d+d'-1}}{\Delta_{d+d'-2}} = -\frac{S_{1,0}(x_0, y_0)}{S_{1,1}(x_0, y_0)}$.

This method allows us to lift a point on $\mathcal{C}$, if there is only one point above (x_0, y_0), but it can be generalized when there are several points above. This generalization is closely related to the subresultant construction of univariate polynomials [330]. Here we want to exploit linear algebra tools from a numerical perspective. The aim is to make the matrix of multiplication by z in the quotient algebra $\mathbb{R}[z]/(P_1(x_0, y_0, z), P_2(x_0, y_0, z))$ appear, in order to compute its eigenvalues which yields z-coordinate of the points above (x_0, y_0) [142].

We proceed as follows: Given a point (x_0, y_0) of the projected curve $\mathcal{C}'$, we construct the Sylvester matrix associated to $P(z), q(z)$. By construction, the columns of this matrix are $P, z\,P, \ldots, z^{d'-1}\,P, q, z\,q, \ldots, z^{p-1}\,q$, written in the basis $1, z, \ldots, z^{d+d'-1}$. Assume that the kernel of the transposed Sylvester matrix $\mathrm{Syl}(x_0, y_0)$ has dimension d and is generated by $\Lambda_1, \ldots, \Lambda_d$.

By transposition, we can interpret the Λ_i ($i = 1 \ldots d$) as linear forms over $\mathbb{Q}_{d+d'-1}[z]$ vanishing on $P, z\,P, \ldots, z^{d'-1}\,p, q, z\,q, \ldots, z^{d-1}\,q$. We can extend the Λ_i over $\mathbb{R}[z]$, considering that these forms vanish over all the ideal generated by p and q. So they can be considered as elements of the dual of $\mathcal{A} = \mathbb{R}[z]/(p(z), q(z))$. As the linear forms Λ_i are independent, they also form a basis of this dual space. The coefficients of Λ_i in the dual basis $(1^*, \ldots, (z^{d-1})^*)$ of the monomial basis $\{1, z, \ldots, z^{d-1}\}$ of $\mathcal{A}$ are $[\Lambda_i(1), \Lambda_i(z), \ldots, \Lambda_i(z^{d-1})]$. By definition of the transposed operator, for any $a \in \mathcal{A}$, $M^t(\Lambda_i)(a) = \Lambda_i(M_z(a)) = \Lambda_i(z\,a)$. Thus we have the relation:

$$
\begin{pmatrix} \Lambda_1(z) & \cdots & \Lambda_d(z) \\ \vdots & & \vdots \\ \Lambda_1(z^d) & \cdots & \Lambda_d(z^d) \end{pmatrix} = M_z^t \begin{pmatrix} \Lambda_1(1) & \cdots & \Lambda_d(1) \\ \vdots & & \vdots \\ \Lambda_1(z^{d-1}) & \cdots & \Lambda_d(z^{d-1}) \end{pmatrix}
$$

where M_z is the operator of multiplication by z in $\mathbb{R}[z]/(p(z), q(z))$. As $d = \dim \ker(\mathrm{Syl}(x_0, y_0)) = \dim \mathcal{A}$, and as $(1, z, \ldots, z^{d-1})$ form a basis of the quotient space, the matrix

$$
\begin{pmatrix} \Lambda_1(1) & \cdots & \Lambda_d(1) \\ \vdots & & \vdots \\ \Lambda_1(z^{d-1}) & \cdots & \Lambda_d(z^{d-1}) \end{pmatrix}
$$

is invertible. We deduce that computing the generalized eigenvalues of the previous matrices yields the eigenvalues of the operator M_z of multiplication by z in $\mathcal{A}$, that is the z-coordinate of the points above (x_0, y_0).

We summarize the algorithm here:

Algorithm 10 Lifting the projection

- Compute the Sylvester matrix $S = \mathrm{Syl}(x_0, y_0)$.
- Compute a basis $\Lambda_1, \ldots, \Lambda_d$ of the kernel of S^t.
- Extract the submatrix A_0 of the coordinates of $\Lambda_1, \ldots, \Lambda_d$ corresponding to the evaluations in $1, \ldots, z^{d-1}$.
- Extract the submatrix A_1 of the coordinates of $\Lambda_1, \ldots, \Lambda_d$ corresponding to the evaluations in $z, \ldots, z^d$.
- Compute the generalized eigenvalues of A_1 and A_0 and output the corresponding z-coordinates of the point above (x_0, y_0).

The last step can be replaced by the computation of $\det(A_1 - z\,A_0)$ and a univariate root finding step.

3.7.4 Computing Points of the Curve above Critical Values

In this section, we are going to describe how we check the genericity condition and how we compute a finite set of points, which will allow us to deduce the topology of $\mathcal{C}$.

First, we check that there is no asymptotic direction parallel to the (y, z)-plane, by testing if the curve $\mathcal{C}$ has a point at infinity in the plane $x = 0$. This is done by checking if the system

$$\frac{P_1^\top}{\Delta}(0, y, z) = \frac{P_2^\top}{\Delta}(0, y, z) = 0$$

has a non-trivial solution, where $P^\top$ is the homogeneous component of highest degree of a polynomial P and $\Delta = \gcd(P_1^\top, P_2^\top)$. It reduces to computing the projective resultant of these two homogeneous polynomials. Since the number of asymptotic directions of $\mathcal{C}$ is finite, by a generic linear change of variables, we can avoid the cases where $\mathcal{C}$ has an asymptotic direction parallel to the (y, z)-plane.

Next, we compute the x-critical points of $\mathcal{C}$ by solving the system (3.1), using Algorithm 8. This computation allows us to check that the system is zero-dimensional and that the x-coordinates of the real solutions are distinct. If this is not the case, we perform a generic change of coordinates.

The cases for which we have to do a change of coordinates are those where a component of $\mathcal{C}$ is in a plane parallel to (y, z) or where a plane parallel to (y, z) is tangent to $\mathcal{C}$ in two distinct points. Such cases are avoided by a generic change of coordinates.

We denote by $\Sigma_0 = \{\sigma_1^0, \ldots, \sigma_{k_0}^0\}$ the x-coordinates of the x-critical points: $\sigma_1^0 < \cdots < \sigma_{k_0}^0$.

Next, we compute the singular points of $\mathcal{C}'$ as (a subset of) the real solutions of the system

$$g(x, y) = 0, \partial_x g(x, y) = 0, \partial_y g(x, y) = 0, \tag{3.2}$$

and of $\mathcal{C}''$, as (a subset of) the real solutions of

$$h(x, z) = 0, \partial_x h(x, z) = 0, \partial_z h(x, z) = 0. \tag{3.3}$$

We denote by $\Sigma_1 = \{\sigma_1^1, \ldots, \sigma_{k_1}^1\}$ the x-coordinates of these singular points: $\sigma_1^1 < \cdots < \sigma_{k_1}^1$.

An important property of the projected curves $\mathcal{C}'$ and $\mathcal{C}''$, that will be used in the algorithm, is the following:

Proposition 9. *The arcs of the curve $\mathcal{C}'$ (resp. $\mathcal{C}''$) above $]\sigma_i, \sigma_{i+1}[$ do not intersect.*

Let $k(x)$ be the square-free part of $\mathrm{Res}_y(g(x, y), \partial_y g(x, y)) * \mathrm{Res}_z(h(x, z), \partial_z h(x, z))$. We consider the zero-dimensional system:

$$\begin{cases} k(x) & = 0 \\ P(x,y,z) = 0 \\ Q(x,y,z) = 0 \end{cases}$$

Computing generalized normal forms and computing a Rational Univariate Resolution of this system (see Section 3.5), we obtain a representation of the solutions of this system as:

$$\begin{cases} f_0(T) = 0 \\ x & = f_1(T) \\ y & = f_2(T) \\ z & = f_3(T) \end{cases}$$

3.7.5 Connecting the Branches

The approach that we are going to describe now for the branch connection, can be seen as an extension of the approach of [191, 185] to the three-dimensional case.

The previous step yields a sequence of strictly increasing values

$$\Sigma = \{\sigma_1, \ldots, \sigma_l\},$$

such that above $]\sigma_i, \sigma_{i+1}[$, the branches of $\mathcal{C}$ are smooth and the arcs of $\mathcal{C}'$, $\mathcal{C}''$ do not intersect. We will use this property to connect the points of $\mathcal{C}$ above the values σ_i. Notice that Proposition 9 is still true if we refine the sequence $\sigma_1, \ldots, \sigma_l$. In particular, it is valid if we consider the x-coordinates of the singular points of a curve, defined by a multiple of the equation of $\mathcal{C}'$ (resp. $\mathcal{C}''$). It is also valid, if we insert new values in between these critical values: $\delta_0 < \sigma_1 < \mu_1 < \cdots < \sigma_l < \delta_1$, where $\mu_i := \frac{\sigma_i + \sigma_{i+1}}{2}$ for $i = 0, \ldots, l-1$, and δ_0, δ_1 are any value such that $]\delta_0, \delta_1[$ contains Σ. We denote by

$$\alpha_0 < \cdots < \alpha_m$$

this new refined sequence of values and by L_i, the set of points on $\mathcal{C}$ above α_i, for $i = 0, \ldots, m$. These points are computed, either

- by substituting $x = \alpha_i$ and solving the 2-dimensional system $P_1(\alpha_i, y, z) = 0$, $P_2(\alpha_i, y, z) = 0$.
- or by computing the points of $\mathcal{C}'$ above α_i and by lifting them to $\mathcal{C}$ (Algorithm 10).

This construction implies the following lemma, which is used in the next theorem, in order to describe how the computed points have to be connected:

Lemma 3. *Two distinct points of a regular section of $\mathcal{C}$ with the same y-coordinate (resp. z-coordinate) are connected to two points of the next section, with the same y-coordinate (resp. z-coordinate) or to a critical point.*

Theorem 8. *Under the genericity condition of Definition 13, the curve $\mathcal{C}$ connects the points L_i to the points L_{i+1}, only in one way.*

To summarize, the connection of the branches from one plane section of $\mathcal{C}$ to the next one, is performed as follows:

Algorithm 11 Connecting the branches

If there is no x-critical point in L_i and possibly an x-critical point c of $\mathcal{C}$ in L_{i+1}, do the following:

1. Decompose L_{i+1} into the subsets $V'_1, \ldots, V'_k$ of the points with the same y-coordinate, listed by increasing y. Let $s_j = |V'_j|$.
2. Compute the index j_0 such that $c \in V'_{j_0}$. Decompose L_i into the subsets $V_1, \ldots, V_k$ in the following way:
   - For $j > j_0$, V_j is the set of s_j greatest points for the lexicographic order with $x > y > z$, among $L_i - \cup_{l>j} V_l$.
   - For $j < j_0$, V_j is the set of s_j smallest points for the lexicographic order, among $L_i - \cup_{l<j} V_l$.
   - V_{j_0} is the remaining set of points $L_i - \cup_{l \neq j_0} V_l$.
3. For $j \neq j_0$ connect the points of V_j to the points of V'_j, according to there z-coordinates, by segments.
4. For $j = j_0$, let A'_{j_0} (resp. B'_{j_0}) be the set of regular points of V'_{j_0}, with z-coordinate $< z(c)$ (resp. $> z(c)$).
   - Connect the $|A'_{j_0}|$ points of smallest z-coordinates in V_{j_0} to the points in A'_{j_0}, according to their z-coordinate, by segments.
   - Connect the $|B'_{j_0}|$ points of greatest z-coordinates in V_{j_0} to the points in B'_{j_0}, according to their z-coordinate, by segments.
   - Connect the remaining points in V_{j_0} to c, by segments.

If there is a x-critical point of $\mathcal{C}$ in L_i, exchange the role of L_i and L_{i+1} in the previous steps.

Proposition 10. *Assume that we are in a generic position. Then the topology of the curve above the segment $[\alpha_i, \alpha_{i+1}]$ is the same as the set of segments produced by the Algorithm 11.*

3.7.6 The Algorithm

We summarize the complete algorithm below:

Remark 1. This algorithm can be easily adapted to the computation of the topology of $\mathcal{C}$ in a box (resp. bounded domain), by considering the points on the border of the box (resp. domain) as x-critical points.

Remark 2. By a generic change of variables, the set of x-coordinates of the x-critical points of $\mathcal{C}'$ will contain those of $\mathcal{C}$ and the resolution of the system

---

Algorithm 12 Topology of the curve $\mathcal{C} \subset \mathbb{R}^3$ defined by two polynomials

INPUT: polynomials $P_1(x, y, z), P_2(x, y, z) \in \mathbb{Q}[x, y, z]$.

- Compute the x-critical points of $\mathcal{C}$ and their x-coordinates $\Sigma := \{\sigma_1^0, \ldots, \sigma_k^0\}$ with $\sigma_1^0 < \cdots < \sigma_k^0$.
- Check the generic position; If the curve is not in a generic position, apply a random change of variables and restart from the first step.
- Compute the square-free part $g(x, y)$ of $\mathrm{Res}_z(P_1, P_2)$.
- Compute the square-free part $h(x, z)$ of $\mathrm{Res}_y(P_1, P_2)$.
- Compute the singular points of the curves $g(x, y) = 0$ and $h(x, z) = 0$ and insert their x-coordinate in Σ.
- Compute the $\mu_i, \delta_0, \delta_1$ and the ordered sequence $\alpha_1 < \cdots < \alpha_l$. Above each α_i for $i = 1, \ldots, l$, compute the set of points L_i on the curve $\mathcal{C}$.
- For each $i = 0, \ldots, l-1$, connect the points L_i to those of L_{i+1} by Algorithm 11.

OUTPUT: the graph of 3D points connected by segments, with the same topology as the curve $\mathcal{C}$.

(3.1) can be replaced by the computation of the x-critical points of $\mathcal{C}'$ and by a lifting operation on $\mathcal{C}$. This allows us to treat unreduced curves, such that $I(\mathcal{C}) \neq (P_1, P_2)$, by using only the square-free part of $G(x, y)$ and $H(x, z)$. However the verification, a posteriori, of the correctness of the result is more delicate.

3.8 Software

The algebraic-geometric computation that we have seen, is based on an exact arithmetic model, which may involve large size numbers. The library GMP [6] provides tools for manipulating integer and rational numbers of arbitrary size. There are also tools for manipulating floating point numbers of arbitrary size. If control over the precision and the rounding mode of each operation are needed, one can also use MPFR, developed in the same spirit [10].

CGAL [2] provides interval arithmetic, as well as a simple multi-precision floating point type able to evaluate signs of polynomials over floating point input exactly, it also interfaces with external libraries such as GMP and LEDA for multi-precision computations. CGAL also provides generic mechanisms to apply filtering techniques to its own kernel of geometric primitives as well as user defined primitives [282, 283, 69]. Finally, CGAL provides some fine-tuned statically filtered predicates [252] for Delaunay triangulations and other classical geometric algorithms.

BOOST [1] provides a generic implementation of interval arithmetic which can be used in many other contexts [71]. Finally, MPFI [9] provides interval arithmetic with multiple-precision bounds based on MPFR [10]. Other resources on interval computations can be found at `http://www.cs.utep.edu/interval-comp/main.html`, together with many other interval arithmetic libraries.

There are several implementations of algebraic numbers which use the separation bound approach for comparison. The LEDA library [76] implements real algebraic root expressions (without the ◇-operator) using the bound given above. The improved separation bound [285, 306] and the diamond operator are implemented in [173]. They will be included in the LEDA library soon.

The CORE library [3] implements real algebraic expression using several different separation bounds.

A package of the library SYNAPS [13] is devoted to the treatment of algebraic numbers represented by an isolating interval and a square-free polynomial of $\mathbb{Z}[x]$. The specialization for small degree developed by E. Tsigaridas and I. Emiris is available there. The isolation and comparison of small degree algebraic numbers are given by precomputed formulas. For higher degrees, isolation and comparison methods based on the Bernstein subdivision method or on the Sturm-Habicht computations are also implemented.

Regarding multivariate polynomial systems solving, one can also find in SYNAPS [13] solvers based on normal form computation (developed by Ph. Trébuchet) or subdivision solvers (developed by J.P. Pavone), which are very interesting for filtering purposes. Implementation of polynomial solvers based on Gröbner computation are also available in GBRS [5], or SINGULAR [12].

Regarding the topology of implicit curves, implementations of the sweeping algorithm and of other methods based on subdivision techniques are available in SYNAPS [13].

4

Differential Geometry on Discrete Surfaces

David Cohen-Steiner and Jean-Marie Morvan*

Point clouds and meshes are ubiquitous in computational geometry and its applications. These subsets of Euclidean space represent in general smooth objects with or without singularities. It is then natural to study their geometry by mimicking the differential geometry techniques adapted for smooth surfaces. The aim of the following pages is to list some geometric quantities (length, area, curvatures) classically defined on smooth curves or surfaces, and to define their analog for discrete objects, justifying our definition by a continuity property: if a sequence of discrete objects tends (in a certain sense) to a smooth object, do the corresponding geometric quantities tend to the ones of the smooth object?

4.1 Geometric Properties of Subsets of Points

Consider a subset S of points of a vector space $\mathbb{E}$, and $\mathcal{G}$ a group acting on $\mathbb{E}$. We say that a quantity Q associated to S is *geometric with respect to $\mathcal{G}$* if the corresponding quantity associated to $g(S)$ is the same, for all $g \in \mathcal{G}$. For instance, if $\mathbb{E}^N$ is the N-dimensional Euclidean space endowed with its standard scalar product, any metric quantity associated to S is geometric with respect to the group of rigid motions: the diameter of S, the area of the convex hull of $\mathcal{P}$ are examples of geometric quantities. But the distance from the origin O to the closest point of S is not, since it is not invariant under translation. It is important to point out that the fact to be geometric *depends on the group*. For instance, let $\mathcal{G}$ be the group of rigid motions of $\mathbb{E}^N$ and let $\mathcal{G}_1$ be the group of projective transformations of $\mathbb{E}^N$. Then, the property of S being a circle is geometric for $\mathcal{G}$ but not for $\mathcal{G}_1$, while the property of being a conic or a straight line is geometric for $\mathcal{G}$ and $\mathcal{G}_1$. In the following, we consider only the group $\mathcal{G}$ of rigid motions, which seems to be the simplest and the most useful one for our purpose. But it is clear that

---

* Chapter coordinator

other interesting studies have been done in the past, and will be done in the future with different groups, as affine group, projective group, quaternionic group, etc...

We are interested in geometric quantities which are continuous in a certain sense. Indeed, if one wants to evaluate a quantity $Q(\mathcal{S})$ defined on $\mathcal{S}$, but if one can only approximate $\mathcal{S}$ by $\mathcal{S}'$, one would like to evaluate the quantity $Q(\mathcal{S}')$, hoping that the result is not too far from $Q(\mathcal{S})$. In other words, we would like to have:

$$\text{if } \lim_{n\to\infty} \mathcal{S}_n = \mathcal{S}, \text{then } \lim_{n\to\infty} Q(\mathcal{S}_n) = Q(\mathcal{S}).$$

Remark that this claim is incomplete since we did not specify the topology on the set of subsets $\mathcal{S}$ of $\mathbb{E}^N$. The simplest one is the Hausdorff topology, but we shall see that it not enough in general. However, if one only considers convex subsets, then a beautiful theorem of Hadwiger [229] states that the space of Hausdorff-continuous additive geometric quantities is spanned by the so-called *intrinsic volumes*. Examples of intrinsic volumes are length for curves, area for surfaces, but also integrals of mean and Gaussian curvature (for smooth convex sets). These examples, which we will study in more detail in what follows, actually exhaust all possibilities for curves and surfaces. Other quantities than intrinsic volumes can be studied along the same lines. For example, recent work by Hildebrandt et. al. [207], which is not covered in this book, proves convergence results for shortest geodesics and Laplace-Beltrami operators.

To classify geometric quantities (as we said, we only consider quantities invariant by rigid motions), we can try to use differential geometry. In fact, some of them can usually be defined when $\mathcal{S}$ is "smooth", or at least "continuous". Involving these particular characteristics, we can try to classify them by the order of differentiability.

Definition 1. *Let Q be a geometric quantity associated to a smooth submanifold M embedded in $\mathbb{E}^N$. We say that the order of Q is k if Q involves the differentials of order k of the embedding of M.*

We must remark that this definition is ambiguous, since, a priori, Q can be computed in many ways involving different degrees of differentiability.

4.2 Length and Curvature of a Curve

4.2.1 The Length of Curves

Any book of differential geometry begins with curves in $\mathbb{E}^2$ or $\mathbb{E}^3$, (see [51], [112] for instance). The length of a segment is classically given by the Euclidean distance between its two endpoints. The length of a polygon is the sum of the length of its edges. Let us now consider the length $l(C)$ of a C^1-curve of $\mathbb{E}^2$

$$C\colon [0,1] \to \mathbb{E}^2.$$

It is well known that $l(C)$ can be computed in two different ways. Using the differential of C, one defines $l(C)$ by

$$l(C) = \int_0^1 \|C'(t)\| dt. \tag{4.1}$$

With the point of view of Sect. (4.1), the length appears as a quantity of order 1. Another definition uses polygons inscribed in C. Take all subdivisions $s = (0, x_1, ..., x_n, 1)$ of the segment $[0, 1]$, and compute the length of the polygon lines

$$(C(0), C(x_1), ..., C(x_n), C(1)).$$

The supremum of these numbers, when s describes the set $\mathbf{S}$ of all subdivisions of $[0, 1]$ is the length of C:

$$l(C) = \sup_{s=(0,x_1,...,x_n,1)\in\mathbf{S}} l((C(0), C(x_1), ..., C(x_n))). \tag{4.2}$$

Using the mean value theorem, it can be proved that these two definitions are equivalent. However, the second one can be given for a more general class of curves, (that is, the class of curves for which (4.2) is finite - the rectifiable curves-). Remark that (4.1) and (4.2) imply immediately a convergence theorem: if P_n is a sequence of polygons inscribed in C, whose Hausdorff limit is C, then

$$\lim_{n \to \infty} l(P_n) = l(C). \tag{4.3}$$

However, this theorem cannot be extended to polygon which are not inscribed in C. Finally, remark that we have defined a (classical) geometric quantity -the length- in two different contexts, (discrete and smooth) and we have studied a continuous property as a bridge between them. This gives a general framework to study other geometric quantities such as curvatures. We shall follow this approach in the next sections.

4.2.2 The Curvature of Curves

The Curvature of a Planar Smooth Curve

A typical example of a geometric property of order 2 is the curvature $k(p)$ of a (regular) curve C at a point p. To compute it, we first need to calculate the unit tangent vector field t, which involves a first derivative, and then take the derivative of the result. Using the arclength s,

$$\left(\frac{dt}{ds}\right)_p = k(p)n(p),$$

where n is the normal vector field of C. If the geometric image of the curve C is smooth enough, it can be locally represented by the graph $(x, f(x))$ of a smooth function f, such that $p = (0,0)$, (*i.e.* $f(0) = 0$), and such that the tangent to C at p is collinear to the x-axis, (*i.e.* $f'(0) = 0$). Then, $k(p) = f''(0)$. Let $\nu = f'''(0)$. Near $p = (0,0)$ we have

$$f(x) = \frac{kx^2}{2} + \frac{\nu x^3}{6} + o(x^3).$$

The Curvature of a Polygonal Line in $\mathbb{E}^2$

By analogy with the smooth case, one can define the curvature $k(p)$ of a polygonal line at one of its interior vertices p (*i.e.* different from the endpoints):

Definition 2. *Let P be an (oriented) polygonal line of $\mathbb{E}^2$ and p be one of its interior vertices. Let $\angle(p) \in [0, \pi)$ be the turning angle between the two consecutive edges e_1, e_2 adjacent to p, of length η_1, η_2. The curvature $\mathbf{k}(p)$ of P at p is defined by the angular defect*

$$\mathbf{k}(p) = \frac{\pi - \angle(p)}{\eta}$$

where $\eta = (\eta_1 + \eta_2)/2$.

So, $\mathbf{k}(p)$ vanishes when the two edges are collinear.

An Approximation Theorem

The previous definition allows us to get an approximation theorem: one can estimate the curvature of a curve from the angular defect α of an inscribed polygon. See [51] and [68] for details.

Theorem 1. *Let p, q and r be three consecutive vertices on C, with η the distance from q to p and η' the distance from q to r. Then the angular defect $\mathbf{k}(p)$ and the curvature $k(p)$ of C at p satisfy:*

- *if $\eta = \eta'$:*
$$\mathbf{k}(p) = k(p) + o(\eta),$$

- *if $\eta \neq \eta'$:*
$$\mathbf{k}(p) = k(p) + o(1).$$

as η and η' go to zero.

Thus, we can conclude that the angular defect is a good approximation of the curvature of the curve provided the edges are short enough.

4.3 The Area of a Surface

4.3.1 Definition of the Area

If T is a triangle, its area is given by: $\mathcal{A}(T) = \frac{1}{2}bh$, where b is a basis of T and h the corresponding height. Summing the areas of all the facets of a triangulated polyhedron, one deduces a definition of the area of any polyhedron, (after having proved that the result does not depend on the chosen triangulation). Let M be a parametric surface, $M = x(U)$, where U is a domain of $\mathbb{E}^2$ (with coordinates (u,v)), and $x : U \to \mathbb{E}^3$ is a smooth embedding. The area $\mathcal{A}(M)$ of M involves the first derivatives of x:

$$\mathcal{A}(M) = \int_U \|x_u \wedge x_v\| du dv.$$

If T is a triangle, the previous definition is consistent with the first one. More generally, area can be defined for a very large class of subsets of Euclidean space, using measure theory. If $\mathcal{S}$ is any subset of points of $\mathbb{E}^3$, the area of $\mathcal{S}$ is the 2-dimensional Hausdorff measure $H^2(\mathcal{S})$ [258]. These definitions are equivalent in the smooth case. An important problem is to know if one has a continuity property similar to equation 4.3. The answer is negative in general, as the following classical example shows. Let us build a 2-parameter family of *generalized Schwarz lanterns* [51] (two of them are displayed in Fig. 4.3.1) Let C be a cylinder of finite height H and radius r in $\mathbb{E}^3$. Let $P(n,m)$ be a triangulation inscribed in C defined as follows: consider $m + 1$ circles on the cylinder C obtained by intersecting C by 2-planes orthogonal to the axis of C. Inscribe on each circle a regular n-gon such that the n-gon on the slice k is deduced from the n-gon of the slice $k - 1$ by a rotation of angle $\frac{\pi}{n}$. Then join each vertex v of the slice $k - 1$ to the two vertices of the slice k which are nearest to v. One obtains a triangulation whose vertices $v_{i,j}$ are defined as follows:

$$\forall i \in \{0, ..n - 1\}, \ v_{i,j} = (r \ \cos(i\alpha), r \ \sin(i\alpha), jh) \text{ if } j \text{ is even,}$$
$$\forall j \in \{0, ..m\}, \quad v_{i,j} = (r \ \cos(i\alpha + \tfrac{\alpha}{2}), r \ \sin(i\alpha + \tfrac{\alpha}{2}), jh) \text{ if } j \text{ is odd,}$$

and whose faces are:

$$v_{i,j} \ v_{i+1,j} \ v_{i,j+1},$$
$$v_{i,j} \ v_{i-1,j+1} \ v_{i,j+1},$$

where $\alpha = \frac{2\pi}{n}$ and $h = \frac{H}{m}$.

Then when n tends to infinity, a simple computation shows that

$$\lim_{n \to \infty} \mathcal{A}(P(n,m)) = 2\pi r H \sqrt{1 + \frac{m^2 \pi^4 r^2}{4n^4 H^2}}.$$

In particular,

Fig. 4.1. Examples of half *Schwarz Lanterns*

$$\lim_{n \to \infty} \mathcal{A}(P(n, n^2)) = 2\pi r H \sqrt{1 + \frac{r^2 \pi^4}{4H^2}} \neq \mathcal{A}(C),$$

and

$$\lim_{n \to \infty} \mathcal{A}(P(n, n^3)) = \infty,$$

although C is the Hausdorff limit of both $P(n, n^2)$ and $P(n, n^3)$. This shows that there exists sequences of triangulations inscribed on a (smooth) surface M whose Hausdorff limit is M, but whose area tends to infinity, or to a limit different from the area of M.

4.3.2 An Approximation Theorem

Our purpose here is to get conditions under which two close surfaces have close areas. This problem is studied in details in [260]. Let M be a compact smooth surface of $\mathbb{E}^3$. There exists a maximal open set U_m of $\mathbb{E}^3$ containing M on which the orthogonal projection pr onto M is well defined and continuous. We need to be more precise to state our results, and use the notion of *reach* introduced by H. Federer, [161], [162]:

Definition 3. *Let A be any subset of $\mathbb{E}^N$ and $a \in A$.*

1. *We denote by reach(A, a) the supremum of the real numbers r such that every point of the ball of radius r and center a has an unique projection on A, (realizing the distance $d_A(m)$ from m to A):*

$$reach(A, a) = \sup\{r \in \mathbb{R} | \{p : |p - a| < r\} \subset U_A\}.$$

2. *The reach of A is defined as the infimum of the values of $reach(A, a)$ for a in A. We denote by $U_r(A)$ the tubular neighborhood of A whose radius is the reach r of A.*

It can be shown that convex sets and C^2-submanifolds are subsets with positive reach. However non-convex polyhedra are not! In [260] is introduced the following:

Definition 4. *Let M be a smooth surface of $\mathbb{E}^3$.*

- *The function $\omega_M : U_r(M) \to \mathbb{R}$ defined by*

$$\omega_M(p) = \|pr(p) - p\| \rho_{pr(p)}$$

(where $\rho_{pr(p)}$ denotes the maximum of the two principal curvatures of M at $pr(p)$) is called the relative curvature function of M on $U_r(M)$, see 4.4.1 for the definitions of the principal curvatures.
- *If P is any subset lying in $U_r(M)$ the real number*

$$\omega_M(P) = \sup_{p \in P} \omega_M(p),$$

is called the relative curvature of P with respect to M.

Following [170] and [260], we introduce the

Definition 5. *Let m be a smooth surface of $\mathbb{E}^3$.*

- *A subset P of $\mathbb{E}^3$ is **nearly parallel** to M if it lies in $U_r(M)$, where r is the reach of M and if the restriction of pr to P is one-to-one.*
- *A polyhedron of $\mathbb{E}^3$ is **closely inscribed** in M if both its vertices lie in M and it is **nearly parallel** to M.*

If P is a surface nearly parallel to M and is differentiable almost everywhere, we define at almost every point p of P the angle $\alpha_m \in \left[0, \frac{\pi}{2}\right]$ between the tangent planes T_pP and $T_{pr(p)}M$ and we call it the *deviation* at p. We put

$$\alpha_{\max} = \max_{p \in P} \alpha_p,$$

$$\alpha_{\min} = \min_{p \in P} \alpha_p.$$

Theorem 2. *Let P be a surface nearly parallel to M. Then,*

$$\frac{\cos \alpha_{\max}}{(1 + \omega_M(P))^2} \mathcal{A}(P) \leq \mathcal{A}(M) \leq \frac{\cos \alpha_{\min}}{(1 - \omega_M(P))^2} \mathcal{A}(P).$$

If the surface M is compact, then ρ is bounded. So we deduce from the previous theorem the following

Corollary 1. *Let M be a compact orientable smooth surface in $\mathbb{E}^3$. Let P_n be a sequence of almost everywhere differentiable surfaces nearly parallel to M. If*

- *the Hausdorff limit of P_n is M, when n goes to infinity, and*
- *the angle between tangent planes T_pP_n and $T_{pr(p)}M$ tends to 0 almost everywhere when n goes to infinity, then:*

$$\lim_{n\to\infty} \mathcal{A}(P_n) = \mathcal{A}(M).$$

Since we only need that the surfaces P_n are diffentiable almost everywhere, we can apply Corollary 1 to a sequence of polyhedra:

Corollary 2. *Let M be a compact orientable C^2 surface in $\mathbb{E}^3$. Let P_n be a sequence of triangulated polyhedra closely inscribed in M. If both*

1. *the Hausdorff limit of P_n is M when n goes to infinity, and*
2. *$cr(P_n)$ tends to 0 when n goes to infinity, (where $cr(P_n)$ denotes the supremum of the circumradii of the triangles of P_n), then:*

$$\lim_{n\to\infty} \mathcal{A}(P_n) = \mathcal{A}(M).$$

Indeed, it can be proved that the angle between the tangent plane to a triangle t inscribed in a smooth surface and the surface tangent plane at any vertex of t is $O(cr(t))$ [260].

4.4 Curvatures of Surfaces

Now we deal with curvatures of surfaces. They are quantities of order 2 in the smooth case. We first give the standard definitions of curvatures in the smooth case. Then we give a pointwise convergence result for the (Gauss) curvature of a sequence of polyhedra approximating a smooth surface. As we will see, the shortcomings of this theorem lead us to adopt a different point of view.

4.4.1 The Smooth Case

Let us first recall some basic definitions and notations in the case of a smooth surface M bounding a compact solid $V \subset \mathbb{E}^3$. The unit normal vector at a point $p \in M$ pointing outward V will be referred to as $n(p)$. Note that M is thereby oriented. Given a vector v in the tangent plane T_pM to M at p, the derivative of $n(p)$ in the direction v is orthogonal to $n(p)$ as $n(q)$ has unit length for any $q \in M$. The derivative D_pn of n at p thus defines an endomorphism of T_pM, known as the Weingarten endomorphism, or shape operator [1]. The Weingarten endomorphism can be shown to be symmetric; the associated quadratic form is called the second fundamental form. Eigenvectors and eigenvalues of the Weingarten endomorphism are respectively called principal

---

[1] for some reason, most authors add a minus sign in the definition of the Weingarten endomorphism.

directions and principal curvatures. Both principal curvatures can be recovered from the trace and determinant of $D_p n$, also called mean [2] and Gaussian curvature at p, respectively denoted by $H(p)$ and $G(p)$. Fig. 4.4.1 shows the geometric meaning of the second fundamental form at a point p : applied to a unit vector v in the tangent plane at p, it yields the signed curvature of the section of the surface by the plane spanned by $n(p)$, v, and passing through p. Principal directions, displayed in bold, correspond to the values of v where the second fundamental form is maximal or minimal. According to the sign of the Gaussian curvature, one gets three different cases, respectively depicted in Fig. 4.4.1: elliptic (positive), parabolic (zero), and hyperbolic (negative).

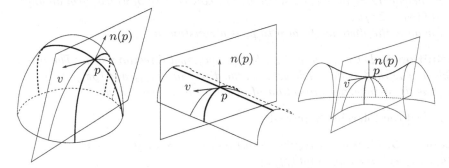

Fig. 4.2. The second fundamental form

4.4.2 Pointwise Approximation of the Gaussian Curvature

Gaussian Curvature of a Polyhedron

Consider now a polyhedral surface P. If p is a vertex of P (which is not on the boundary of P), one can assign to p the angular defect $2\pi - \sum_i \gamma_i$ at p, where the γ_i's stand for the angles at p of the facets incident to p. Using these notations, it is classical to define a notion of *discrete Gauss curvature* [42]:

Definition 6. *Let P be a polyhedral surface and p be a vertex that is not on the boundary. The Gauss curvature $\mathbf{G}(p)$ of P at p is defined by*

$$\mathbf{G}(p) = \frac{2\pi - \sum_i \gamma_i}{\mathcal{A}}$$

where $\mathcal{A}$ denotes the sum of the areas of the triangles adjacent to p.

Remark that in this formula, the Gauss curvature of a vertex of a polyhedron scales like the inverse of a surface area, like the Gauss curvature of a smooth surface at one of its point.

---

[2] the mean curvature is usually defined as the half trace of the Weingarten endomorphism.

An Approximation Theorem

We shall now compare the curvatures of a smooth surface M at a point p with the angular defect of a polyhedron P inscribed in M having this point p as an interior vertex. Our problem is local around p, so we consider the set of triangles incident to p, (which we call *the one ring of P*). A *normal section* through p is the intersection of M with a plane containing p and the normal vector to p.

Definition 7. *Let p be a point of a smooth surface M and let $p_i, i = 1, \ldots, n$ be its neighbors. Point p is called a* **regular vertex** *if*

- *its neighbors lie in normal sections, two consecutive of which form an angle of $\theta(n) = 2\pi/n$,*
- *for all i, the distance from p to p_i is a constant η.*

Definition 8. *Let v_{max} and v_{min} be the principal directions at p. The offset angle a is defined as the angle between the directions v_{max} and $(p\pi(p_1))$, where $\pi(p_1)$ is the orthogonal projection of p_1 onto the tangent plane.*

In [68] the following is proved:

Theorem 3. *Consider a regular vertex p of valence n. If γ_i denotes the angle between directions (pp_i) and (pp_{i+1}):*

- *There exist two functions $A(a, n)$ and $B(a, n)$ such that*

$$2\pi - \sum_{i=1}^{n} \gamma_i = \left[A(a,n)G(p) + B(a,n) \ (k_{max}^2 + k_{min}^2) \right] \eta^2 + o(\eta^2).$$

- *The only value of n such that the functions $A(a, n)$ and $B(a, n)$ depend upon a is $n = 4$, and then*

$$2\pi - \sum_{i=1}^{n} \gamma_i = \left[(1 - 2\cos^2 a \sin^2 a)G(p) + \cos^2 a \sin^2 a \ (k_{max}^2 + k_{min}^2) \right] \eta^2 + o(\eta^2).$$

- *If $n \neq 4$:*

$$2\pi - \sum_{i=1}^{n} \gamma_i = \frac{n}{16 \sin 2\pi/n} \Big[\Big(2 - \cos \frac{4\pi}{n} - \cos \frac{2\pi}{n} \Big) \ G(p) +$$

$$\Big(1 + \frac{1}{2} \cos \frac{4\pi}{n} - \frac{3}{2} \cos \frac{2\pi}{n} \Big) \ (k_{max}^2 + k_{min}^2) \Big] \eta^2 + o(\eta^2).$$

In particular, the only value of n such that $B(a, n) = 0$ is $n = 6$, and then $A(a, 6) = \sqrt{3}/2$, that is

$$2\pi - \sum_{i=1}^{n} \gamma_i = \frac{\sqrt{3}}{2} G(p)\eta^2 + o(\eta^2).$$

Remark that the principal curvatures can be estimated from two different meshes of valences n_1 and n_2 such that $n_1 \neq 4, n_2 \neq 4, n_1 \neq n_2$, by solving a system of equations deduced from Theorem 3.

Finally, the previous result shows that in general, at a point p of a smooth surface endowed with a triangulated polyhedron, the defect angle $\mathbf{G}(p)$ is not a good approximation of the (pointwise) Gauss curvature at p. The approximation is "good" in very special cases, for instance when the one ring around p is regular and the valence is 6.

4.4.3 From Pointwise to Local

In the previous section, we have seen the difficulty of approximating the pointwise Gaussian curvature of a smooth surface by the one of a polyhedron. As shown in Fig. 4.3 in the case of curves, the very concept of pointwise curvature does not even make sense for the class of piecewise linear objects. Indeed, at any point lying in the interior of an edge, the curvature is 0, whereas at a vertex, it seems infinite. This problematic situation is easily overcome by shifting from the pointwise point of view to the *measure theoretic* one : instead of considering curvatures at a given point, one should consider integrals of curvature over a given region. For instance, even if the curvature at a given point of a polygonal line is not geometrically relevant, it is intuitively clear that the total amount of curvature in the region B (in bold in Fig. 4.3) is β, the angle between the normals at the two endpoints of B.

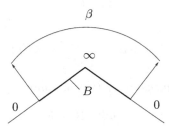

Fig. 4.3. What is the curvature of a polygonal line?

The function that associates to each region B the number β is the simplest example of what are called *curvature measures*. We now define curvature measures for smooth or polyhedral surfaces.

Notion of Curvature Measures

We denote by $\mathcal{B}$ the set of Borel subsets[3] of $\mathbb{E}^3$. If P is a triangulated poly-hedron, we denote by $\mathbf{V}$, $\mathbf{E}$, $\mathbf{T}$ the set of its vertices, edges, and triangles, respectively.

Definition 9. *Let M be an oriented smooth surface and P be an oriented polyhedron of $\mathbb{E}^3$. We respectively denote mean and Gaussian curvatures at a point $p \in M$ by $H(p)$ and $G(p)$.*

- *The Gaussian curvature measure of M is the map*

$$\Phi_M^G : \mathcal{B} \to \mathbb{R},$$

  defined by

$$\Phi_M^G(B) = \int_{B \cap M} G(p) dp.$$

- *The mean curvature measure of M is the map*

$$\Phi_M^H : \mathcal{B} \to \mathbb{R},$$

  defined by

$$\Phi_M^H(B) = \int_{B \cap M} H(p) dp.$$

- *The Gaussian curvature measure of P is the map*

$$\Phi_P^G : \mathcal{B} \to \mathbb{R},$$

  defined by

$$\Phi_P^G(B) = \sum_{p \in \mathbf{V} \cap B \cap P} \mathbf{G}(p),$$

- *The mean curvature measure of P is the map*

$$\Phi_P^H : \mathcal{B} \to \mathbb{R},$$

  defined by

$$\Phi_P^H(B) = \sum_{e \in E} length\ (e \cap B) \beta(e),$$

where $|\beta(e)| \in [0, \pi]$ is the angle between the normals to the triangles incident on e. The sign of $\beta(e)$ is positive if e is convex (i.e. salient) and negative if e is concave.

---

[3]Borel subsets are the building blocks of measure theory. They include in particular all open and closed subsets.

Up to now, these definitions of curvature measures of polyhedra are completely arbitrary. In fact, there is a strong consistency in these definitions. In [170], J. Fu proved a convergence theorem of curvatures. Here we mention a convergence theorem based on homotopic deformation of integral currents [161]. Details on the proof and extensions can be found in [97], [95], [96]:

Theorem 4. *Let P be a triangulated polyhedron closely inscribed in M. Let B be the relative interior of a union of triangles. Then, assuming P and M are consistently oriented:*

- $|\Phi_P^G(B) - \Phi_M^G(pr(B))| \leq C_M \mathbf{K}\epsilon;$
- $|\Phi_P^H(B) - \Phi_M^H(pr(B))| \leq C_M \mathbf{K}\epsilon;$

where pr denotes the orthogonal projection on M and

- C_M *is a real number depending only on the maximum curvature of M,*
- $\mathbf{K} = \sum_{t \in \mathbf{T}, t \subset \overline{B}} cr(t)^2 + \sum_{t \in \mathbf{T}, t \subset \overline{B}, t \cap \partial B \neq \emptyset} cr(t);$
- $\epsilon = \max\{cr(t), t \in \mathbf{T}, t \subset \overline{B}\};$

where $cr(t)$ is the circumradius of the triangle t.

We will see in Sect. 4.4.5 that in several practically important cases, the number $\mathbf{K}$ can be bounded from above, implying that the curvature measures of a sequence of increasingly fine triangulations of a smooth surface converge to the ones of the smooth surface.

Exterior Calculus

Before we explain where these theorems come from, we need some background on calculus on manifolds (see [80] for further information). Let S be a smooth manifold of dimension at least two embedded in some euclidean space $\mathbb{E}^k$. If f is a vector field on S, we denote by $f_x \in T_x S$ the vector associated with a point $x \in S$. differential 2-forms are, in a certain sense, 2-dimensional analogs of (co-)vector fields:

Definition 10. *A differential 2-form ω on S associates with every point $x \in S$ a skew-symmetric bilinear form on $T_x S$, denoted by ω_x.*

The following definition shows how a differential 2-form can be built from two vector fields:

Definition 11. *The exterior product $f \wedge g$ of two vector fields f and g on S, is the differential 2-form defined by:*

$$(f \wedge g)_x(u, v) = (f_x \wedge g_x)(u, v) = \begin{vmatrix} f_x.u & g_x.u \\ f_x.v & g_x.v \end{vmatrix}$$

for all x in S and $u, v \in T_x S$. Here . is the inner product of $\mathbb{E}^k$.

Exterior products are special cases of differential 2-forms. However, they provide a good intuition of the general case: any 2-differential form can actually be written as a linear combination of exterior products of vector fields. It can be seen from the definition of an exterior product that if A is a linear transformation of the plane P spanned by u and v, then $(f \wedge g)_x(Au, Av) = det(A)(f \wedge g)_x(u, v)$. In particular, $(f \wedge g)_x(u, v) = (f \wedge g)_x(u', v')$ for any two orthonormal frames (u, v) and (u', v') of P with the same orientation. Note that this property extend to general differential 2-forms by linearity. Similarly, we have $f_x \wedge g_x(u, v) = f'_x \wedge g'_x(u, v)$ for any couple of orthonormal frames (f_x, g_x) and (f'_x, g'_x) spanning the same oriented plane. Important examples of exterior products are *area forms*. Area forms are a way to represent oriented surfaces as 2-differential forms. If $T \subset S$ is an oriented surface, then the area form of T is constructed as follows: for each point $x \in T$, pick an orthonormal frame of the tangent plane $T_x T$ compatible with the orientation of the surface, say (a_x, b_x). We will call such frames positively oriented orthonormal frames. For $x \notin T$, set $a_x = b_x = 0$. The area form of T, denoted by a_T, is the differential 2-form $a \wedge b$. Intuitively, area forms can be thought of as fields of surface elements: when applied to two vectors u and v in $T_x S$, a_{T_x} yields the signed area of the parallelogram spanned by the projections of u and v on $T_x T$.

Integration.

Differential 2-forms can be integrated on oriented surfaces, in the same way vector fields can be integrated on oriented curves. To see how, let T be an oriented surface in S and, for each $x \in T$, let (u_x, v_x) be a positively oriented orthonormal frame of the tangent plane $T_x T$. The integral of a differential 2-form ω on T is defined to be:

$$\int_T \omega = \int_T \omega_x(u_x, v_x)dx$$

whenever the right-hand side is defined. For instance, one has $\int_T a_T = area(T)$, which is why area forms are called this way.

Change of variable.

A change of variable is merely a diffeomorphism $\phi: S' \longrightarrow S$ where S' is the manifold where the new variables live. Using such a map, a differential 2-form ω on S can be transformed into a differential 2-form on S', by a process called *pullback*:

Definition 12. *The pullback of ω by ϕ, denoted by $\phi^*\omega$ is given by:*

$$\phi^*\omega_x(u, v) = \omega_{\phi(x)}(D_x\phi(u), D_x\phi(v))$$

for all $x \in S'$ and $u, v \in T_x S'$.

In a certain sense, pulling a differential 2-form back amounts to expressing it in terms of the new variables. The change of variable formula relates the integral of a differential 2-form with the one of its pullback. The result turns out to be particularly simple:

$$\int_{S'} \phi^* \omega = \int_{\phi(S')} \omega \tag{4.4}$$

For example, if $S = S' = \mathbb{E}^2$ and h is an integrable function from S to $\mathbb{R}$, applying (4.4) to $\omega = h a_T$ yields $\phi^* \omega = Jac(\phi)(h \circ \phi) a_T$, where $Jac(\phi)$ denotes the determinant of the Jacobian matrix of ϕ. Equation (4.4) thus generalizes the classical change of variable formula. For this formula to hold, ϕ need actually not be a diffeomorphism from S' to S; the only requirement is that ϕ should be a diffeomorphism from S' to $\phi(S')$.

Integral 2-currents.

Integral 2-currents generalize oriented surfaces [258]. They can be formally defined as linear combinations of oriented surfaces with integral coefficients. In particular, any oriented surface T can be considered as an integral 2-current, which we will abusively also denote T. Integration of differential 2-forms is extended to integral 2-currents by linearity:

$$\int_{nT+pT'} \omega = n \int_T \omega + p \int_{T'} \omega$$

The surface that is setwise the same as T but with reverse orientation thus corresponds to the current $-T$. Geometrically, integral 2-currents can be thought of as oriented surfaces with *multiplicities*. For instance, if T and T' are two oriented surfaces such that orientations of T and T' agree on $T \cap T'$, $T + T'$ can be represented as $T \cup T'$ endowed with the same orientation as T and T', points in $T \cap T'$ having a multiplicity equal to 2. If orientations of T and T' do not agree, then summing T and T' yields a cancellation on $T \cap T'$. Finally, integration of differential forms on integral currents is traditionally not denoted by an integration sign, but rather by a bracket, namely one writes $\langle C, \omega \rangle$ instead of $\int_C \omega$.

A Unified Framework, the Theory of Normal Cycles

The proof of Theorem 4 relies on the theory of the normal cycle. This theory has been developped by P. Wintgen and M. Zähle to give a general method to define curvatures of a large class of objects, [349], [171], [170]. The main observation is that the curvature measures of smooth surfaces can be interpreted as integrals of particular differential forms on the unit normal bundle of surfaces. So, the crucial point is to define a generalization of this unit normal bundle for non-smooth objects, called the normal cycle. In this section, we

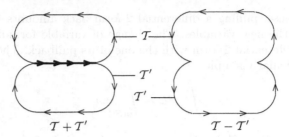

Fig. 4.4. Sum of integral currents

describe the differential forms ω_G and ω_H which induce the Gauss curvature measure and mean curvature measure, when integrated on the normal cycles associated to the studied object, and we shall describe the construction of the normal cycle of a polyhedron.

Invariant 2-forms.

Now set $\mathcal{S} = \mathbb{E}^3 \times S^2$. $\mathcal{S}$ is obviously a subset of $\mathbb{E}^3 \times \mathbb{E}^3$. We will call the first factor of the latter product the *point space*, E_p, and the second one the *normal space*, E_n. The reason for this is that an element of $\mathcal{S}$ can be thought of as a point in space together with a unit normal vector. If u is a 3-vector, u^n will denote the vector $(0, u) \in E_p \times E_n$, and u^p the vector $(u, 0) \in E_p \times E_n$. Rigid motions of $\mathbb{E}^3$ can be naturally extended to $\mathcal{S}$: if g is such a motion, one can set $\hat{g}(p, n) = (g(p), \bar{g}(n))$, where $\bar{g}$ is the rotation associated with g. We now define two particular differential 2-forms on $\mathcal{S}$. Let $(p, n) \in \mathcal{S}$ and $x, y \in \mathbb{E}^3$ such that (x, y, n) is a positively oriented orthonormal frame of $\mathbb{E}^3$. We set:

$$\omega^H_{(p,n)} = x^p \wedge y^n + x^n \wedge y^p$$
$$\omega^G_{(p,n)} = x^n \wedge y^n$$

One can actually check that these 2-forms do not depend on the choice of x and y. Moreover, they are *invariant under rigid motions*, that is satisfy $\hat{g}^* \omega = \omega$ for any rigid motion g.

Normal cycle of a smooth surface and a polyhedron.

Let us give a short overview of the theory of *normal cycle*.

- In the smooth case, the normal cycle is a generalization of the unit normal bundle. Recall that if M is an oriented smooth surface in $\mathbb{E}^3$, the unit normal bundle of M is the manifold

$$ST^{\perp}M = \{(p, n(p)), p \in M\}$$

endowed with the orientation induced by the one of M. Now if V is the solid enclosed by a closed smooth surface in $\mathbb{E}^3$, the normal cycle $N(V)$ of V is nothing but the current canonically associated to $ST^\perp \partial V$: If ω is any 2-differential form defined on $\mathbb{E}^3 \times \mathbb{E}^3$, the duality bracket $\langle , \rangle$ is given by

$$\langle N(V), \omega \rangle = \int_{ST^\perp \partial V} \omega.$$

- This definition can be easily generalized to convex bodies C of $\mathbb{E}^3$: We begin by defining the normal cone $\mathcal{C}_p(C)$ of a point p of C as the set of unit vectors n such that

$$\forall q \in C, \; \overrightarrow{pq}.n \leq 0.$$

The normal cone $\mathcal{C}(C)$ of C is the union of the sets $\{p\} \times \mathcal{C}_p(C)$, when p runs over C. Now the normal cycle of C is nothing but the current associated to $\mathcal{C}(C)$ endowed with its canonical orientation. A crucial property of the normal cycle is its additivity: if C_1 and C_2 are two convex subsets of $\mathbb{E}^3$ such that $C_1 \cup C_2$ is convex, then,

$$N(C_1 \cup C_2) = N(C_1) + N(C_2) - N(C_1 \cap C_2). \tag{4.5}$$

- Remarking that a polyhedron is the union of (convex) tetrahedra, triangles, edges and vertices, we define the normal cycle of a polyhedron by applying (4.5) recursively. If polyhedron V is triangulated using tetrahedra $(t_i)_{i=1..n}$, this leads to the following formula

$$N(V) = \sum_{n=1}^{\infty} (-1)^{n+1} \sum_{1 \leq i_1 < .. < i_n \leq n} N(\cap_{j=1}^{n} t_{i_j})$$

by inclusion-exclusion. Of course one must check that the definition is independent of the decomposition of the polyhedron, but it is not too difficult to prove.

Curvature measures.

To recover the curvature measures in a unified way, we use the theory of normal cycles. If M is smooth and bounds a solid V, let i be its Gauss map:

$$i : M \to ST^\perp M$$

defined by

$$i(p) = (p, n(p)).$$

One has the following (we refer to [97], [95], [96]):

Theorem 5. *Let V be a solid whose boundary M is either a smooth surface or a polyhedron. If B is a Borel subset of $\mathbb{E}^3$, then:*

- $\Phi_M^G(B) = \langle N(V), \chi(i(B \cap V))\omega_G \rangle,$
- $\Phi_M^H(B) = \langle N(V), \chi(i(B \cap V))\omega_H \rangle$

where χ denotes the characteristic function.

Let us prove this theorem for Gaussian curvature in the smooth case. By definition we have:

$$\int_{N(V)} \omega^G \chi(i(B \cap M)) = \int_{i(B \cap M)} \omega^G$$

The change of variable formula now states that:

$$\int_{i(B \cap M)} \omega^G = \int_{B \cap M} i^* \omega^G$$

To prove the first claim, it is thus sufficient to show that:

$$i^* \omega^G = G a_M$$

Let (u, v) be a positively oriented orthonormal frame of $T_x M$, where $x \in M$. By definition, the derivative of i is given by:

$$\forall w \in T_x M \ D_x i(w) = w^p + D_x n(w)^n$$

Expressing $\omega_{i(x)}^G$ in the frame (u^n, v^n, n_x^n), we get

$$(i^* \omega^G)_x(u, v) = \begin{vmatrix} u^n.(u^p + D_x n(u)^n) & v^n.(u^p + D_x n(u)^n) \\ u^n.(v^p + D_x n(v)^n) & v^n.(v^p + D_x n(v)^n) \end{vmatrix}$$

$$= \begin{vmatrix} u.D_x n(u) & v.D_x n(u) \\ u.D_x n(v) & v.D_x n(v) \end{vmatrix} = G(x)$$

The proof of the second equality is similar.

This interpretation of curvature measures in terms of differential forms leads to the approximation theorem 4 by applying classical results on deformation of integral currents [162].

4.4.4 Anisotropic Curvature Measures

In the previous section, we have seen how to define Gauss and mean curvature measures on polyhedra, which generalize Gauss and mean curvatures on smooth surfaces. From the Gauss and mean curvature, the principal curvatures are easily recovered by solving a quadratic equation. However, this does not give any information on the principal directions. In this section, we solve this problem by giving an extension of the concept of curvature tensor to polyhedra, based on the normal cycle. At each point p of M, we denote by $\tilde{h}_p$ the bilinear form

$$\tilde{h}_p \colon \mathbb{E}_p^3 \times \mathbb{E}_p^3 \to \mathbb{R}$$

defined as the composition of the second fundamental form at p, h_p, with the projection pr_{T_pM} on T_pM:

$$\tilde{h}_p = h_p \circ (pr_{T_pM}, pr_{T_pM}).$$

$\tilde{h}_p$ is clearly null on the normal space of M at p.

We give the following [98]:

Definition 13. *Let X, Y be two (constant) vectors of $\mathbb{E}^3$, and (p, n) be a point of $\mathbb{E}^3 \times \mathbb{E}^3$. The 2-form*

$$\omega^{X,Y}_{(p,n)} = (n \times X)^p \wedge Y^n,$$

is called the anisotropic 2-form.

In this definition, $\times$ is the standard cross product in $\mathbb{E}^3$. The relation between $\omega^{X,Y}$ and $\tilde{h}_p(X, Y)$ is given by [98]:

Theorem 6. *Let M be a smooth surface bounding a solid V, and B a Borel subset of $\mathbb{E}^3$. Then,*

$$\langle N(V), \chi(\pi^{-1}(B \cap V))\omega^{X,Y}\rangle = \int_{B \cap M} \tilde{h}_p(X, Y)dp.$$

We put

$$h_M(B)(X, Y) = \langle N(V), \chi(\pi^{-1}(B \cap M)\omega^{X,Y}\rangle. \tag{4.6}$$

Remark that one could also introduce the other form

$$\overline{\omega}^{X,Y}_{(p,n)} = X^p \wedge (n^n \times Y^n)$$

and define

$$\bar{h}_M(B)(X, Y) = \langle N(V), \chi(\pi^{-1}(B \cap M)\bar{\omega}^{X,Y}\rangle.$$

This possibility leads to a similar result, namely

$$\bar{h}_M(B) = \int_{B \cap M} (\text{Trace}(\tilde{h}_p)Id - \tilde{h}_p)dp$$

By analogy with (4.6), we define the anisotropic curvature measure $\overline{h}_P$ associated to a polyhedron of $\mathbb{E}^3$:

Definition 14. *Let P be a polyhedron of $\mathbb{E}^3$, bounding a solid V, and B be a Borel subset of $\mathbb{E}^3$. We put*

$$\overline{h}_P(B)(X, Y) = \langle N(V), \chi(\pi^{-1}(B \cap V)\overline{\omega}^{X,Y}\rangle.$$

The reason why we use $\overline{\omega}$ instead of ω in this definition is that it leads to simpler expressions. Indeed, a computation gives

Theorem 7. *Let P be a polyhedron and B a Borel subset of* $\mathbb{E}^3$. *Then,*

$$\overline{h}_P(B)(X,Y) = \sum_{e \in E} \beta(e) \ length \ (e \cap B)\overrightarrow{e} \otimes \overrightarrow{e}.$$

where $\overrightarrow{e} \otimes \overrightarrow{e}$ *denotes the bilinear form with matrix* $\overrightarrow{e}.transpose(\overrightarrow{e})$.

Using the same techniques as in Sect. 4.4.3, we have an approximation result similar to Theorem 4: with the same notations,

Theorem 8. *Let P be a triangulated polyhedron closely inscribed in M. Let B be the relative interior of a union of triangles. Then,*

$$\|\tilde{h}_P(B) - \tilde{h}_M(B)\| \leq C_M\mathbf{K}\epsilon,$$

where $C_M, \mathbf{K}, \epsilon$ *are defined in Theorem 4.*

Classically, $\|\overline{h}_P(B) - \overline{h}_M(B)\| = \sup_{\|X\|=1,\|Y\|=1} |\overline{h}_P(B) - \overline{h}_M(B)(X,Y)|$.

It must be remarked that $h_P(B)(X,Y)$ is symmetric in (X,Y). This implies that one can diagonalize $h_P(B)$. One gets three eigendirections: the one associated to the smallest eigenvalue can be called the *estimated normal* of $B \cap P$, and the two others can be called its *principal directions*, by analogy with the smooth case. We note that even if the estimated curvature tensor always converges to the actual one, it might not be the case for the eigendirections. Indeed, on a cylinder, the curvature tensor has rank one, so the estimated normal may lie anywhere in the plane orthogonal to the axis of the cylinder. For that reason, this method should not be used strictly for normal estimation purposes, for which simpler techniques exist.

In practice, one may take for B a small number of rings of triangles around the considered vertex. Experimental results are shown in Fig. 4.5 and Fig. 4.7.

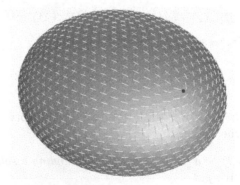

Fig. 4.5. The principal directions estimated on an ellipsoid with 1442 vertices, look very similar to the actual ones, whose integral lines are shown in Fig. 4.6

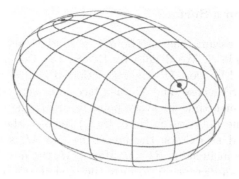

Fig. 4.6. Lines of curvature of an ellipsoid [206]

Fig. 4.7. Directions of minimal curvature estimated on a mesh of Michelangelo's David. For each vertex, the averaging domain used for computations is the 2-ring of that vertex

4.4.5 ϵ-samples on a Surface

In this paragraph, assume that M bounds a solid V and is endowed with a finite point sample S. We denote by $\mathrm{Del}(S)$ the Delaunay triangulation associated to S, and by $\mathrm{Del}_M(S)$ the restriction of $\mathrm{Del}(S)$ to M, that is, the set of faces of $\mathrm{Del}(S)$ whose dual Voronoi face intersects M. An ϵ-sample S on M is a sample such that for every point m of M, the ball $B(m, \epsilon \mathrm{lfs}(m))$ encloses at least one point of S. Here, $\mathrm{lfs}(m)$ denotes the *local feature size* of M at m, defined as follows: the *medial axis* of M is the center of the maximal open balls included in $\mathbb{E}^3 \setminus M$. If $m \in M$, the *local feature size* of M at m is the Euclidean distance from m to the medial axis of M. This notion, also used in Chap. 6 Sect. 6.2.2, actually coincides with the notion of *reach* introduced before (Definition 3).

In [21] and [25], N. Amenta *et al.* proved that if S is an ϵ-sampling of M with $\epsilon < 0.06$, then $\mathrm{Del}_M(S)$ is closely inscribed in M. Moreover they proved

Lemma 1. *Let S be an ϵ-sample on M, with $\epsilon < 0.06$. If B be any Borel subset of $\mathrm{Del}_M(S)$, then the Hausdorff distance $\delta_B = \delta(B, pr(B))$ between B and its projection onto M satisfies:*

$$\delta_B \leq \epsilon \sup_{m \in pr(B)} lfs(m).$$

Moreover,

$$cr(B) \leq \epsilon \sup_{m \in pr(B)} lfs(m),$$

where $cr(B)$ is the maximum of the circumradius of all triangles lying in B.

When M is approximated by the restricted Delaunay triangulation of an ϵ-sample, one gets a "good" approximation of the normal:

Theorem 9. *Let S be an ϵ-sample on M, with $\epsilon < 0.06$. Then, the maximum angle between the normal of a triangle of the restricted Delaunay triangulation and the corresponding tangent plane of M at any of the vertices of the triangle is $O(\epsilon)$.*

However, in general, one cannot deduce that the difference between the curvature measures of $B \cap \mathrm{Del}_M(S)$ and $pr(B) \cap M$ is $O(\epsilon)$. For that we need a stronger assumption. If we assume that the ϵ-sample on M is κ-*light* [93], namely that for every point m of M, the ball $B(m, \epsilon \mathrm{lfs}(m))$ encloses at most κ points of $\mathcal{S}$, then it can be shown that the parameter $\mathbf{K}$ in Theorem 4 satisfies

$$\mathbf{K} = O(\mathrm{area}(M))$$

in Theorem 4 and Theorem 8. So one has:

Theorem 10. *Let S be a κ-light ϵ-sample on M. Then,*

$$\|\varPhi^G_{Del_M(S)} - \varPhi^G_M\| = O(\epsilon);$$

$$\|\varPhi^H_{Del_M(\mathcal{S})} - \varPhi^H_M\| = O(\epsilon);$$

$$\|\tilde{h}_{Del_M(S)} - \tilde{h}_M\| = O(\epsilon).$$

4.4.6 Application

Estimating the curvature tensor from a triangulated approximation of a surface is an important building block for several applications in geometric modeling and computer graphics. To end this chapter, we briefly mention one of them: anisotropic remeshing of surfaces [19]. The goal here is to approximate a given triangulated surface by a surface (mostly) made of quadrilaterals, in such a way that the approximation error is as small as possible. As shown by approximation theory [319], in such an optimal quadrangulation, the quadrilaterals axes should be aligned with the principal directions of the surface, and their aspect ratio should be proportional the square root of the principal curvatures ratios. A strategy to build such a quadrangulation (see figure 4.8) is to estimate the curvature tensor at each vertex of the surface, then numerically integrate two networks of curvature lines (one for the maximum curvature, and one for the minimum curvature), choosing the spacing between consecutive lines as a function of principal curvatures (left). Finally, overlaying the two networks (center) and straightening the curves between intersection points (right) provides the desired quadrangulation.

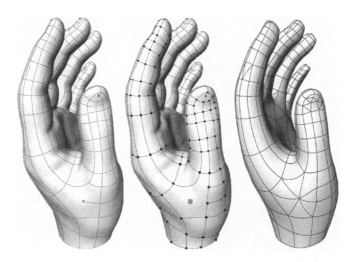

Fig. 4.8. Anisotropic remeshing of a hand

5

Meshing of Surfaces

Jean-Daniel Boissonnat, David Cohen-Steiner, Bernard Mourrain, Günter Rote*, and Gert Vegter

5.1 Introduction: What is Meshing?

Meshing is the process of computing, for a given surface, a representation consisting of pieces of simple surface patches. In the easiest case, the result will be a triangulated polygonal surface. More general meshes also include quadrilateral (not necessarily planar) patches or more complicated pieces, but they will not be discussed here. For example, Fig. 5.1a shows a meshed

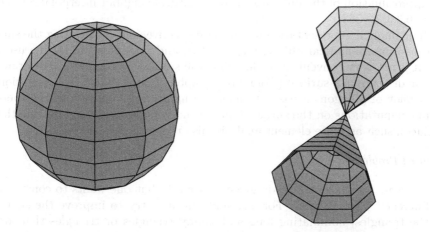

Fig. 5.1. (a) A meshed sphere. (b) A meshed double-cone

sphere $x^2 + y^2 + z^2 = 1$. Fig. 5.1b shows a good mesh of a double-cone $x^2 + y^2 = z^2$. Note that the cone has pinching point at the apex, which is represented correctly in this mesh. The automatic construction of good

---

* Chapter coordinator

meshes for surfaces with singularities is still an open research area. Here we will mainly concentrate on smooth surfaces, with the exception of Sect. 5.4, where surfaces with singularities are also treated.

Why Meshing?

The meshing problem occurs in different settings, depending on the way how a surface is given, and on the purpose of meshing.

Usually we assume that a surface is given *implicitly*, as the solution of an equation

$$f(x, y, z) = 0.$$

The function f comes from various sources. It can be explicitly given, often as a polynomial, like in the examples above, or as a sum of exponential "blob" functions like $\exp(-\|A(x-b)\|) - 1$ for a matrix A and a center b, which allow flexible modelling of shapes.

One can also try to fit data by defining an appropriate function f from the data. For example, in scattered data interpolation, a scanned image may be given a two-dimensional (or sometimes three-dimensional) grid of grey-level values. If f is a function that interpolates these values on each grid square, the equation $f(x, y) = \text{const}$ extracts a level curve (iso-curve, or iso-surface in higher dimensions). In the area of surface reconstruction from scattered data points, there are procedures for defining a function f whose zero set is an approximation of the unknown surface (natural neighbor interpolation, see Sect. 6.3.3, p. 263.)

The implicit representation of a surface is convenient for defining the surface as a mathematical object, for modeling and manipulation by the user, but it is not very convenient for handling the surface by computer. For drawing or displaying a surface (Computer Graphics), an *explicit* representation as a union of polygons is easier to handle. Engineering applications that perform computations on the surface (and on the volume inside and outside the surface), such as finite element analysis, also require a meshed surface.

Related Problems.

Sometimes, a surface is already given as a mesh, but one wants to construct a different, "better" mesh. For example, one may try to improve the shape of the triangles, eliminating long and skinny triangles or triangles that are too large, or one may want to produce a coarser mesh, eliminating areas that are meshed too densely, with the purpose of reducing the amount of data for storage or transmission (data compression). These are the problems of *remeshing*, *mesh refinement* and *mesh simplification*. These problems are also applicable for plane meshes, where the given "surface" is a region of the plane [301, 317]. Some of the methods that we will discuss below have been applied to this setting, and to the meshing problem for polyhedral surfaces in general, but we will only mention this briefly.

As mentioned, engineering applications also require three-dimensional *volume meshes* or even higher-dimensional meshes. Extending a given boundary mesh of a surface to a mesh of the enclosed volume is a difficult problem of its own.

In this chapter, we will concentrate on surface meshing. The other problems described above are not covered in this book. For simplicity, we restrict our attention to surfaces without boundary. Some algorithms can clip a surface by some bounding box or by intersecting it with some other surface, but we will not discuss this.

Meshing of curves, by comparison, is a much easier problem: Here we look for a polygonal chain approximating a curve. We will often discuss curve meshing because the main ideas of many meshing algorithms can be illustrated in this setting.

Meshing is related to surface reconstruction, which is the subject of Chapter 6: in both cases, the desired output is a meshed surface. However, surface reconstruction starts from a set of sample points of the surface that is given as input, and which is usually the result of some measurement process. In meshing, one also constructs a point sample of the surface, but the selection of these points is under the control of the algorithm.

There is some overlap in the techniques applied, in particular in the area of *Delaunay meshing* (Sect. 5.3). As mentioned above, one way of reconstructing a surface is by defining a function f whose zero set is the reconstructed surface.

Goals of Meshing Algorithms—Correctness.

There is a vast literature on meshing with many practical algorithms in the areas of Computer-Aided Geometric Design (CAGD) and Computer Graphics. In this book we concentrate on methods with proven correctness and quality guarantees. Correctness means that the result should be *topologically correct* and *geometrically close*.

The definition of what it means for a mesh to be topologically correct has evolved over the past few years. It is not sufficient to require that a surface S and its mesh S' are homeomorphic. A torus and a knotted torus are homeomorphic when viewed as surfaces in isolation, but one would certainly not accept one as a topologically correct representation of the other, see Fig. 5.2. The following definition combines the strongest notions of having the correct topology with the requirement of geometric closeness.

Definition 1. *An* ambient isotopy *between two surfaces* $S, S' \subset \mathbb{R}^3$ *is a continuous mapping*

$$\gamma \colon \mathbb{R}^3 \times [0,1] \to \mathbb{R}^3$$

which, for any fixed $t \subseteq [0,1]$, *is a homeomorphism* $\gamma(\cdot, t)$ *from* $\mathbb{R}^3$ *to itself, and which continuously deforms* S *into the mesh* S': $\gamma(S, 1) = S'$.

In addition, the approximation error D *is the largest distance by which a point is moved by this homeomorphism:*

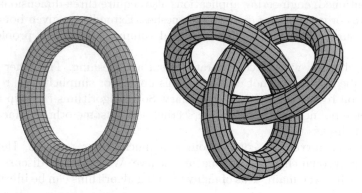

Fig. 5.2. Two homeomorphic surfaces which are not isotopic

$$\|x - \gamma(x, 1)\| \leq D \text{ for all } x \in S$$

This implies that the surface S and the meshed surface S' are homeomorphic to each other, and their Hausdorff distance (see the definition in Sect. 6.2.3 on p. 251) is at most D. There is also the concept of isotopy between two surfaces, which only deforms S without deforming the ambient space $\mathbb{R}^3$, see Sect. 7.2 (p. 281):

Definition 2. *An* isotopy *between two surfaces* $S, S' \subset \mathbb{R}^3$ *is a continuous mapping*

$$\gamma \colon S \times [0, 1] \to \mathbb{R}^3$$

which, for any fixed $t \subseteq [0, 1]$*, is a homeomorphism* $\gamma(\cdot, t)$ *from* S *onto its image, and which continuously deforms* S *into the mesh* S'*:* $\gamma(S, 1) = S'$*.*

Formally, isotopy is weaker than ambient isotopy. However, for our purposes, there is no difference between between isotopy and ambient isotopy: The isotopy extension lemma ensures that an isotopy between two smooth surfaces (of class C^1) embedded in $\mathbb{R}^3$ can always be extended to an ambient isotopy [208, Theorem 1.3 of Chapter 8, p. 180]. This does not directly apply to a piecewise linear surface mesh S', but it is easy to show that a piecewise linear surface is ambient isotopic to an approximating smooth surface, to which the theorem applies. The isotopy extension lemma cannot be used in the algorithm of Sect. 5.4, which deals with singular surfaces, but in this case, the ambient isotopy will be constructed explicitly.

Theorems in earlier papers have only made claims about the Hausdorff distance or about the existence of a homeomorphism, but it is often not difficult to obtain also isotopy. Typically, the mapping constructed in the proofs moves points along fibers that sweep out the space between the surface S

and its approximation S'. These fibers move each point of the mesh S' to its closest neighbor on the surface S', as in Fig. 5.3a. (For a curve in the plane, one can also map each curve point to its closest neighbor on the mesh, as in Fig. 5.3b; this does not work in higher dimensions because it may lead to a discontinuous mapping.) Each point x on the mesh is mapped to the "correct"

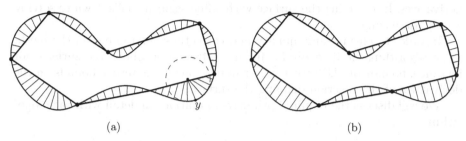

(a)   (b)

Fig. 5.3. The isotopy between a smooth curve and the approximating polygon is indicated by fibers perpendicular to the curve (a) or to the polygon edges (b). The curve can be continuously deformed into the polygon by moving each point along its fiber

corresponding closest point y on the surface, if x is not too far from S, in particular, if the distance from y to x is not larger than the radius of the medial sphere, see Fig. 5.3a, which shows the medial circle at $y \in S$. (See p. 109 in Sect. 2.7 for the definition of medial sphere.)

A common tool for establishing isotopy is a *tubular neighborhood* $\hat{S}$ of a surface S. It is a thickening of the surface such that within the volume of $\hat{S}$, the projection of a point x to the nearest point $\pi_S(x)$ on S is well-defined. The points x which have the same nearest neighbor $\pi_S(x) = p$ form a segment through p normal to the surface. These segments are called *fibers* of the tubular neighborhood, and they form a partition of $\hat{S}$.

Lemma 1 (see [302, Theorem 4.1]). *Let S be a compact closed surface of class C^2 in $\mathbb{R}^3$ with a tubular neighborhood $\hat{S}$. Let T be a closed surface (not necessarily smooth) contained in $\hat{S}$ such that every fiber intersects T in exactly one point.*

Then $\pi_S \colon T \to S$ induces an ambient isotopy that maps T to S.   □

The isotopy interpolates between T and S along the fibers, and it does not move points by more than the length of the longest fiber.

Principal Approach and Primitive Operations.

Although other approaches are conceivable, all methods basically select vertices *on the surface* and connect them appropriately. The fundamental operation is to find the intersection point of a line segment with the surface.

For a few algorithms, it is sufficient to compute these intersections only approximately, and thus the mesh vertices will lie only *close* to the surface. (In particular, this holds for the marching cubes algorithms in Sect. marching-cubes and piecewise-linear interpolation in Sect. 5.5.2).

Some algorithms also compute certain critical points of the surface. All these operations require access to the function f defining the surface and its derivatives. Intersecting the surface with a line segment boils down to solving a univariate equation.

In order to ensure some quality and correctness guarantees on the mesh, some algorithms need to obtain further information about the surface, for instance, bounds on the curvature, or in more algebraic terms, bounds on the derivatives of the function defining the surface.

We will discuss the required primitive operations in detail with each algorithm.

Other Quality Criteria.

Besides topological correctness and geometric closeness to the original surface, we may want to achieve other criteria.

1. Normals: The normals of the mesh should not deviate too much from the normals of the surface. Note that a "wiggly" approximation of a given surface or a curve can be isotopic and have arbitrarily small Hausdorff distance, but still have normals deviating very badly from the original normals.
2. Smoothness: Adjacent facets should not form a sharp angle.
3. Desired density. We may impose an upper bound on the size of the mesh triangles. This bound may depend on the location. For example, in fluid mechanics calculations, a region of turbulence will require a finer mesh than a region of smooth flow.
4. Regularity and Shape. We want to avoid skinny triangles with sharp angles.

There are many other criteria for individual mesh elements. All of these criteria, except the first two in the above list, also apply to plane meshes, and they have been studied extensively in the literature. The algorithms in this chapter concentrate on achieving correctness; geometric quality criteria are often considered in a secondary refinement step.

Basic Assumptions about Smoothness.

The basic assumption for most part of this chapter is that f is a smooth function and the surface has no singularities.

Assumption 1 NONSINGULARITY.
The function f and its gradient ∇f are never simultaneously zero.

This implies that the equation $f(x) = 0$ defines a collection of smooth surfaces without boundary. As mentioned in the introduction, meshing in the vicinity of singularities is a difficult open problem and an active area of research.

Section: Algorithm	Strategy	Topological Correctness	Scaffolding
5.2.3: Snyder [321, 322]	refinement	global parameterizability	cubes
5.2.4: Plantinga and Vegter [286]	refinement	Small Normal Variation	cubes
5.3.1: Boissonnat and Oudot [64]	refinement	sample density, local feature size	Voronoi diagram
5.3.2: Cheng, Dey, Ramos and Ray [90]	refinement	topological ball property*	Voronoi diagram
5.4: Mourrain and Técourt [263, 327]	space sweep, vertical projection	treatment of critical points	vertical planes
5.5.1: Stander and Hart [324]	parameter sweep	Morse theory	—
5.5.2: Boissonnat, Cohen-Steiner, and Vegter [61]	refinement	(Morse theory)*[†]	boxes[†]

Table 5.1. A rough taxonomy of the meshing algorithms described in this chapter, according to the overall strategy, the way how topological correctness is achieved, and the kind of spatial ground structure ("scaffolding") that is used.
* These two algorithms employ a hierarchy of conditions, which are successively tested, in order to achieve correctness.
[†] This algorithm uses Morse theory in a more indirect way. It works with boxes which are subdivided into tetrahedra.

5.1.1 Overview

Meshing algorithms can be roughly characterized as (i) continuation-based methods, that grow a mesh following the surface, and (ii) mesh-based methods, which build some sort of three-dimensional scaffolding around the surface. Although continuation-based methods are often used in practice, it is not easy to achieve correctness guarantees for them. Thus, all algorithms discussed in this chapter fall into the second category. There are three types of adaptive "grid structures" which are used: axis-aligned cubes, vertical planes, and the Voronoi diagram. The algorithms use different algorithmic strategies and a variety of conditions to ensure topological correctness. Table 5.1 summarizes these characteristics. One can see that the algorithms are related in various different ways. In the remainder of this chapter, we have chosen to group the algorithms mainly by the mathematical idea that underlies their correctness, but a different organization might be equally reasonable. It is perhaps rewarding to return to Table 5.1 after reading the chapter.

5.2 Marching Cubes and Cube-Based Algorithms

The *marching cubes algorithm* [342, 244] conceptually covers space by a *grid* of small cubes, and locally computes a mesh for each cube which is intersected by the surface. The algorithm computes f at all grid points. The surface must pass between the grid points with positive f and the grid points with negative f. The algorithm computes intersection points between the surface and the grid edges whose endpoints have opposite signs, and uses them as the vertices of the mesh. Depending on the desired accuracy, these intersection points can be computed by linear interpolation of f between the endpoints or some more elaborate method, or one can simply choose the midpoints of the edges.

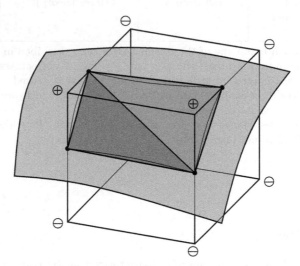

Fig. 5.4. A cube intersected by the surface $f(x, y, z) = 0$. The sign of f at the vertices is shown. Inside the cube, the surface will be represented by two triangles connecting the points where the surface intersects the edges of the cube

These vertices have to be connected, inside each cube, by a triangular mesh that separates the positive and the negative vertices, as shown in Fig. 5.4. There are different possible patterns of positive and negative cube vertices, and the triangulation can be chosen according to a precomputed table of cases.

(In three dimensions, the $2^8 = 256$ cases are reduced to 15 different cases by taking into account symmetries.)

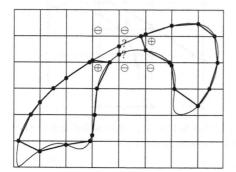 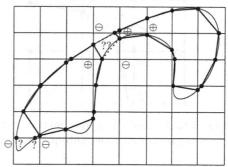

Fig. 5.5. The intersections marked with ? are not found by checking the signs of f at grid points. In the left picture, this leads to a curve with two components instead of one. The right figure shows a shifted copy of the same curve. When all 4 sides of a square are intersected, as in the square marked ??, there are two ways of connecting them pairwise. Applying a local rule to decide between the two possibilities may not always give the correct result. In addition, the meshed curve misses a large part of the protrusion in the lower left corner

Problems with this method appear when the grid is not sufficiently fine to capture the features of the surface. In some cases, the connecting triangulation between the mesh vertices inside a cube is ambiguous, and some arbitrary decision has to be made. Fig. 5.5 shows a two-dimensional instance of a curve whose constructed mesh has an incorrect topology (2 cycles instead of a single cycle). A slightly shifted grid leads to an ambiguous situation: the sign pattern of f at the vertices is not sufficient to decide how the points should be connected.

It can also happen that components of the implicit surface or curve which are smaller than a grid cell may be completely missed.

Originally, the marching cubes algorithm was designed to find isosurfaces of the form $f(x) = \text{const}$ in a scalar field f resulting from medical imaging procedures, with one data value (gray-level) $f(p)$ for each point p in a pixel or voxel grid. In this case, one has to live with ambiguities, trying to make the best out of the available data. One can resolve ambiguities by any rule which ensures that triangles in different cubes fit together to form a closed surface. (In fact, the rules as given originally in [244] may resolve ambiguities inconsistently, leading to meshes which are not watertight surfaces.)

In our setting, however, the function f is defined everywhere, and it is not sufficient to achieve consistency: we want a topologically correct mesh; on the other hand, we are free to compute additional values at any point we like. Thus, the obvious way to solve ambiguities is to refine the grid, see Fig. 5.6.

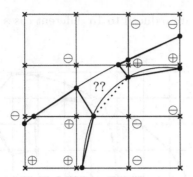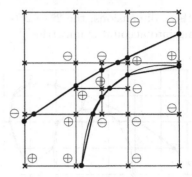

Fig. 5.6. A section from the right part of Fig. 5.5 before and after subdividing the offending square into four subsquares. The ambiguity is resolved

5.2.1 Criteria for a Correct Mesh Inside a Cube

The basic strategy is to subdivide the cells of the grid until sufficient information is available for forming a correct mesh. The question is now: How can we tell that a mesh within some grid cube is correct and the cube need not be further subdivided.

We present two criteria, an older one due to Snyder [322, 321], and a recent one, due to Plantinga and Vegter [286].

5.2.2 Interval Arithmetic for Estimating the Range of a Function

Both methods are based on estimating the range of values of a function $\{ g(p) \mid p \in B \}$ when the parameter p ranges over some domain B, which is typically a box. In most cases, g will be the function f or one of its derivatives, and one tries to establish that the range of g does not contain 0, i.e., $g(p) \neq 0$ for all $p \in B$.

For example, if B is a grid cube and $f(p) \neq 0$ for all $p \in B$, we know that the surface S contains no point of B, and no further processing of B is necessary.

The standard technique for estimating the range of a function is *interval arithmetic*. The idea of using interval arithmetic in connection with meshing was pioneered by Snyder [322, 321].

Interval arithmetic has been introduced in Sect. 3.2.2 as a method to cope with the limited precision of floating-point computation. Ideally, one would like to have the exact result x of a computation. By interval arithmetic, one obtains an interval $[a, b]$ which definitely contains the correct result x. By repeating the calculations with increased accuracy, one can decrease the size $b - a$ of the interval until it becomes smaller than any prespecified limit $\varepsilon > 0$.

Interval arithmetic is also suited to get the range of a function whose inputs $x_1, x_2, x_3, \ldots$ vary in some given intervals. Instead of starting with zero-length intervals for exact inputs, one takes the given starting intervals. Note that this approach may suffer from a systematic overestimation of the resulting interval, for example when subtracting quantities that are close, like in $f(x) = x - \sin x$. For $x = [0, 0.3]$, we get the interval approximation $\Box f([0, 0.3]) = [0, 0.3] \boxminus (\Box \sin)([0, 0.3]) = [0, 0.3] \boxminus [0, 0.2956] = [-0.2956, 0.3]$, whereas the true range of f is contained in $[0, 0.0045]$. There are techniques such as Fast Automatic Differentiation [193] which can circumvent this problem.

For the remainder of this chapter, we assume, for a given function g the availability of some operation $\Box g$ such that $\Box g([a_1, b_1], [a_2, b_2], \ldots)$ returns an interval that contains the range $\{ g(x_1, x_2, \ldots) \mid a_1 \leq x_1 \leq b_1, \ a_2 \leq x_2 \leq b_2, \ldots \}$. We require that the resulting interval can be made arbitrarily small if the input intervals are small enough:

Assumption 2 CONVERGENCE *(of Interval Arithmetic).*
The size of the interval $\Box g([a_1, b_1], [a_2, b_2], \ldots)$ goes to zero if the sizes of the parameter intervals $[a_1, b_1]$, $[a_2, b_2]$, $\ldots$ tend to zero.

This requirement is fulfilled for ordinary interval arithmetic in the vicinity of every point for which g is continuous, provided that the accuracy of the calculations can be increased arbitrarily. If the computation of g does not involve operations that may be undefined, like division by zero, and in particular, if g is a polynomial, interval arithmetic converges.

If a continuous function g satisfies $g(x) \neq 0$ for all x in a box B, convergence implies that interval arithmetic will establish this fact by subdividing B into sufficiently many sub-boxes.

5.2.3 Global Parameterizability: Snyder's Algorithm

We want to compute a mesh for a surface $f(x) = 0$ inside some box $X = [x_{min}, x_{max}] \times [y_{min}, y_{max}] \times [z_{min}, z_{max}]$. The method works by induction on the dimension. Thus, as a subroutine, the algorithm will need to compute a mesh for a curve $f(x) = 0$ (an approximating polygonal chain) inside some rectangle $X = [x_{min}, x_{max}] \times [y_{min}, y_{max}]$.

Snyder's criterion for telling when a cell need not be further subdivided is *global parameterizability* of f inside a box X in two of the variables x, y, or z. We say that $\{ (x, y, z) \in X \mid f(x, y, z) = 0 \}$ is *globally parameterizable* in the parameters x and y, if, for each pair (x, y), the equation $f(x, y, z) = 0$ has at most one solution z in the box X. One way to establish this property is to ensure that the derivative with respect to the third variable is nowhere zero:

$$\frac{\partial}{\partial z} f(x, y, z) \neq 0 \text{ for all } (x, y, z) \in X \tag{5.1}$$

An analogous definition holds for a curve $\{ (x, y) \in X \mid f(x, y) = 0 \}$ in two dimensions: it is globally parameterizable in the parameter x, if, for each value x, the equation $f(x, y) = 0$ has at most one solution y in the box X.

Global parameterizability of a curve in the parameter x means that the solution consists of a sequence of curves which can be written in the parameterized form $y = C(x)$, over sequence of disjoint intervals for the parameter x. Similarly, global parameterizability of a surface in the parameters x and y means that the solution consists of parameterized surface patches $z = S(x, y)$ over a set of disjoint domains for (x, y).

Suppose that a curve in a two-dimensional rectangle X is globally parameterizable in x. The curve has at most one intersection with the left and right edge, and an arbitrary number of intersections with the bottom and top edge. Let $x_1, x_2, x_3, \ldots$ denote the sequence of intersections, sorted from left to right, see Fig. 5.7a.

Between the first two successive intersections x_1 and x_2, there can either be no solution inside X, or the solution can be an x-monotone curve in X. These two possibilities can be distinguished by intersecting the curve with a vertical line segment ℓ half-way between x_1 and x_2. More precisely, we just need to compute the signs of f at the endpoints of ℓ. Given this information, we can draw polygonal connections between the points x_i which are a topologically correct representation of the curve pieces inside X, as in Fig. 5.7b. To connect two points x_i and x_{i+1} on the same edge, we can for example draw two 45° segments. Points on different edges can be connected by straight lines.

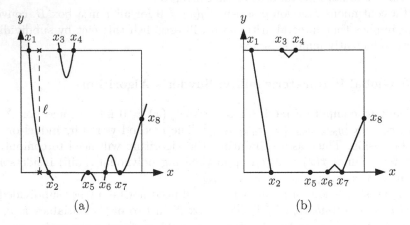

Fig. 5.7. Finding a correct mesh for a curve in a square

The following lemma summarizes this procedure, and it also formulates the three-dimensional version.

Lemma 2. *1. If a curve $f(x, y) = 0$ is globally parameterizable in x in a two-dimensional box X, and if one can find the zeros of f on the edges of the box, then one can construct a topologically correct mesh for the curve inside X.*

2. • *If a surface $f(x, y, z) = 0$ is globally parameterizable in x and y in a three-dimensional box X, and*
   • *if, on the top and bottom face of X, each of the two functions $f(x, y, z_{max})$ and $f(x, y, z_{min})$ is everywhere nonzero or globally parameterizable in x or y, (not necessarily both in the same variable)*
   then one can construct a topologically correct mesh for the surface inside X, provided that one can find the zeros of f on the edges of the box.

In each case, the only additional information required is the sign of f at a few points on the edges of the box.

We call a function f *well-behaved* with respect to the box X if the conditions of part 2 of the lemma are satisfied, possibly after permuting the coordinates x, y, and z.

Proof (Proof of part 2). Suppose the surface intersects the bottom face as in Fig. 5.7a and the top face as in Fig. 5.8a. Fig. 5.8b shows the overlay of the two intersection patterns, like in a top view onto the box. By global parameterizability in x and y, the intersections with the top face and the bottom face cannot cross. In each region which is delimited by these intersection curves, there is either no intersection of the surface with X or there is a single surface patch. These two cases can be distinguished by checking whether an appropriate vertical line segment in the boundary of X intersects the surface, i.e., whether f has opposite signs at the endpoints of this segment.

Fig. 5.8c shows the polygonal mesh for the intersection with the top face, and Fig. 5.8d shows the overlay with Fig. 5.7b. The shaded areas in Fig. 5.8b and 5.8d represent the existing patches of the surface in X and the corresponding patches of the mesh that are to be constructed. Such a mesh can be constructed easily: we may need to find intersection points at the vertical edges of X, and we may need to add $45°$ segments on the vertical sides of X. On each vertical side, the mesh edges will look exactly as the ones that would be produced by part 1 of the lemma. This is important to ensure that the surface patches in adjacent boxes fit together across box boundaries, even if the adjacent box is parameterizable in y and z or in x and z.

Any triangulation of the grey polygons in Fig. 5.8d will now lead to appropriate triangulated surface patches, as shown in Fig. 5.8e. We leave it as exercise to construct an isotopy that shows topological correctness in the sense of Definition 1.

We can now present the overall algorithm. We suppose we are given an initial box containing the part of the surface in which we are interested.

The algorithm maintains a list of boxes that are to be processed. We select a box X from the list and process it as follows. First we try to establish that $f(x) \neq 0$ in X, using interval arithmetic. If this is the case, we can discard the box without further processing. Otherwise, we check if Lemma 2 can be applied. Using interval arithmetic, we try to show that one of the partial derivatives is nonzero in X, see (5.1), which implies that f is globally

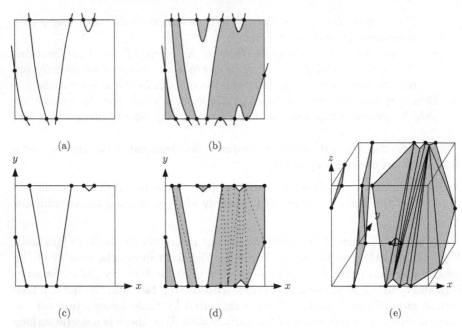

Fig. 5.8. Finding a correct mesh for a surface in a cube

parameterizable in two of the parameters x, y, and z. We then also have to check the "top" and "bottom" faces of X, in a completely analogous way: Either f or one of its partial derivatives must be nonzero on the face. If this test succeeds, we know that we can mesh the surface in X. Otherwise, we subdivide X into smaller boxes and put them on the list for further processing.

The approximation error is trivially bounded by the diameter of the box, regardless of how we construct the mesh in each box. Thus, if we want to guarantee a small error, we can achieve this by subdividing boxes that are too large, before checking global parameterizability.

In the end, we have a bunch of boxes of different sizes in which we have to construct meshes. Cubes of different sizes may touch, and therefore the method of Lemma 2 must be adapted: The surface is first meshed inside the smallest boxes. The pattern of intersection with the boundary is transmitted to larger adjacent boxes, and thus the mesh boundary on the sides of the boxes may look more involved than in Fig. 5.8e. The largest boxes are meshed last.

We still have to discuss the assumption of Lemma 2 that the intersections of the surface with the cube edges can be found. Snyder [322, 321] proposed to use interval arithmetic also for this task. In fact, this problem is just the meshing problem in one dimension: finding zeros $f(x) = 0$ of a univariate function f. (The two- and three-dimensional versions are treated in Lemma 2.) Global parameterizability in this setting boils down to requiring $f' \neq 0$.

The basic algorithm successively subdivides a starting interval until $f \neq 0$ or $f' \neq 0$ can be established to hold throughout each subinterval, by interval arithmetic. Then one can establish the existence of a unique zero or the absence of a zero by computing the sign of f at all interval endpoints. The results is a sequence of disjoint isolating intervals $[u_1, v_1]$, $[u_2, v_2]$, ..., where each interval is known to contain a unique zero (cf. Sect. 3.3.2).

Note that the sequence of points x_1, x_2, ... need not be exact for the algorithm in part 1 of the lemma to work. All that is required is that they have the correct order. There will be a sequence of disjoint isolating intervals for the intersections with the upper edge and another sequence of disjoint isolating intervals for the intersections with the lower edge. If any two of these intervals overlap, the intervals must be refined until they become disjoint. This will eventually happen, since, by global parameterizability, the zeros on the upper edge and on the lower edge are distinct. Then any point from the respective interval can be used to construct the mesh.

Note however, that interval arithmetic fails to converge for zeros where the function only touches the zero line or does not cross it transversally such as the points x_5 and x_7 in Fig. 5.7a, or generally when $f(x) = f'(x) = 0$ (*grazing intersections*). No amount of subdivision will suffice to show the presence or absence of a zero in this case.

Thus, to cope with these cases, one has to resort to the exact methods of Chap. 3. In practice, one could of course simply stop the subdivision when the size of the intervals become smaller than some threshold and "declare" the presence of a zero in this interval, giving up any correctness claims below the precision threshold.

Another issue is termination of Snyder's algorithm. It turns out that this question is closely related to the problem of grazing intersections: the algorithm may fail to terminate in certain special situations.

Consider the elliptic paraboloid $z = x^2 - xy + y^2$. In a cube X in the first orthant with the corner at the origin, the surface is globally parameterizable in x and y, but not in any other pair of variables. However, on the bottom face $z = 0$, the partial derivatives with respect to x and to y are $2x - y$ and $2y - x$, and neither of them has a uniform sign in X. This box will never satisfy the condition of Lemma 2, and subdivision will produce a smaller box of the same type. Thus the algorithm will not terminate. Note that this surface is not in any way difficult to mesh; it presents no problems for the algorithm if the origin is inside some (small enough) cube: the surface will simply intersect the four vertical edges, and the mesh will consist of two triangles.

In both cases discussed above, the difficulty results from a special position of the grid relative to the surface: The surface is tangent to an edge (in the case of grazing intersections for the one-dimensional problem) or a face (in the case of non-termination) of a grid cube. In fact, one can show that this is the only source of difficulties: If no face of a cube that is created during the algorithm is tangent to the surface, the algorithm will terminate. Thus, a translation and rotation of the initial grid to a "generic" position guarantees termination: All cube faces are parallel to one of three given directions. A smooth surface has only finitely many points with a specified randomly chosen normal direction, the grid must be translated such that no grid plane will go through one of these critical points. This is ensured, for example, if the coordinates of the critical points are not multiplies of powers of 2, in the grid coordinate system.

In all likelihood, such a translation and rotation of the initial grid to a generic position should also remove the problem or grazing intersections with grid edges, but this has not been analyzed.

Exercise 1. If $\{\,(x,y,z) \in X \mid f(x,y,z) = 0\,\}$ is globally parameterizable in the parameters x and y, then the intersection curves on the vertical sides of X (parallel to the z-axis) are globally parameterizable in the parameters x or y, respectively.

Exercise 2. Construct an explicit isotopy between the original curve pieces and the polygonal approximating curve in the case of Lemma 2, part 1. (Fig.s 5.7a and 5.7b). Assume first that the intersections $x_1, x_2, \ldots$ with the boundary are given exactly. Then extend the construction to the case when the intersections are replaced by approximate values $x'_1, x'_2, \ldots$ etc. The only property that can be assumed is that they are ordered in the same way as the true values $x_1, x_2, \ldots$

Exercise 3. Extend the previous exercise to part 2 of Lemma 2. Assume that an isotopy from the true intersection curves to the polygonal approximation is given on each face of the cube X.

Exercise 4. Examine termination of Snyder's algorithm for a parabolic cylinder $(y - x)^2 - z = 0$ and for the hyperbolic paraboloids $x^2 - y^2 - z = 0$ and $xy - z = 0$, starting with eight unit cubes meeting at the origin. Assume that the range of f and its derivatives over any box (a) can be calculated exactly, or (b) is evaluated using interval arithmetic.

5.2.4 Small Normal Variation

Plantinga and Vegter [286] used a stronger condition than global parameterizability to guide the subdivision process, the SMALL NORMAL VARIATION condition:

$$\langle \nabla f(x_1), \nabla f(x_2) \rangle \geq 0, \text{ for all } x_1, x_2 \in X \tag{5.2}$$

In other words, there is an upper bound of 90° on the angle between two gradient vectors, and in particular, between two normal vectors of the surface.

Exercises 5–7 below explore the relation to global parameterizability and Lemma 2. In particular, Small Normal Variation implies that the function is monotone in some coordinate direction, and therefore the surface (or curve) is globally parameterizable.

Condition (5.2) can be checked by interval arithmetic. We compute an interval representation $\Box\nabla f(X) = (\Box\partial f/\partial x, \Box\partial f/\partial y, \Box\partial f/\partial z)$ of the gradient and take the interval scalar product of this vector with itself. If the resulting interval does not contain 0, we have established the Small Normal Variation condition.

The algorithm starts with a given box and recursively subdivides it until, in every box X, the following *termination condition* is satisfied:

Either $f(x) \neq 0$ for all $x \in X$, or the Small Normal Variation condition (5.2) holds.

Both conditions are checked with interval arithmetic.

Theorem 1. *If the surface* $S = \{\, x \mid f(x) = 0 \,\}$ *has no singular points and interval arithmetic converges, this subdivision procedure terminates.*

Proof. By the nonsingularity assumption and since f and ∇f are continuous, there is a positive minimum distance ε between the solution sets of $f(x) = 0$ and $\nabla f(x) = 0$ inside the starting box. This means that every box X which is smaller than ε has either $f(x) \neq 0$ or $\nabla f(x) \neq 0$ for all $x \in X$. Convergence implies that interval arithmetic will establish $f(x) \neq 0$ or $\|\nabla f(x)\|^2 \neq 0$, respectively, after finitely many subdivision steps. However, the interval computation of $\|\nabla f(x)\|^2 = \langle \nabla f(x), \nabla f(x)\rangle$ is identical to the calculation of $\langle \Box\nabla f(X), \Box\nabla f(X)\rangle$ that is used to check the Small Normal Variation condition.

One can see that the granularity of the subdivision adapts to the properties of the function f. In places where f and ∇f have a large variation and f is close to 0, the algorithm will have to subdivide the cubes a lot, but in regions where f is "well-behaved", not much refinement will be necessary.

We still have to show that the signs of f at the vertices of all boxes give sufficient information to construct a correct mesh. We first discuss the case of a curve in the plane, for illustration. For simplicity, let us ignore the case when f is zero at some box vertex. The algorithm will simply insert a vertex on every edge for which f has opposite signs at the endpoints. Now, the ambiguous case that caused so much headache in Fig. 5.5 is excluded: If the signs alternate in the four corners, then f is neither monotone in the x-direction nor in the y-direction, contradicting the Small Normal Variation condition, see Fig. 5.9a.

It may happen that the curve intersects an edge twice, and these intersections go unnoticed, as in Fig. 5.9b. However, the Small Normal Variation condition ensures that the curve cannot escape too far before coming back, see Fig. 5.9c.

Before trying to mesh the curve inside the boxes, the algorithm refines the subdivision until it becomes *balanced*: The size of two boxes that are adjacent via an edge differs at most by a factor of 2. As long as two adjacent boxes differ by a larger factor, the bigger box is subdivided into four boxes. (Boxes in which $f(x) \neq 0$ need not be subdivided, of course.) At this stage, we need

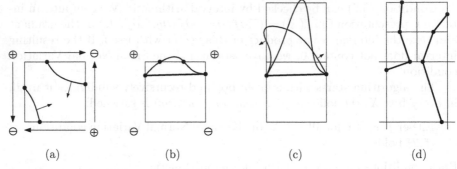

(a)   (b)   (c)   (d)

Fig. 5.9. (a) The ambiguous sign pattern cannot arise. The arrows along the sides indicate the direction in which f cannot be increasing. The little arrows indicate two normals of a hypothetical solution, which form an angle larger than $\pi/2$. (b) The two intersections with the upper edge are missed, but the straight segment between the endpoints is isotopic to the true curve. (c) In particular, the curve cannot leave the adjacent cube without violating the Small Normal Variation condition in the adjacent cube. Thus, the approximating segment is not only isotopic, it is even geometrically close. (d) The connections between endpoints in a square can be chosen by simple local rules

not worry about the termination condition inside the boxes, because they are automatically fulfilled.

Finally, we insert a mesh vertex on every edge whose endpoints have different signs. We have to decide how to connect these vertices inside each square. Due to the balancing operation, there is only a small number of cases to analyze. It turns out that there can be zero, two, or four vertices on the boundary of a square. If there are two vertices, we simply connect them by a straight line. If there are four vertices, two of them must lie on the same side, since the case of Fig. 5.9a is excluded. We connect each of them to one of the other vertices, see Fig. 5.9d for an example.

Theorem 2 ([286]). *The polygonal approximation constructed by the algorithm is isotopic to the curve S.* □

The algorithm works similarly for surfaces in three dimensions: The refinement step has the same termination criterion as in the plane. After balancing the subdivision, a vertex is inserted at every edge with endpoints of opposite signs. The analysis of the possible cases is now more involved. In particular, there can be an ambiguity on the *face* of a box without contradicting the Small Normal Variation condition, as in Fig. 5.10a. This is because this condition does not carry over from a cube to a face: The gradient of the restricted function $f(x, y, z_{\max})$ on a face of the cube is the projection of the three-dimensional gradient vector ∇f, and two gradient vectors with angles less than π may form a larger angle after projection.

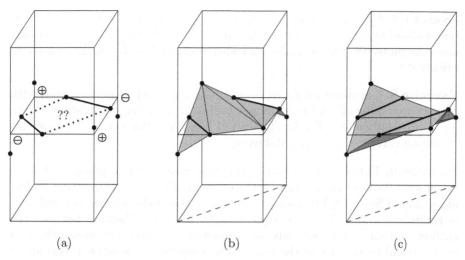

Fig. 5.10. (a) The ambiguous sign pattern can arise on a face of a cube. (b-c) The ambiguity can be resolved in two possible ways. The two resulting meshes cross the boundary face in different patterns, but they are isotopic to each other

However, in this case one can insert the two edges arbitrarily in the ambiguous face. Each choice leads to a different mesh in the two boxes, see Fig. 5.10b–c. But when the boxes are combined, the two choices lead to isotopic meshes.

Fig. 5.10 is representative of the different cases that can arise. One just has to ensure that the choice of edges is done in a consistent manner for adjacent boxes, for example, by always favoring the edges which do not cross the diagonal in direction $(1,1,0)$ (Fig. 5.10c) over the alternate choice, or by consulting the value of f in the middle of the square. The algorithm constructs a mesh that is isotopic to the surface S.

Comparison with Snyder's algorithm.

Looking at Fig. 5.10, we can see why Snyder's algorithm of Sect. 5.2.3 has a harder time to terminate: it insists on topological correctness *within each single cube* separately. The example of Fig. 5.10 shows that this is not necessary to get the correct topology in a global level. Snyder's algorithm may refine the grid to some unneeded precision when surface interacts with the grid in an unfavorable way.

Exercise 5. If all three partial derivatives are nonzero in a box X (and hence f is globally parameterizable in each pair of parameters out of x, y, and z), then f has Small Normal Variation.

Exercise 6. If f satisfies Small Normal Variation, then it monotone in x, y, or z, and in particular, it is globally parameterizable in some pair of parameters out of x, y, and z.

Exercise 7. Construct a function f with Small Normal Variation which is not well-behaved in the sense of Lemma 2, part 2. The function f should have the property that Small Normal Variation can be established by interval arithmetic.

Exercise 8. Construct an example of a function which is well-behaved with respect to a cube X, but is no longer well-behaved after subdividing X into eight equal subcubes. (For Small Normal Variation, this cannot happen: this condition carries over to all sub-boxes.)

Exercise 9. This exercise explores the properties of interval arithmetic for estimating the maximum angle between two gradient vectors in a box X. The algorithm of Sect. 5.2.4 terminates as soon as the angle between two different normals is less than $90°$. Consequently, the geometric distance between the surface and the mesh can only be estimated very crudely; essentially it is proportional to the size of the box. If the angle bound is smaller, one might derive better bounds (see Research Problem 3, p. 227).

The standard way to estimate the angle is by the formula

$$\cos\alpha = \frac{\langle x, y\rangle}{\|x\| \cdot \|y\|}$$

where $x, y \in \Box\nabla f(X)$. If $\langle\Box\nabla f(X), \Box\nabla f(X)\rangle = [a, b]$ for some interval with $0 < a < b$, the standard interval arithmetic calculation leads to a bound of $\alpha \leq \arccos\frac{b}{a}$. Assuming that $\Box\nabla f(X) = ([a_1, b_1], [a_2, b_2], [a_3, b_3])$, can one derive a better bound on α by tackling the problem more directly? By how much can one improve the crude bound $\alpha \leq \arccos\frac{b}{a}$? Are there instances where the crude bound cannot be improved?

Exercise 10. Suppose that f satisfies the Small Normal Variation condition. Then there is an infinite circular double-cone C (like in Fig. 5.1b) of opening angle $\alpha = 2\arcsin\sqrt{1/3} \approx 70°$ with the following property: When the apex of C is translated to any point x on the surface S, the two cones lie on different sides of S and intersect S only in x.
(Hint: α is the opening angle of the largest cone that fits into the first orthant. On the unit sphere S^2 of directions, a set of diameter π (measured in angles) is contained in a spherical disc of radius $(\pi - \alpha)/2$.)

Exercise 11. Prove that the sign pattern on the vertices shown in Fig. 5.11 cannot arise, for a function with Small Normal Variation. (This pattern is configuration 13 in [286, Fig. 5].)
(This exercise seems to require some geometric arguments which are not straightforward. The previous exercise may be useful.)

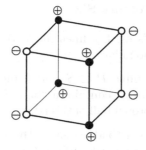

Fig. 5.11. Is this sign pattern possible when Small Normal Variation holds?

5.3 Delaunay Refinement Algorithms

The Restricted Delaunay Triangulation.

Given a set of points P and a surface S, the *Delaunay triangulation restricted by S* is formed by all faces of the three-dimensional Delaunay triangulation whose dual Voronoi faces intersect S. In particular, it consists of those triangles whose dual Voronoi edges intersect S. In the applications, the points of P will always lie on S.

Generically, Voronoi vertices will not happen to lie on S; thus, the restricted Delaunay triangulation T contains no tetrahedra; it is at most two-dimensional. If P is a sufficiently good sample of S, then T will form a surface that is isotopic to S. A restricted Delaunay triangle xyz is characterized by the existence of an empty ball through the vertices xyz whose center p lies on the surface. We call this ball the *surface Delaunay ball*. It may happen that a vertex $p \in P$ is incident to no edge and no triangle of the restricted Delaunay triangulation T: then p must lie on a small component of S that is completely contained in the Voronoi cell $V(p)$. It can also happen that p is incident to some edges but to no triangle of T.

We will present two algorithms that use the restricted Delaunay triangulation as a mesh. They start with some initial point sample and add points until it is guaranteed that the restricted Delaunay triangulation forms a polyhedral surface that is isotopic to the given surface. The algorithms are adaptations of the greedy "farthest point" technique of Chew [91] in which the points that are added are centers of surface Delaunay balls.

The algorithms differ in the way how the correct topology is ensured, and they differ in the primitive operations that are used to obtain information about the surface. We will present an algorithm of Boissonnat and Oudot, which is based on the local feature size, and an algorithm of Cheng et al. that works towards establishing the so-called topological ball property.

5.3.1 Using the Local Feature Size

The generic form of Chew's mesh refinement algorithm is as follows, for the case of a surface without boundary.

> Select some starting sample $P \subset S$, and compute its restricted Delaunay triangulation T. If T contains a "bad" triangle xyz, insert the center of the surface Voronoi ball of this triangle into P, and update T.

Depending on the definition of "bad" triangle, the algorithm will give different results. The algorithm can also treat surfaces (and plane regions) with boundary, and in fact, this is the real challenge in the design of the algorithm. For simplicity, we will discuss only the case of smooth surfaces without boundary.

The *local feature size* of a point $x \in S$, denoted by $\mathrm{lfs}(x)$, is the distance from x to the closest point on the medial axis, see Fig. 5.12. (See p. 109 in Sect. 2.7 for the definition of the medial axis; see also Sect. 6.2.2, pp. 244–247, for a more extensive discussion about the medial axis and the local feature size.)

In the case of a curve, the local feature size is small where the curve makes sharp bends, as in the region C of Fig. 5.12. More generally, for a surface, the curvature radius corresponding to the maximum principal curvature (see Sect. 4.4.1) at x is an upper bound on $\mathrm{lfs}(x)$. The local feature size is also small when a different part of the curve comes close, as in the region around A of Fig. 5.12. It is therefore a somewhat natural measure for specifying the necessary density of the mesh. There are however instances where the local feature size overestimates the density: for example, two parallel flat sheets of a surface that approach each other very closely have a small local feature size, but they can be meshed with few vertices. The local feature size is also related to the length of the fibers in a tubular neighborhood of the surface, see Lemma 1, Sect. 5.1.

The local feature size is nonzero if S is smooth at x. It is zero at edges or other singular points of S. The local feature size is a Lipschitz-continuous function with constant 1:

$$\mathrm{lfs}(x) - \mathrm{lfs}(y) \leq \|x - y\|$$

For a smooth compact surface, the local feature size is therefore bounded from below by a positive constant $\mathrm{lfs}_{\min} > 0$.

ε-Samples and Weak ε-Samples.

A fundamental concept is the notion of an *ε-sample* $P \subset S$ of a surface S, introduced by Amenta and Bern [22]. It is defined by the following condition: For every point $x \in S$, there is a point $p \in P$, such that $\|p - x\| \leq \varepsilon \cdot \mathrm{lfs}(x)$.

Since $\mathrm{lfs}(x)$ is difficult to obtain in practice, one replaces it by some other function ψ: A ψ-sample $P \subset S$ for a function $\psi \colon S \to \mathbb{R}^+$ is a subset with the following property: For every point $x \in S$, there is a point $p \in P$, such that

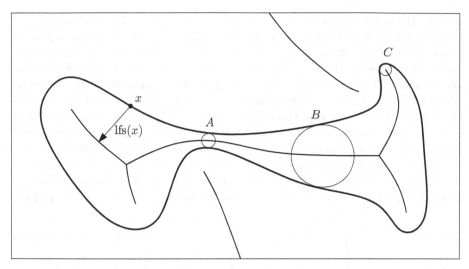

Fig. 5.12. A curve, its medial axis, and the local feature at a point x on the curve

$\|p - x\| \le \psi(x)$. Thus, an ε-sample is the same as an $(\varepsilon \cdot \mathrm{lfs})$-sample. It will always be clear from the context which definition is meant.

Both of these notions are still difficult to check because the definition involves a condition for infinitely many points $x \in S$. The following concept relaxes this condition to a finite set of points.

For every surface Delaunay ball with center x and radius r, $r \le \varepsilon \cdot \mathrm{lfs}(x)$, or $r \le \psi(x)$, respectively.

A point sample with this property is called a *weak ε-sample* (a weak ψ-sample, respectively). (Originally, this was called a *loose ε-sample* [64].)

The difference between ε-samples and weak ε-samples is not too big, however. It can be shown that every weak ε-sample, for small enough ε is also an ε'-sample, with $\varepsilon' = O(\varepsilon)$ [64, Theorem 1]. To exclude trivial counterexamples, one has to assume that, for each connected component C of S, the restricted Delaunay triangulation of P has a triangle with at least one vertex on C.

Theorem 3. *If $P \subset S$ is a weak ε-sample of S for $\varepsilon < 0.1$, and, for every connected component C of S, the restricted Delaunay triangulation T of P contains a triangle incident to a sample point on C, then T is ambient isotopic to S. The isotopy moves each point $x \in S$ by a distance at most $O(\varepsilon^2 \mathrm{lfs}(x))$.*

The theorem was first formulated for ε-samples (with the same bound of 0.1) by Amenta and Bern [22], see also Theorem 6 in Chap. 6 (p. 248) for a related theorem. The extension to weak ε-samples is due to Boissonnat and Oudot [65].

We give a rough sketch of the proof, showing the geometric ideas, but omitting the calculations. Let us consider a ball tangent to S in some point

$x \in S$. The definition of local feature size implies that such a ball contains no other point of S as long as its radius is smaller than $\mathrm{lfs}(x)$. Thus, when we draw the two balls of radius $\mathrm{lfs}(x)$ tangent to S, it follows that the the surface must pass between them, and hence, in the neighborhood of x, S must be more or less "flat", see Fig. 5.13. By the Lipschitz continuity of lfs, the same property (with slightly smaller balls) must hold for all other surface points in the vicinity of x. Consider three points $a, b, c \in S$ at distance $r = \varepsilon \cdot \mathrm{lfs}(x)$ from x. Then the normal of the plane through these points differs from the surface normal at x only by a small angle, which can be shown to be $O(\varepsilon)$.

We now apply this observation to the situation when x is a center of a surface Voronoi ball through the vertices a, b, c of a restricted Delaunay triangle Δ. Since the variation of surface normals is bounded, one can show

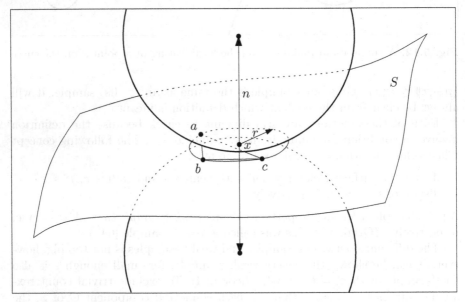

Fig. 5.13. The surface must squeeze between two tangent balls of radius $\mathrm{lfs}(x)$

that the maximum distance from a point of Δ to the closest point on S is $O(\varepsilon^2 \mathrm{lfs}(x))$. It follows that the projection π_S that maps every point of Δ to the closest point on S is injective: if we extend an open segment of length $\mathrm{lfs}(y)$ from every surface point y to both sides of S in the direction normal to S, these segments do not intersect, and they can be used as the fibers of a tubular neighborhood $\hat{S}$ of S. Each point of such a segment has y as its unique closest neighbor on S. For small enough ε, the triangle Δ is contained in $\hat{S}$. Thus, the mapping π_S defines an isotopy between Δ and a corresponding surface patch, as in Lemma 1 and Fig. 5.3a.

One can show that two triangles of T that share an edge or a vertex have normals that differ by at most $O(\varepsilon)$, and the mapping π_S extends continuously across the edge or the vertex. It follows that the projection π_S is a homeomorphism that is invertible locally. (In topological terms, $\pi_S: T \to S$ is a covering map, if we can establish that it is surjective.) By assumption, on every component, there is at least one vertex contained in a triangle of T. This ensures that $\pi_S(T)$ contains that vertex, and since the mapping can be continued locally, it follows that every component of S is covered at least once. It is now still possible that some component is covered more than once by π_S. This would imply that some sample point $p \in P$ is covered more than once. However, one can show quite easily that no point of T except p itself has p as its closest neighbor on S.

With the help of Lemma 1, one can obtain the desired ambient isotopy between T and S.   □

The Delaunay refinement algorithm of Boissonnat and Oudot [64] applies this theorem to obtain a topologically correct mesh. It requires some function $\psi(x)$ for determining the necessary degree of refinement. The function ψ must be a lower estimate for the local feature size. To obtain a bound on the running time, it should be a Lipschitz function with constant 1:

$$\psi(x) - \psi(y) \le \|x - y\|$$

The refinement algorithm is an instance of the general Delaunay refinement algorithm described on p. 202. A triangle with surface Delaunay ball of radius r centered at x is declared "bad" if $r > \psi(x)$.

Thus the main loop of the algorithm runs as follows. We have to intersect every Voronoi edge with the surface S. If there is an intersection point x, it is the center of a surface Delaunay ball, and it is the witness for a corresponding triangle in the restricted Delaunay triangulation. If the radius r of the ball is larger than $\psi(x)$, we insert x into P and update the Voronoi diagram. It may happen that a Voronoi edge intersects the surface more than once. Then we carry out the test for all intersection points. If follows from the arguments in the proof of Theorem 3 that at least one point x has $r > \phi(x)$ and can be inserted into P.

The following lemma helps to prove termination of this algorithm.

Lemma 3. *Let ψ be a Lipschitz function with $0 < \psi(x) \le \mathrm{lfs}(x)$. Consider an algorithm which successively inserts points into a sample P with the property that every point p has distance at least $\psi(p)$ from all previously inserted points. Then the number of points is at most $O(H(\psi, S))$, with*

$$H(\psi, S) := \int_{x \in S} \frac{1}{\psi(x)^2}\, dx.$$

Proof. This is proved by a packing argument: Let L be the Lipschitz constant of ψ. First we prove that

$$\frac{1}{L+2}\psi(p) + \frac{1}{L+2}\psi(q) \leq \|p - q\| \tag{5.3}$$

for any two points $p, q \in P$. Assume that p was inserted before q. Then, by assumption, $\|p - q\| \geq \psi(q)$. By the Lipschitz property, we have

$$\psi(p) \leq \psi(q) + L \cdot \|p - q\| \leq (L + 1) \cdot \|p - q\|,$$

which implies $\psi(p) + \psi(q) \leq (L + 2) \cdot \|p - q\|$ and hence (5.3). It follows that we can draw disjoint disks D_p of radius $\psi(p)/(L+2)$ around all points $p \in P$. It is not difficult to show that the integral over these disks is bounded from below:

$$\int_{x \in D_p} \frac{1}{\psi(x)^2} \, dx \geq \Omega\left(\frac{1}{(L+2)^2}\right) \tag{5.4}$$

The argument is a follows. The area of D_p is $\Omega(\psi(p)/(L+2))^2$: it can be somewhat smaller than a plane disk of radius $\psi(p)/(L+2)$, since D_p lies on the curved surface S, but since the radius is bounded in terms of the local feature size, it cannot be smaller by more than a constant factor. By the Lipschitz property, the integrand cannot deviate too much from the value $1/\psi(p)^2$ at the center of the disk. Multiplying the integrand by the area of integration yields the lower bound (5.4). Since the disks D_p are disjoint, it follows that the number of points is $O((L+2)^2 H(\psi, S))$.

In a similar way, but using a covering argument instead of a packing argument, one can prove a lower bound on the size of a weak ε-sample (an hence on an ε-sample), cf. [150]:

Lemma 4. Let ψ be a Lipschitz function with $0 < \psi(p) \leq \text{lfs}(p)$. Any ψ-sample of S with respect to ψ must have at least $\Omega(H(\psi, S))$ points.   □

We still have to ensure that the restricted Delaunay triangulation contains a triangle on every component. A *seed triangle* is a triangle in the restricted Delaunay triangulation with a surface Delaunay ball of radius $r \leq \psi(x)/3$, where x is the center. Since this triangle is so small, one can show that the refinement algorithm will never insert a point in its surface Delaunay ball.

Lemma 5. If the sample P contains a seed triangle, this triangle will remain in the restricted Delaunay triangulation.   □

The algorithm starts with sample P that consists of a seed triangle on every component of the surface. The lemma ensures that this triangle remains in the restricted Delaunay triangulation till the end, thus fulfilling the last assumption of Theorem 3. Since $\int 1/\text{lfs}(x)^2 \, dx \geq \Omega(1)$ on every closed surface, the extra seed points fall withing the asymptotic bound of Lemma 3.

The following theorem summarizes the conclusions from the above theorems and lemmas about the Delaunay refinement algorithm.

Theorem 4. Let ψ be a Lipschitz-continuous function with Lipschitz constant 1 and $0 < \psi(x) \leq \text{lfs}(x)$. Suppose that P is initialized with seed triangle

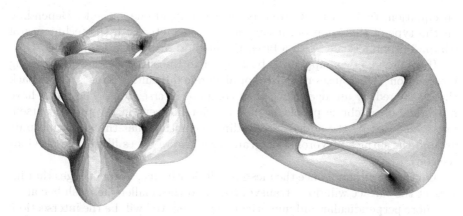

Fig. 5.14. Two meshes constructed by the Delaunay refinement algorithm of Boissonnat and Oudot [64]

on every connected component of S. Then the Delaunay refinement algorithm computes a sample of $\Theta(H(\psi, S))$ points. The resulting mesh is a weak ψ-sample. If $\psi(x) \leq \varepsilon \cdot \mathrm{lfs}(x)$ for $\varepsilon \leq 0.1$, it is ambient isotopic to S. The isotopy maps every point to a point of distance at most $O(\varepsilon^2 \mathrm{lfs}_{max})$.   □

Geometry Improvement.

After obtaining the correct topology, one can improve the geometry of the mesh by eliminating triangles with small angles (with bad "aspect ratio").

A triangle is declared "bad" if the minimum angle is below some bound. This can be done concurrently with the size criterion specified by ψ. If the bound is not too strict (less than $\pi/6$), termination of the algorithm is guaranteed. For further details, we refer to [64].

Primitive Operations.

The algorithm needs some lower estimate ψ on the local feature size as external input. In some applications, such information can be known in advance. As a heuristic without correctness guarantee, one can also use the distance to the poles of the Voronoi diagram (roughly, the largest distance to a Voronoi vertex in the Voronoi cell of each sample point on each side of the surface (see the definition on p. 236 in Sect. 6.2.1). This is a suitable substitute for the local feature size, once a reasonably fine starting mesh has been obtained (Theorem 7 in Chap. 6, p. 249). The distance to the closest pole should be scaled down by a constant factor, to obtain a lower bound on the local feature size with some safety margin.

The essential primitive operation during the algorithm is the intersection of the surface with a Voronoi edge. If the surface is given as a solution set of an equation, this can be written as an equation in one variable. Depending on the type of the equation, it can be solved exactly with the techniques of Chap. 3 or numerically with a bisection method or interval arithmetic.

If a Voronoi edge happens to be tangent to the surface or have a "grazing intersection", this causes numerical or algebraic difficulties. This is a drawback that is shared by all algorithms that compute intersections: They may have numerical or algebraic problems that are not inherent in the problem itself, but are caused by the added "scaffolding" structure that the algorithm puts around the surface (a Voronoi diagram, or a grid of cubes like in the previous section).

On the other hand, once the mesh is sufficiently fine, Voronoi edges that intersect the surface will do so transversally, and the smaller the mesh becomes, the more perpendicular and numerically well-behaved will be the intersection. Also, in connection with appropriate conditions like Small Normal Variation, one can avoid attempts to compute difficult intersections.

Besides the estimate ψ and the computation of intersection points, the algorithm requires only a small initial seed triangle on every component. One can construct examples with some tiny components that are missed completely unless they are specified initially. For instances which are not so complicated, it is often sufficient to intersect a few random lines with the surface to provide a starting sample. An alternative approach is to look for critical points on the surface in some arbitrary direction d, by solving for points where ∇f is parallel to d. This will insert at least two seed vertices on every component of S. Then one can test the incidence criterion of Theorem 3 at the end, by checking whether the seed points are incident to some triangles of the restricted Delaunay triangulation, and if necessary, restart the algorithm after a few additional refinements. The surface mesher of Boissonnat and Oudot is implemented in the CGAL library. Fig. 5.14 shows meshes that were obtained by this algorithm.

The algorithm is restricted to smooth surfaces, because the local feature size is 0 at non-smooth points. However, by introducing a new sampling condition, it has been shown recently that the algorithm also works for some non-smooth surfaces provided that the normal deviation is not too large at the singular points [66]. Experimental results on polyhedral surfaces, piecewise smooth surfaces, and algebraic surfaces with singularities can be found in [65].

Exercise 12. Assume we know a Lipschitz continuous function $\phi(x)$ which is a lower estimate on the local feature size lfs(x), with $\phi(x) \geq \phi_{\min} > 0$. Refine the analysis leading to Theorem 4 to show that an approximating mesh with Hausdorff error $O(D)$, for $D \leq \phi_{\min}$, can be obtained by running the Delaunay refinement of Sect. 5.3 with threshold function $\psi(x) := \sqrt{D\phi(x)}$. Show that ψ satisfies the assumptions of Theorem 4.

Exercise 13. 1. Take a set P of at least 4 points from a sphere S, but otherwise in general position. Show that the Delaunay triangulation of P restricted by S is homeomorphic to S.
  2. Find a point set P which lies very close to the sphere S, whose restricted Delaunay triangulation is not homeomorphic to S.

5.3.2 Using Critical Points

Another meshing algorithm of Cheng, Dey, Ramos and Ray [90] uses a different criterion for topological correctness of the restricted Delaunay triangulation. We say that a point set P on a surface S satisfies the *topological ball property* if every Voronoi face F of dimension k in the Voronoi diagram of P intersects S in a closed topological $(k - 1)$-ball or in the empty set, for every $k \geq 0$. (For example, the non-empty intersection of a 2-dimensional Voronoi face F with the surface must be a curve segment.) Moreover, the intersection must be nondegenerate, in the following sense: the (relative) boundary of the intersection must coincide with the intersection of S with the boundary ∂F. (For example, it is forbidden that an interior point of the curve segment in the above example touches the boundary of F.)

Theorem 5 ([138]). *Let P be a non-empty finite point set. The Delaunay triangulation of P restricted by a surface S is homeomorphic to S if the topological ball property is satisfied.*

In the paper [138], where this property was introduced, it was called the *closed ball property*, see also Theorem 1 in Chap. 6 (p. 241).

The topological ball property is by itself not restricted to smooth surfaces. However, the termination proof of the algorithm and the methods that are used to establish the topological ball property are restricted to surfaces without singularities, defined by a smooth function f.

Instead of using some quantitative argument as in Sect. 5.3.1, the algorithm that we are going to describe tries to establish the topological ball property directly. For this purpose, we need some computable conditions which imply the topological ball property for a given surface $f(x) = 0$. These conditions will be given in the sequel. In each case, testing the condition involves finding a point on the surface with certain criticality properties, which reduces to the task of solving a system of equations involving f and its derivatives. There will be three equations in three unknowns, and thus generically a zero-dimensional set of solutions. If f is a polynomial, the number of solutions will be finite and the techniques of Chap. 3 can be applied to solve the problem exactly, thereby leading to a provably correct algorithm. In this section, however, we will restrict ourselves to the task of setting up the geometric and topological conditions and deriving the corresponding systems of equations.

The following lemma gives a sufficient condition for a two-dimensional surface patch to be isotopic to a disk.

Lemma 6 (Silhouette Lemma). *Let $M \subset S$ be a connected, compact 2-manifold whose boundary is a single cycle, and let $d \in \mathbb{S}^2$ be an arbitrary direction. If M contains no point whose normal is perpendicular to d, then it is a topological disk.*

See Exercise 14 for a proof. The set of points whose normal is perpendicular to d is called the *silhouette* of the surface S in direction d. Algebraically, it is defined by the condition $\langle \nabla f, d \rangle = 0$ (and $f = 0$), which specifies a one-dimensional family of solutions. Generically, it is a set of smooth curves.

The algorithm performs a sequence of four tests to establish the topological ball property. If any test fails, a new sample point is identified. This point is inserted into the sample, and the restricted Delaunay triangulation is updated. We assume throughout that no Voronoi vertex lies on S. (If this case should happen, it can be resolved by perturbing the set P slightly, either by a small random amount, or conceptually with symbolic perturbation [136, 344]. In this way, one can also ensure that P contains no 5 co-spherical points, and thus the Delaunay triangulation is indeed a proper triangulation.)

(1) For a Voronoi edge e, the topological ball property amounts to requiring that e intersects S in at most one point (but not in one of its endpoints).

Thus, we look at each Voronoi edge e in turn, and intersect e with S. If there is more than one intersection, then insert the intersection point p^* which is farthest from p, where p is the sample point in any Voronoi cell to which e belongs. (The choice of p has no influence on the distances to p^*.) The possibility that the intersection point is an endpoint of e, and thus a Voronoi vertex, has been excluded above.

(2) The second check is some topological consistency check for the restricted Delaunay triangulation T. (The purpose of this test will become apparent later.) We check if T is a two-dimensional manifold: each edge must be shared by exactly two triangles, and the triangles incident to each vertex p must form a single cycle around p. If this holds, these triangles form a *topological disk* around p. If the test fails for some vertex p or for some edge pq, we intersect all edges of the Voronoi cell of p with S and insert into P the intersection point p^* which is farthest from p.

(3) Next, we look at each Voronoi facet F. The topological ball property requires that F intersects S in a single curve connecting two boundary points. In general, the intersection $F \cap S$ could consist of several closed curves or open curves that end at the boundary of F.

Suppose that F is the intersection of the Voronoi cells $V(p)$ and $V(q)$. First, we exclude the possibility of some closed loop. If there is a closed loop, then it must have an extreme point in some direction. Thus, we choose some arbitrary direction d in F and compute the points of $F \cap S$ with tangent direction d. Algebraically, this amounts to finding a point x on $F \cap S$ where the surface normal ∇f is perpendicular to d in $\mathbb{R}^3$. We take the line l in F that is perpendicular to d and goes through x, and we intersect l it with S. If x is part of a cycle, then l must intersect S in a point x^* of F different from x.

Among all such points x^*, we choose the one with largest distance from p and insert it into P. If necessary, we have to repeat this test for each critical point x which lies in F.

If no point p^* is found in this way, we have excluded the possibility of a closed cycle in $F \cap S$. $F \cap S$ might still consist of more than one topological interval. However, since no Voronoi edge intersects S in more than one point, S must intersect more than two Voronoi edges of F. This means the dual Delaunay edge pq of F is incident to more that two triangles in the restricted Delaunay triangulation, violating the topological disk condition for p and q that was established in (2).

Thus, we have established the topological ball property for all Voronoi edges and for all two-dimensional Voronoi faces.

(4) Finally, we have to check that the surface forms a topological disk in each Voronoi cell $V(p)$. Because of the test (2), we already know that S intersects the boundary of $V(p)$ in a single cycle. However S could still contain a handle or some more involved topological structure inside $V(p)$. To exclude this possibility, we use the Silhouette Lemma (Lemma 6). We take the normal direction $d = \nabla f(p)$ in the point p and look at the silhouette in that direction, defined as the set of points whose normal is perpendicular to d. Arguing in the same was as in (3), a silhouette can either form a cycle inside $V(p)$, or it must intersect the boundary. We choose an arbitrary direction $d' \perp d$, and we look for a critical point in direction d' on the silhouette curve; we also intersect the silhouette with all boundary facets of $V(p)$. If we find any point in $V(p)$ in this way, we insert it into the sample. (It is not necessary to choose the point farthest from p.) Otherwise, we know that no silhouette exists in $V(p)$, and the surface must be a topological disk.

For intersection of the silhouette with a plane, we have the two equations characterizing the silhouette:

$$f(x) = 0, \ \langle \nabla f(x), d \rangle = 0,$$

and the plane equation, leading to a 0-dimensional solution set. The third equation for defining the critical point in direction d' can be worked out as the condition that the three vectors $H_f(x) \cdot d$, d', and the gradient $\nabla f(x)$ should be coplanar:

$$\det(H_f(x)d, d', \nabla f(x)) = 0, \tag{5.5}$$

where $H_f(x)$ is the Hessian matrix at x.

If we start the algorithm with an empty sample set $P = \emptyset$, the algorithm will directly proceed to step (4) to find a critical point on some silhouette (in an arbitrary direction). Any component of S without sample points has a silhouette with critical points (for any directions d and d') and will thus eventually be detected. Of course, any other method to find some seed points on the surface is also appropriate for starting the algorithm. Cheng et al. [90] propose to initialize P with all critical points of S in the vertical direction.

Here is a summary of the *Topology Refinement Algorithm* to obtain a topologically correct mesh.

Check the conditions given in (1)–(4), in this order. If any condition fails, it defines a new point. Insert it into P and update the Voronoi diagram. Continue the process till no new point is inserted.

As discussed above, this process ensures that the topological ball property holds when the algorithm terminates, and therefore, the restricted Delaunay triangulation is a homeomorphic mesh for S. It is however, not known whether the mesh is actually isotopic to S, see Research Problem 4 on p. 228. The proof of termination assumes that f is smooth and defines a smooth compact surface S, and it is based on the local feature size. One can show that a new point p^* that is inserted has distance at least $k \cdot \mathrm{lfs}(p)$ from all previous points, for $k = 0.06$. Therefore, Lemma 3 applies and gives the explicit bound $|P| = O(H(S, \mathrm{lfs}))$. (Note that the algorithm does not have to know the local feature size.)

The tests in (1) and (2) insert centers of surface Delaunay balls and fall within the framework of Delaunay refinement. The tests in (3) and (4) are different and require more involved computations.

The above discussion has ignored a few degenerate cases which do not happen in practice, but which need to be excluded nevertheless, to obtain a certified method. For example, in step (3), the intersection $S \cap F$ with a Voronoi face might be tangent to a third edge between its two endpoints. This would violate the topological ball property for F because the boundary of the intersection $S \cap F$ does not coincide with the intersection $S \cap \partial F$ with the boundary. However, this situation would lead to an extra triangle incident to the Delaunay edge pq and be detected in step (2). It is possible that the intersection $S \cap F$ is formed by S being tangent to F from one side, either in a point, in a curve, or even in a two-dimensional area. This would be in violation of the topological ball property for one of the incident cells $V(p)$ or $V(q)$. However, such a point of tangency is detected by the geometric condition of having a surface normal ∇f perpendicular to d, in step (3), and can be inserted into the sample.

Improving the Geometry.

The mesh obtained in this way has the correct topology, but it may still be a very rough approximation. As in the Delaunay refinement algorithm of the previous section, one can refine the mesh in order meet various geometric quality criteria. One can eliminate triangles with small angles or edges with sharp face angles between adjacent triangles. This refinement may destroy the correct topology, hence it is necessary to go back to steps (1)–(4) and then come back for another round of geometry improvement, and so on, until the process stabilizes. It can be shown that the asymptotic upper bound $|P| = O(H(S, \psi))$ from Lemma 3 still applies (with a different constant), and hence, the method terminates, for smooth surfaces.

Primitive Operations.

The only primitive operation is the solution of systems of equations involving the function f and its derivatives, and various plane or line equations that come from the Voronoi diagram. These equations have generically a zero-dimensional set of solutions, and thus are amenable to techniques of Chap. 3 and to software such as SYNAPS, in the case of an algebraic function f. In contrast to the Delaunay refinement algorithm of the previous section, the equations go beyond intersecting a line segment with the surface: they involve derivatives of f up to second order.

On the other hand, no a-priori knowledge or estimate about the local feature size or the location of different connected components is required by this algorithm.

Polyhedral Input.

This Delaunay refinement algorithm has been extended to the case when the input is already a polygonal surface S whose dihedral angles are not too sharp (bigger than π) [126]. For example, S can come from a sufficiently fine sample of a smooth surface. Actually this falls into the area of remeshing, which is not within the scope of this chapter, but the algorithm is based on the same ideas as for the smooth case. It is is implemented in SURFREMESH software by Tamal K Dey and Tathagata Ray[1]. Of course, for a polyhedral input surface, some simplifications are possible. The algorithm refines the surface with Delaunay triangles that have bounded aspect ratio, and it achieves a small approximation error.

Exercise 14. Prove the Silhouette Lemma (Lemma 6) by projecting M in direction d onto a plane.

Exercise 15. Prove that the points which are critical in direction d' on the silhouette in direction d are defined by (5.5).

5.4 A Sweep Algorithm

Mourrain and Técourt [263, 327] have proposed a meshing algorithm for algebraic surfaces that is based on sweeping a vertical plane over the surface. We have already seen in previous sections that critical points play a crucial role in determining the topological structure of a surface. Accordingly, the algorithm uses certain critical points to guide the sweep. In contrast to previous methods discussed in this chapter, this algorithm makes no smoothness or regularity assumptions about the input surface (other than those which follow from being an algebraic surface). The algorithm works for surfaces with self-intersections, fold lines, or other singularities. On the other hand, it makes

---

[1] http://www.cse.ohio-state.edu/~tamaldey/surfremesh.html

no guarantees about the geometric accuracy of the mesh, and it cannot be extended in a straightforward way to provide a more accurate mesh.

We first give a rough overview, concentrating on the geometric ideas. We will then discuss the primitive geometric operations that are necessary. In the case of algebraic surfaces, these operations can be carried out exactly.

The algorithm cuts the surface M into vertical slabs by a series of planes parallel to the y-z-plane in such a way that, between two successive planes, the surface has a "trivial" structure that can be constructed easily. Fig. 5.15 shows a mesh for a torus shape that is constructed in this way. Before meshing

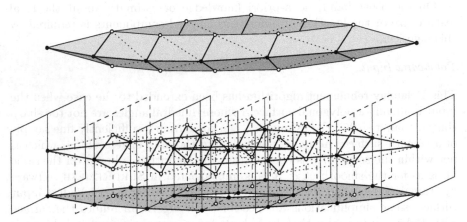

Fig. 5.15. A triangular mesh for a torus constructed by connecting a sequence of vertical cuts. The mesh is shown both as a surface and as a transparent skeleton. The vertices and edges are marked according to the conventions of Fig. 5.16 and Fig. 5.17 below

the surface between the planes, we first have to construct the intersection of M with each plane. This is the meshing problem for a plane curve. The algorithm solves this problem in an analogous way, cutting it by vertical lines through certain critical points, finding the intersections with these lines, and connecting the points between them. Thus, the algorithm proceeds by induction on the dimension.

We first discuss the problem of finding a topologically correct mesh for a planar curve C, in the form of a planar straight-line graph which is ambient isotopic to C. This algorithm is due to González-Vega and Necula [185]. We will assume throughout that C and M contains no straight line segment which is vertical. (This can most conveniently be achieved by a sufficiently random rotation of the coordinate system; alternatively, one can handle vertical parts explicitly.) For simplicity of exposition, we will also assume that the curve or surface is bounded. The algorithm can also deal with curves and surfaces that are clipped by a bounding box or sphere, and it can be extended to handle algebraic surfaces with infinite parts.

5.4.1 Meshing a Curve

We want to construct an embedded planar straight-line graph which is isotopic to the solution set of the equation $f(x, y) = 0$. First we compute the x-coordinates of all potential "critical points": points where the curve has a vertical tangent, or where it crosses itself or has some other singularity. These points are characterized by the equation

$$f_y(x, y) = 0, \quad f(x, y) = 0. \tag{5.6}$$

Let X be the finite set of x-coordinates of the solutions of this system of equations. In the x-interval between two successive critical points, the solution consists of a constant number of x-monotone curves, since $f_y(x, y) \neq 0$, and by the implicit function theorem, the curve can be locally written as a graph of a function $y = h(x)$, see Fig. 5.16. To make life easier, we insert an intermediate

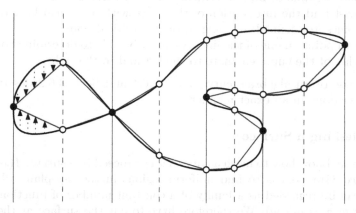

Fig. 5.16. A plane curve and its critical points (full circles). The vertical lines through the critical points and the additional lines between the critical points (dotted) define the polygonal mesh. In the left part, the isotopy that maps the curve to the meshed curve is indicated

x-value between each pair of successive values of X. Now, for each value x in this enlarged set, we intersect the curve with the vertical line at x. (This is nothing but the *one-dimensional* version of the meshing problem.) We get a discrete set of points on each vertical line. By construction, the intersection equation has only simple roots at each intermediate value x, whereas it has at least one multiple root at each value of X.

Now that we have the points on each vertical line, we need to figure out how to connect them. For each point, we need to know how many parts of the curve emanate on each side. For a simple root (including all points on the intermediate lines) there is one piece emanating on each side. For the critical points (multiple roots), we have to use algebraic tools to find this information. This is described in detail in Sect. 3.6.1 and Sect. 3.6.2. In each vertical slab, we now scan the points from bottom to top and connect them simply by a straight segment, taking into account the multiplicity of emanating arcs, see Fig. 5.16. By adding the intermediate lines, we have ensured that these line segments are disjoint. (In the example of Fig. 5.16 we could not connect points on the first and third line directly by straight segments if the second line were not there.)

The resulting graph is ambient isotopic to the given curve. We can even explicitly construct an isotopy of a very restricted kind, because all we have to do is to move points vertically upward or downward from each curve piece to the corresponding segment, as indicated in Fig. 5.16. The points on the vertical lines remain fixed.

Finally, we summarize the geometric primitives that we need to provide in order to make this algorithm work:

- We must be able to find all critical points by solving (5.6).
- We must find the intersection of the curve with a vertical line. (Some of these lines are defined by going through a critical point.)
- For each point on one of the lines, we must be able to determine how many branches of the curve emanate to the left and to the right.

For the case of an algebraic curve, Sect. 3.6.1 and Sect. 3.6.2 describe how this can be done in an exact manner.

5.4.2 Meshing a Surface

Now that we know how to mesh a curve, let us proceed to a surface $f(x, y, z) = 0$ in space. Our goal is to find uniform regions in the x-y-plane where the surface can be regarded as a family of a constant number of function graphs of the form $z = h(x, y)$. We therefore have to cut the surface at the points of self-intersection and the points that have vertical tangents (the *silhouette*). These points form the *polar variety* C, which is defined by

$$f_z(x, y, z) = 0, \quad f(x, y, z) = 0. \tag{5.7}$$

This system of two equations in three variables will in general have a one-dimensional solution space, consisting of curves on the surface M. (The points of self-intersection, as well as other sorts of critical surface points, satisfy in fact the stronger condition $f_x(x, y, z) = f_y(x, y, z) = f_z(x, y, z) = 0$, and hence they are included in the solution of (5.7).) When we cut the surface at the polar variety, we obtain x-y-monotone surface patches that can be parameterized in x and y.

As discussed in the beginning, we partition the surface into vertical slabs by planes perpendicular to the x-axis, in such a way that, between two sections, the topology of the surface "does not change". The points at which we cut will be called *slab points*. These points include all x-critical points of the polar variety, as well as all points where the projection of the polar variety on the x-y-plane intersects itself.

The system of equations that characterize the x-critical points has been given in (5.5) for two general directions d and d'. In our case, d is the z-direction and d' is the x-direction. Thus, the critical points are given by the system

$$(f_z \cdot f_{yz} - f_y \cdot f_{zz})(x,y,z) = 0, \quad f_z(x,y,z) = 0, \quad f(x,y,z) = 0. \quad (5.8)$$

This includes the x-critical points of the surface itself, i. e., the points where x has a local extremum: these points have a tangent plane perpendicular to the x-axis, and a fortiori a vertical tangent line, and therefore they lie on the silhouette. There are cases when the system (5.8) does not have a zero-dimensional solution set, and therefore it cannot be used to define slab points. (The example of Fig. 5.20 below is an instance of this.) In these cases, one must modify the system to obtain a finite set of slab points, as described in [263, 327].

The points where the vertical projection of the polar variety onto the x-y-plane crosses itself are the points (x,y) for which (5.7) has more than one solution z. For a polynomial f, these points can be found by computing the resultant of the polynomials in (5.7), see Chap. 3 for details. A slab point (x,y) of this type will be called a *multiple slab point* if more than two curves of the polar variety pass through the vertical line at (x,y) without going through the same point in space.

We make the following important *nondegeneracy assumption*:

> There is a finite set of slab points, there are no multiple slab points, and no two slab points have the same x-coordinate.

This assumption excludes for example a surface which consists of two equal spheres vertically above each other. The two silhouettes (equators) would coincide in the projection. It also excludes a torus with a horizontal axis, or a vertical cylinder (for which the polar variety would be two-dimensional), for the same reason. Such cases are very special, and they can easily be avoided by a random transformation of the coordinate system. Still, any number of curves of the polar variety may go through the same point in space, and in particular, the surface can have self-intersections of arbitrary order. Thus, the nondegeneracy assumption is no restriction on the generality of the surface M.

Now we proceed as in the planar case. We take the x-coordinates of all slab points, we add intermediate "regular" x-values between them, and we compute all vertical cross-sections at these values, using the algorithm of Sect. 5.4.1 for plane curve meshing. Note that the intersections of the polar variety with the vertical planes become critical points for the two-dimensional meshing problem. This can be seen by comparing (5.7) with (5.6), noting that the z-coordinate of our three-dimensional problem becomes the y direction of the two-dimensional problem. The algorithm produces in each vertical plane a planar graph that is ambient isotopic to the cross-section. The isotopy has only deformed the curves vertically.

Now, as we look at a slab from the top, the polar variety will form x-monotone non-crossing curves from one plane to the next, as in Fig. 5.17a. The strip between the boundaries is divided into triangular and quadrangular *regions* that are bounded by two curves of the polar variety C, and one or two straight pieces from the boundary walls. (In addition, there are the unbounded regions at the extremes, but by the boundedness assumption on the surface M, there cannot be any part of M in these areas.) We must find the correct assignment between the critical points on the two planes that have to be connected by the polar variety in the projection. By construction, one of the planes is an "intermediate" plane without a slab point; so each critical point is incident to one piece of C. By assumption, the other plane contains at most one slab point, and we know which one it is. We can therefore find the correct connections by assigning critical points in a one-to-one manner, with the projected slab point absorbing the difference between the number of critical points on the two sides. In the mesh, these pieces of C will be replaced by straight line segments, see Fig. 5.17b. Fig. 5.17 shows an example where the critical points in the regular cross-section outnumber the critical points on the other side, and thus s has to accept two connections. A different case arises if s is a local x-minimum of the surface, or in the situation of Fig. 5.19: s receives no connections at all from the left.

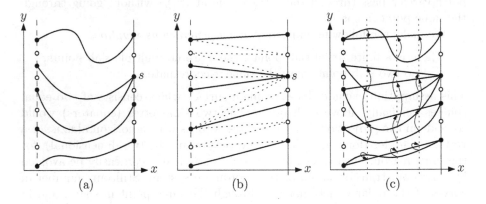

$$(a) \qquad\qquad (b) \qquad\qquad (c)$$

Fig. 5.17. (a) Vertical projection of the polar variety between two planes. The critical points in each vertical plane are marked by full circles. The plane on the right contains a slab point s, the plane on the left is a "regular" cross-section. (b) Vertical projection of the resulting mesh. (c) The horizontal component of the isotopy

Now we have to construct the surface pieces. *Above each region of the projected picture, the surface M consists of a constant number of x-y-monotone*

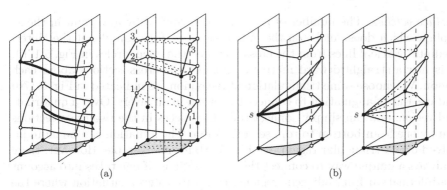

(a)   (b)

Fig. 5.18. Connecting a region in several layers: (a) A simple situation with three layers above the projection and a one-to-one assignment between two successive cross-sections. The pieces of the polar variety are shown in thick lines. The figure on the left includes a piece of the surface from an adjacent region, to show how the segment projected polar variety in the projection arises. This part of the surface will be meshed as part of the adjacent region. (b) Four layers over a triangular region. Three parts of the surface intersect in the point s, which is therefore a slab point

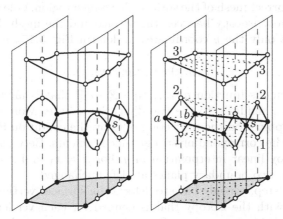

Fig. 5.19. Connecting a quadrilateral region in several layers: The triangulation of the region must avoid to connect the critical point s with the boundary points a and b on the other side, because otherwise the first and second layer of the surface would touch along this diagonal. This situation occurs for example in the second slab for the torus of Fig. 5.15

surface patches. The number of patches is determined by any point in the x-y-plane which does not lie on the projection of C, for example, at the "intermediate" vertical lines from the cross-sections (the open circles in Fig. 5.17b). It is now a straightforward matter to connect the cross-sections above each region. We choose some triangulation of the region (as indicated in Fig. 5.17b) and use this triangulation to connect the pieces in all layers. Over a quadrilateral region, one can simply connect the curve pieces in the two cross-sections one by one from bottom to top, see Fig. 5.18a. The situation can be more involved over a triangular region, see Fig. 5.18b for an example. However, there is always a unique way to connect the cross-sections, if one takes into account the information from adjacent regions. Fig. 5.19 shows a situation where the triangulation of the region cannot be chosen arbitrarily. There are degenerate situations which are more complicated, for example when more than three surface patches intersect in the same point, or when an x-minimal point on a self-intersection curve has at the same time a vertical tangent plane. Since we know that there is only one slab point on every vertical line and we know which point it is, these cases can also be resolved.

It is clear that the resulting triangles do not cross, and hence form a topologically correct mesh of the surface above each region. One can even write down the ambient isotopy between the surface and the mesh: In a first step, one transforms only the y coordinates to deform Fig. 5.17a into Fig. 5.17b, see Fig. 5.17c:

$$(x, y, z) \mapsto (x, g(x, y), z),$$

for some continuous function $g \colon [x_1, x_2] \times \mathbb{R} \to \mathbb{R}$ that is monotone in y for each value of x, similarly to the two-dimensional case. More explicitly, g is defined for all points on the projection of C by the condition that they must be mapped to the corresponding straight line segments. Between these points, g is extended by linear interpolation in y. For $x = x_1$ and $x = x_2$, we have $g(x, y) = y$: the two boundary planes are left unchanged.

In a second step, we only have to deform the surfaces vertically. Note that this coincides with the isotopy that is defined for each vertical slab by the planar curve meshing procedure. Thus, by concatenating the two isotopies (first in the y-direction and then in the z-direction) and gluing them together across all slabs, we get the isotopy between M and the mesh.

Theorem 6. *The mesh constructed by this algorithm is ambient isotopic to the surface M.*

For an algebraic surface, one can analyze the number of solutions that the equations arising in the course of the solution might have [263, 327]:

Theorem 7. *For an algebraic surface of degree d, the algorithm constructs a mesh with at most $O(d^7)$ vertices.*

Note that the solution set M of the equation $f(x, y, z) = 0$ may not be a surface at all. Of course, without any smoothness requirements whatsoever,

M could be some "wild" set. But even when f is a polynomial (the case of an *algebraic "surface"*), M can be a space curve or a set of isolated points. It can even be a mixture of parts of different dimensions, for example the union of a sphere and a line through the sphere, plus a few isolated points. The algorithm can be extended to handle these cases.

In particular, if the set M contains a space curve C, then all points on that curve will automatically form part of the polar variety. Figs. 5.20–5.21 show an example of a sphere and a line that are defined by the equation

$$(x^2 + y^2 + z^2 - 1)\left((x+z)^2 + (y+z)^2\right) = 0.$$

In such cases, the connection between two vertical sections will contain edges with no incident triangles.

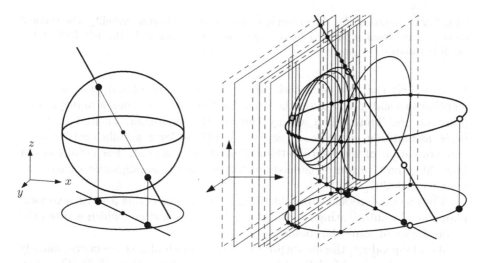

Fig. 5.20. The union of a sphere and a line, and the first half of the vertical cross-sections. The cross-sections in the right half are symmetric. The slab points are marked white

In fact, when the curve meshing problem (Sect. 5.4.1) is used as a subroutine for the surface meshing problem, degenerate cases of this type *will* occur. For example, an x-critical point p of M which is a local minimum or maximum in the x-direction will become an isolated point in the vertical plane through p. A saddle point in the x-direction will become a double point of the curve.

Finally, let us recall the geometric primitive that is needed, in addition to those that are necessary for the curves in the two-dimensional vertical cross-sections:

- We must be able find all slab points.

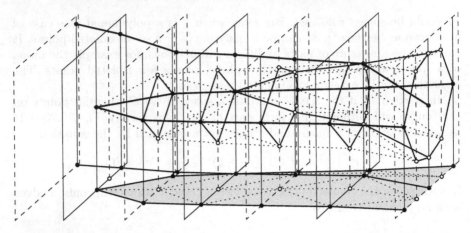

Fig. 5.21. The mesh for the example of Fig. 5.20. For better visibility, the vertical sections have been separated by a large amount. Again, only the left half of the mesh is shown

It is implicit that we can check whether a finite set of slab points exists, whether two slab points have the same x-coordinate, or when a multiple slab point occurs. Thus, when at any time in the algorithm, we find that our basic assumption is violated, we can simply perform a sufficiently generic transformation of the coordinates and start from scratch. For details about how this primitive can be carried out for the case of an algebraic surface, we refer to [263, 327].

The two-dimensional subproblems arise from intersecting M with a vertical plane, i.e., by substituting the variable x by some constant (which is often the x-coordinate of some slab point).

As a by-product, the algorithm produces a mesh of a space curve, namely the polar variety on M, defined by two polynomial equations (5.7). The algorithm can be extended to construct a topologically correct polygonal approximation for a space curve that is defined by two *arbitrary* polynomials [177].

Finally, let us step back and look at the algorithm from a broader perspective. Some ideas recur that we have already seen in connection with Snyder's algorithm (Sect. 5.2.3): the algorithm proceeds by induction on the dimension, and the condition when it is safe to construct a mesh is very similar to global parameterizability, except that there are several curve pieces (a constant number of them), each of which is parameterizable.

Silhouettes and the polar variety, which play an important part in this algorithm, are also used in the algorithm of Cheng, Dey, Ramos and Ray [90] of Sect. 5.3.2 to avoid complicated topological situations.

Exercise 16. By applying a random transformation of coordinates, one can assume in the meshing algorithm for an algebraic curve (Sect. 5.4.1) that no two critical points have the same x-coordinate. Is this statement still true when the curve meshing algorithm is used as a subroutine for the vertical sections of the surface meshing algorithm (Sect. 5.4.2)?

5.5 Obtaining a Correct Mesh by Morse Theory

5.5.1 Sweeping through Parameter Space

Stander and Hart [324] proposed a method for obtaining a topologically correct mesh that is based on sweeping through the family of surfaces $f(x, y, z) = a$ for varying parameters a and watching the critical points where the topology changes. Morse theory (see Sect. 7.4.2 on p. 300) classifies these changes. This method works theoretically, but there is no completely analyzed guaranteed finite algorithm to implement it. We sketch the main idea of this method.

For a given parameter a, the surface $f(x, y, z) = a$ can be interpreted as the *level set* of a trivariate function $f \colon \mathbb{R}^3 \to \mathbb{R}$. The idea is to start with a very small (or very large) value a for which $f(x, y, z) = a$ has no solution, and to gradually increase a until $a = 0$ and the surface in which we are interested is at hand. This is related to the space sweep method of Sect. 5.4, except that it works in one dimension higher: It sweeps a hyperplane $a = \mathrm{const}$ through the four-dimensional space of points (x, y, z, a) and maintains the intersection with the hypersurface $f(x, y, z) = a$.

As a varies, the surface "expands" continuously, except when a passes a *critical value* of f, where the topology changes. A critical value is the value of f at a *critical point*, i. e., at a point x where $\nabla f(x) = 0$. (These are precisely the values that we have avoided in the discussion so far, by assuming that the surface has no critical points.) At a *non-degenerate* critical point x, the Hessian H_f has full rank, and the number of its negative eigenvalues (the Morse index) gives information about the type of topology change. A critical value of Morse index 0 or 3 is a local minimum or maximum of f, and it corresponds to the situation when a small sphere-like component of the surface appears or disappears as a increases. The more interesting cases are the *saddle points*, the critical points of Morse index 1 and 2. Generically, they look like a hyperboloid $x^2 + y^2 - z^2 = a$ in the vicinity of the origin, for $a \approx 0$. For $a > 0$, we have a hyperboloid of one sheet, and for $a < 0$, we have a hyperboloid of two sheets, see Fig. 5.22. The transition occurs at $a = 0$, where the surface is a cone. Depending on the Morse index (1 or 2), the transition in Fig. 5.22 takes place from left to right or from right to left as a increases. The eigenvectors of the Hessian give the coordinate frame for rotating and scaling the picture such that it looks like the standard situation in Fig. 5.22.

Degenerate critical points, where the Hessian H_f does not have full rank, would pose a difficulty for this approach. They can be avoided by multiplying f by some suitably generic positive function g like $g(x) = a + \|x - b\|$ for some arbitrarily chosen scalar point b and scalar $a > 0$.

The algorithm of Stander and Hart [324] proceeds as follows: First we compute all critical points and critical values. This amounts to solving a 0-

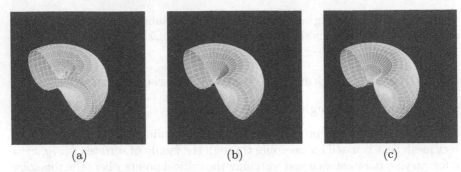

    (a)   (b)   (c)

Fig. 5.22. The change of the surface at a saddle point of f. Two separate pieces of the surface (a) come together in a pinching point (b) and form a tunnel (c)

dimensional system of equations. Then we let a vary from $a = a_{\min}$, where the surface is empty, to $a = 0$ in small steps. At each step, we maintain a mesh of the surface $f(x, y, z) = a$. Between critical values, we simply update the mesh. We know that the surface has no singularities, and we know that the topology is unchanged from the previous step. Any standard continuation method that builds a mesh on each component of the surface, taking into account Lipschitz constants for ∇f, can be applied.

At a critical point, we have to implement the appropriate topological change in the surface. A critical point of index 0 is easy to handle: One just has to generate a small spherical component of the surface. A critical point of index 3 is even easier: a small spherical component is simply deleted.

At a critical point, we have to implement the topological change indicated in Fig. 5.22. Going from left to right, two surface patches meet, forming a tunnel. We shoot rays from the origin in the positive and negative z direction (which is given by one of the eigenvectors of the Hessian), and remove the two mesh triangles that we hit first. Connecting the two triangles by a cylindrical ring establishes the new topology.

Going from right to left corresponds to closing a tunnel and separating the surface into two pieces which are locally disconnected. We intersect the x-y-plane with the surface and remove the ring of intersected triangles. By triangulating the two polygonal boundaries that are formed in the upper and in the lower half-plane, the two holes are closed.

To make a rigorous and robust method, one has to analyze the required step length that makes the approximations work, but this has not been done so far. Also, the complexity of the resulting mesh has not been analyzed.

5.5.2 Piecewise-Linear Interpolation of the Defining Function

The method of Boissonnat, Cohen-Steiner, and Vegter [61] also uses Morse theory, but in a more indirect way. The basic idea is to output the zero-set of a piecewise-linear interpolation of the defining function f. More precisely, let $S = f^{-1}(0)$ denote the surface that we want to mesh, and assume S is contained in some bounding box. Let T denote a tetrahedral mesh of this bounding box, $\hat{f}$ be the function obtained by linear interpolation of f on T, and set $\hat{S} = \hat{f}^{-1}(0)$. The algorithm consists in building a tetrahedral mesh T such that the output mesh $\hat{S}$ is isotopic to S.

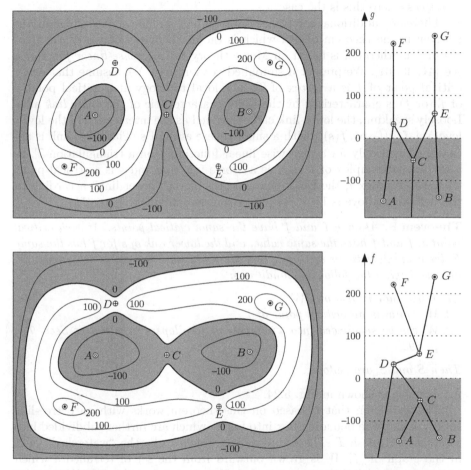

Fig. 5.23. Critical points do not determine the topology of level sets. The two functions have the same critical points of the same types at the same heights, but different level sets at level 0. Minima and maxima are indicated by empty and full circles, and crosses denote saddle points. On the right, the corresponding contour trees (Sect. 7.4.2) are shown

To ensure that this is the case, the mesh T must of course satisfy certain conditions. From Morse theory, one might require that f and $\hat{f}$ have the same critical points, the same value at critical points, and the same types of critical points. Unfortunately, this is not sufficient even for implicit curves in the plane. Indeed, the situation in figure 5.23 is a two-dimensional example of two zero-sets $S = f^{-1}(0)$ and $S' = g^{-1}(0)$ (boundaries of the grey regions) which are not homeomorphic, though their defining functions have the same critical points, with the same values and indices. In this example, g cannot be obtained from f by piecewise-linear interpolation, but it is possible to design examples where this is the case.

Therefore, additional conditions are required. A sufficient set of conditions is given in the theorem below, which is the mathematical basis of the algorithm. The theorem is based on Morse theory for *piecewise-linear functions*, see [41, 42, 61]. We present a simplified version here. We assume that every critical point of f is a vertex of T. The local topology at a critical point s of f (or $\hat{f}$) is characterized by the Euler characteristic of the *lower link* at s. Loosely speaking, the lower link can be defined as the intersection of the lower level set $f^{-1}((-\infty, f(s)])$ with a small sphere around s. The lower link is actually defined only for a piecewise linear function $\hat{f}$ on a triangulation T, as a certain subcomplex of T. If f is a Morse function and s is a critical point with Morse index i, the Euler characteristic if the "lower link" according to the definition above is $1 - (-1)^i$, see Exercise 3 in Chap. 7 (p. 311).

Theorem 8. *Assume f and $\hat{f}$ have the same critical points. At each critical point s, f and $\hat{f}$ have the same value, and the lower link of s for f has the same Euler characteristic as the lower link for $\hat{f}$. Suppose there is a subcomplex W of T satisfying the following conditions:*

1. *f does not vanish on ∂W.*
2. *W contains no critical point of f.*
3. *W can be subdivided into a complex that collapses onto $\hat{S}$ (see Sect. 7.3, p. 292).*

Then S and $\hat{S}$ are isotopic.

An example is shown in Fig. 5.24.

The algorithm that is based on this theorem works with an octree-like subdivision of the bounding box into boxes, which are further subdivided into a tetrahedral mesh T. The complex W is taken to be the "watershed" of $\hat{S}$ in the graph of $|\hat{f}|$: W is grown outward from the set of tetrahedra which have vertices with different signs of f. Tetrahedra are added to W in order to fulfill Condition 1, while trying to avoid the inclusion of critical points (Condition 2). If a set W cannot be found, the mesh T is refined. Note that fulfilling the conditions requires to compute all critical points of f exactly, which is difficult, in particular in the case of nearly degenerate critical points. This is why the algorithm actually uses a relaxed (but still sufficient) set of conditions that permits an implementation within the framework of interval analysis. This algorithm is not meant to provide a geometrically accurate approximation of S, but rather to build a topologically correct approximation using as few elements as possible.

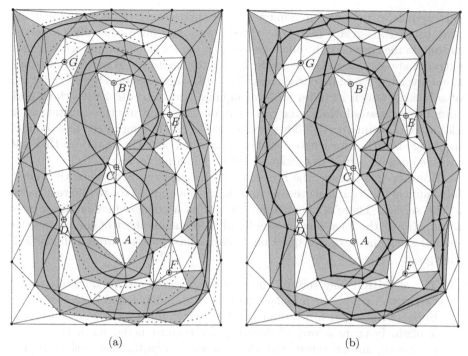

(a)   (b)

Fig. 5.24. (a) a triangulation T for the function f of Fig. 5.23, rotated by 90°. The subcomplex W is shaded. Since W must collapse to S, it must form two bands that enclose the two components of S, without common vertices. (b) the zero-set of the piecewise linear function $\hat{f}$

5.6 Research Problems.

1. It was mentioned in Sect. 5.2.4 that the behavior of the Small Normal Variation refinement algorithm Plantinga and Vegter [286] adapts the refinement to the properties of f. Estimate the number of cubes generated by the algorithm in terms of properties of the function f, like total variation of f and ∇f, etc.
2. Can the balancing operation be eliminated in the algorithm of Sect. 5.2.4? Try to define rules for constructing a mesh when a cube may have an arbitrary number of small neighboring boxes.
3. The algorithm in Sect. 5.2.4 stops as soon as the angle between two surface normals inside a cube is bounded by $\pi/2$. If we impose some smaller bound α on the angle, what can be said about the distance between the surface and the approximating mesh? How should the mesh be chosen to obtain a good approximation?

The Delaunay refinement algorithm requires the knowledge of a lower estimate $\psi(p)$ on the local feature size. The *minimum* local feature size lfs$_{\min}$

corresponds to points of maximum principal curvature or to medial spheres that touch the surface in two or three points and have a locally minimum radius. In the case of an implicit surface $f(x, y, z) = 0$, the points where these extrema are attained can be found by solving appropriate systems of equations involving f and its derivatives. Generally, these systems have a finite set of solutions, which includes all *local* minima and maxima. By checking and comparing these solutions, one can compute lfs$_{min}$ and use this constant as a global lower estimate $\psi(p)$. This yields a theoretically guaranteed and reliable meshing algorithm for smooth surfaces, provided that the equations that are involved can be solved (for example, when f is a polynomial).

However, in this case, the necessary mesh density is dictated by the global minimum of the local feature size, and thus it does not adapt to different parts of the surface. There is no reliable way to find a good *individual* lower estimate $\psi(p)$ on the local feature size lfs(p) beforehand, short of computing the medial axis. The next two questions address this question from a theoretical viewpoint.

1. The algorithm may rely on the user to specify the function $\psi(p)$, which can as well simply be a global constant ψ_{min} independent of the location. Suppose the algorithm terminates, for a given function ψ, and constructs a mesh. Is there a way of deciding if the constructed mesh is at least *consistent*, in the sense that *there exists* a hypothetical surface S' for which ψ is a lower bound on the local feature size, and for which the same mesh would be obtained? (This idea of having a "certificate" of consistency is similar to the approach of [127] for curve reconstruction.) Note that in practice, one may apply the algorithm to a non-smooth surface and be perfectly happy with the resulting mesh; however, the hypothetical surface S' in the above question would necessarily have to be smooth. Otherwise it would contain points with lfs $= 0$.

2. For an implicit surface $f(x, y, z) = 0$, is there a way of estimating the local feature size within some given range for the variables x, y, z, by looking at the function f and its derivatives? Can one use interval arithmetic to obtain a conservative lower bound ψ?

3. The test (5.5) for critical points in a silhouette involves second derivatives (cf. Exercise 15). Is there a zero-dimensional system of equations for establishing the topological ball property that only involves f and its first derivatives?

4. **The topological ball property and isotopy.**
   The topological ball property only guarantees a homeomorphism between the original surface and the reconstruction, it does not provide an isotopy. In fact, Theorem 5 can be extended to manifolds in arbitrary dimension k (and even to non-manifolds [138]): For a k-dimensional manifold $M \subset \mathbb{R}^n$, the topological ball property means that every Voronoi face F of dimension d intersects M in a closed topological $(d - n + k)$-ball or in the empty set.

For manifolds of codimension at least 2, the topological ball property is not sufficient to establish isotopy. For example, the topological ball property for a point sample P on a curve C in $\mathbb{R}^3$ ($k = 1$, $n = 3$) will not detect whether C is knotted inside a Voronoi cell, and thus the restricted Delaunay will not always be isotopic to C.

a) Does the topological ball property for a surface S in $\mathbb{R}^3$ (or more generally, for an $(n - 1)$-manifold embedded in $\mathbb{R}^n$) imply that the restricted Delaunay triangulation is isotopic to S?

b) Find an appropriate strengthening of the topological ball property that ensures isotopy of the restricted Delaunay triangulation.

5. For a curve $f(x, y) = 0$, the critical points in direction $\binom{u}{v}$ are given by the equation

$$u \cdot f_x(x, y) + v \cdot f_y(x, y) = 0, \quad f(x, y) = 0.$$

If f is a polynomial of degree d, give an upper bound on the number of directions for which two distinct critical points lie on a line parallel to $\binom{u}{v}$.

6

Delaunay Triangulation Based Surface Reconstruction

Frédéric Cazals and Joachim Giesen

6.1 Introduction

6.1.1 Surface Reconstruction

The surfaces considered in surface reconstruction are 2-manifolds that might have boundaries and are embedded in some Euclidean space $\mathbb{R}^d$. In the surface reconstruction problem we are given only a finite sample $P \subset \mathbb{R}^d$ of an unknown surface S. The task is to compute a model of S from P. This model is referred to as the reconstruction of S from P. It is generally represented as a triangulated surface that can be directly used by downstream computer programs for further processing. The reconstruction should match the original surface in terms of geometric and topological properties. In general surface reconstruction is an ill-posed problem since there are several triangulated surfaces that might fulfill these criteria. Note, that this is in contrast to the curve reconstruction problem where the optimal reconstruction is a polygon that connects the sample points in exactly the same way as they are connected along the original curve. The difficulty of meeting geometric or topological criteria depends on properties of the sample and on properties of the sampled surface. In particular, sparsity, redundancy, noisiness of the sample or non-smoothness and boundaries of the surface make surface reconstruction a challenging problem.

Notation. The surface that has to be reconstructed is always denoted by S and a finite sample of S is denoted by P. The size of P is denoted by n, i.e., $n = |P|$.

6.1.2 Applications

The surface reconstruction problem naturally arises in computer aided geometric design where it is often referred to as reverse engineering. Typically, the surface of some solid, e.g., a clay mock-up of a new car, has to be turned into a computer model. This modeling stage consists of (i) acquiring data points on the surface of the solid using a scanner (ii) reconstructing the surface from these points. Notice that the previous step is usually decomposed into two stages. First a piece-wise linear surface is reconstructed, and second, a piecewise-smooth surface is built upon the mesh.

Surface reconstruction is also ubiquitous in medical applications and natural sciences, e.g., geology. In most of these applications the embedding space of the original surface is $\mathbb{R}^3$. That is why we restrict ourselves in the following to the reconstruction of surfaces embedded in $\mathbb{R}^3$.

6.1.3 Reconstruction Using the Delaunay Triangulation

Because reconstruction boils down to establishing neighborhood connections between samples, any geometric construction defining a simplicial complex on these samples is a candidate auxiliary data structure for reconstruction. One such data structure is the Delaunay triangulation of the sample points. The intuition that it might be extremely well suited for reconstruction was first raised in [54] and is illustrated in Fig. 6.1 which features a sampled curve and the Delaunay triangulation of the samples. It seems that the Delaunay triangulation explores the neighborhood of a sample point in all relevant directions in a way that even accommodates non-uniform samples.

The Delaunay triangulation is a cell complex that subdivides the convex hull of the sample. If the sample fulfills certain non-degeneracy conditions then all faces in the Delaunay triangulation are simplices and the Delaunay triangulation is unique. The combinatorial and algorithmic worst case complexity of the Delaunay triangulation grow exponentially with the dimension of the embedding space of the original surface. In $\mathbb{R}^3$ the combinatorial as well as the algorithmic complexity of the Delaunay triangulation is $\Theta(n^2)$, where $n = |P|$ is the size of the sample. However, it has been shown [33] that the Delaunay triangulation of points that are well distributed on a smooth surface has complexity $O(n \log n)$. Robust and efficient methods to compute the Delaunay triangulation in $\mathbb{R}^3$ exist [2]. Also important for the reconstruction problem is the Voronoi diagram which is dual to the Delaunay triangulation. The Voronoi diagram subdivides the whole space into convex cells where each cell is associated with exactly one sample point.

There are also approaches toward the surface reconstruction problem that are not based on the Delaunay triangulation, e.g., level set methods [350], radial basis function based methods [79] and moving least squares methods [16]. That we do not cover these approaches in this chapter does not mean that they are less suited or worse. On the practical side, many of them are very successfully applied in daily practice. On the theoretical side though, these algorithms often involve non-local constructions making a theoretical analysis difficult. As opposed to these, algorithms elaborating upon Delaunay are more

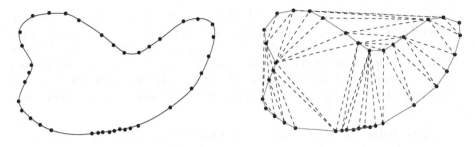

Fig. 6.1. Left: a sampled curve. Right: Delaunay contains a piece-wise linear approximation of the curve. Notice the Delaunay triangulation has neighbors in *all* directions, no matter how non-uniform the sample

prone to such an analysis, and one of the goals of this survey is to outline the key geometric features involved in these analysis.

6.1.4 A Classification of Delaunay Based Surface Reconstruction Methods

Using the Delaunay triangulation still leaves room for quite different approaches to solve the reconstruction problem. But all these approaches, that we sketch below, benefit from the structure of the Delaunay triangulation and the Voronoi diagram, respectively, of the sample points. We should note already here that many of the algorithms combine features of different approaches and as such are not easy to classify. We did the classification by what we consider the dominant idea behind a specific algorithm.

Tangent plane methods. If one considers a smooth surface with a sufficiently dense sample, the *neighbors* of a point in the point cloud should not deviate too much from the tangent plane of the surface at that point. It turns out that this tangent plane can be well approximated by exploiting the fact that under the condition of sufficiently dense sample the Voronoi cell of the sample point is elongated in the direction of the surface normal at the sample point. This normal or tangent plane information, respectively, can be used to derive a local triangulation around each point.

Restricted Delaunay based methods. It is possible to define subcomplexes of the Delaunay triangulation by restricting it to some given subset of $\mathbb{R}^3$. Restricted Delaunay based methods compute such a subset from the Delaunay triangulation of the sample. This subset should contain the unknown surface S provided the sample is dense enough. The reconstruction basically is the Delaunay triangulation of P restricted to the computed subset.

Inside / outside labeling. Given a closed surface S one can attempt to classify the tetrahedra in the Delaunay triangulation as either inside or outside with respect to S. The interface between the inside and outside tetrahedra should provide a good reconstruction of S. Algorithms that follow the inside / outside labeling paradigm often shell simplices from the outside of the Delaunay triangulation of the sample points in order to discover the surface to be reconstructed. A subclass of the shelling algorithms guide the shelling by topological information like the critical points of some function which can be derived from the sample.

Empty balls methods. When reconstructing a surface, the simplices reported should be *local* according to some definition. One such definition consists of requiring the existence of a sphere that circumscribes the simplex and does not contain any sample point on its bounded side. The ball bounded by such a sphere is called an empty ball. All Delaunay simplices are local in this sense. This property can be used to filter simplices from the Delaunay triangulation, e.g., by considering the radii of the empty balls.

6.1.5 Organization of the Chapter

The rest of this chapter is subdivided into two sections. Sect. 6.2 contains mathematical pre-requisites that are necessary to understand the ideas and guarantees behind the algorithms that are detailed in Sect. 6.3.

6.2 Prerequisites

Some prerequisites that we introduce here in order to describe the various reconstruction algorithms and the guarantees they come with are also described in other chapters. Voronoi diagrams are introduced in much more generality in Chap. 2, the restricted Delaunay triangulation, ε-samples and the topological concepts of homeomorphy and isotopy also play a dominant role in Chap. 5 on meshing, most differential geometric concepts are more detailed in Chap. 4 and all topological concepts appear in more detail in Chap. 7. The reason for this redundancy is mostly to make this chapter self contained and to provide a reader only interested in reconstruction with the minimally needed background.

6.2.1 Delaunay Triangulations, Voronoi Diagrams and Related Concepts

General Position.

The sample P is said to be in general position if there are no degeneracies of the following kind: no three points on a common line, no four points on a common circle or hyperplane and no five points on a common sphere. In the following we always assume that the sample P is in general position. But note that the case that P is not in general position can also be dealt with algorithmically [136]. We make the general position assumption only to keep the exposition simple.

Voronoi Diagram.

The Voronoi diagram $V(P)$ of P is a cell decomposition of $\mathbb{R}^3$ in convex polyhedra. Every *Voronoi cell* corresponds to exactly one sample point and contains all points of $\mathbb{R}^3$ that do not have a smaller distance to any other sample point, i.e. the Voronoi cell corresponding to $p \in P$ is given as follows

$$V_p = \{x \in \mathbb{R}^3 : \forall q \in P \quad \|x - p\| \leq \|x - q\|\}.$$

Closed facets shared by two Voronoi cells are called *Voronoi facets*, closed edges shared by three Voronoi cells are called *Voronoi edges* and the points shared by four Voronoi cells are called *Voronoi vertices*. The term *Voronoi face* can denote either a Voronoi cell, facet, edge or vertex. The Voronoi diagram is the collection of all Voronoi faces. See Fig. 6.2 for a two-dimensional example of a Voronoi diagram.

Delaunay Triangulation.

The Delaunay triangulation $D(P)$ of P is the dual of the Voronoi diagram, in the following sense. Whenever a collection $V_1, \ldots, V_k$ of Voronoi cells corresponding to points $p_1, \ldots, p_k$ have a non-empty intersection, the simplex whose vertices are $p_1, \ldots, p_k$ belongs to the Delaunay triangulation. It is a simplicial complex that decomposes the convex hull of the points in P. That is, the convex hull of four points in P defines a *Delaunay cell (tetrahedron)* if the common intersection of the corresponding Voronoi cells is not empty. Analogously, the convex hull of three or two points defines a *Delaunay facet* or *Delaunay edge*, respectively, if the intersection of their corresponding Voronoi cells is not empty. Every point in P is a *Delaunay vertex*. The term*Delaunay simplex* can denote either a Delaunay cell, facet, edge or vertex. See Fig. 6.2 for a two-dimensional example of a Delaunay triangulation.

Flat Tetrahedra.

In surface reconstruction *flat tetrahedra* may cause problems for some algorithms. The most notorious flat tetrahedra are *slivers*. These are Delaunay tetrahedra that have a small volume but do not have a large circumscribing ball and do not have a small edge. Here all comparisons in size are made with respect to the length of the longest edge of the tetrahedron. See Fig. 6.3 for an illustration of a sliver and a cap, and refer to [89] for a classification of *baldly* shaped tetrahedra.

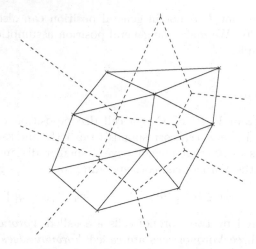

Fig. 6.2. Voronoi and Delaunay diagrams in the plane

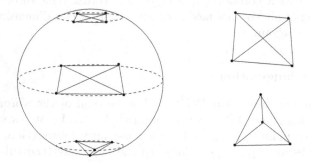

Fig. 6.3. A nearly flat tetrahedron can be located near the equatorial plane or the north pole of its circumscribing sphere. The tetrahedra near the poles have a large circumscribing ball. Only the tetrahedron near the equatorial plane is a sliver (also shown on the top right, the bottom right tetrahedron being a cap.)

Pole.

There are *positive* and *negative poles* associated with a Voronoi cell V_p. If V_p is bounded then the positive pole is the Voronoi vertex in V_p with the largest distance to the sample point p. Let $\mathbf{u}$ be the vector from p to the positive pole. If V_p is unbounded then there is no positive pole. In this case let $\mathbf{u}$ be a vector in the average direction of all unbounded Voronoi edges incident to V_p. The negative pole is the Voronoi vertex v in V_p with the largest distance to p such that the vector $\mathbf{u}$ and the vector from p to v make an angle larger than $\pi/2$.

Empty-ball Property.

It follows from the definitions of Voronoi diagrams and Delaunay triangulations that the relative interior of a Voronoi face of dimension k, which is dual to a Delaunay simplex of dimension $3 - k$, consists of the set of points having exactly $3 - k + 1$ nearest neighbors. Therefore, for any point in such a Voronoi face, there exists a ball empty of sample points containing the vertices of the dual simplex on its boundary. The simplex is said to have the *empty ball property*. See also Fig. 6.4 for a two-dimensional example. For Delaunay tetrahedra there is only one empty ball whereas there is a continuum of empty balls for Delaunay triangles and edges.

The empty ball property can be used to define sub-complexes of the Delaunay triangulation by imposing additional constraints on the empty balls. Here we discuss two such restrictions that lead to Gabriel simplices and α-shapes, respectively.

Gabriel Simplex.

A simplex of dimension less then 3 is called *Gabriel* if its smallest circumscribing ball is empty. Obviously all Gabriel simplices are contained in the Delaunay triangulation. Gabriel simplices also have a dual characterization: a Delaunay simplex is Gabriel iff its dual Voronoi face intersects the affine hull of the simplex.

Well known and heavily used is the Gabriel graph which is the geometric graph that contains all one dimensional Gabriel simplices.

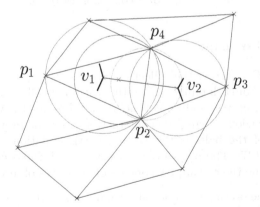

Fig. 6.4. Empty balls centered on Voronoi faces

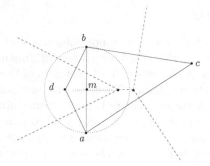

Fig. 6.5. All edges but edge ab are Gabriel edges

Restricted Voronoi Diagram and Restricted Delaunay Triangulation.

Given a subset $X \subset \mathbb{R}^3$ we can restrict the Voronoi diagram of P to X by replacing every Voronoi face with its intersection with X. The restricted Voronoi diagram is denoted as $V_X(P)$. The Delaunay triangulation $D_X(P)$ of P restricted to X is defined similarly as the Delaunay triangulation of P. The only difference is that instead of taking the common intersection of Voronoi cells now the common intersection of restricted Voronoi cells is taken. That is, whenever a collection $V_1 \cap X, \ldots, V_k \cap X$ of Voronoi cells corresponding to points $p_1, \ldots, p_k$ restricted to X have a non-empty intersection, the simplex whose vertices are $p_1, \ldots, p_k$ belongs to the restricted Delaunay triangulation. The restricted Delaunay triangulation of a plane curve is illustrated in Fig. 6.6. The restricted Delaunay triangulation is also most convenient to introduce the so-called α-complex and α-shape of a collection of balls.

α-complex and α-shape.

Given a sample P, consider the collection of balls of square radius α centered at these points [1]. For each ball, consider the *restricted* ball, i.e., the intersection of the ball with its corresponding Voronoi region. Finally, let X be the union of these restricted regions. Using the construction from the previous paragraph, the α-complex of the balls is the Delaunay triangulation restricted to the domain X [131, 137]. The polytope [2] associated with the α-complex is called the α-shape. While the α-complex consists of simplices of any dimension, i.e.,

---

[1]We present the α-complex for a collection of balls of the same radius $\sqrt{\alpha}$. The variable α stands for the square radius rather than the radius, a constraint stemming from the construction of the α-complex for a collection a balls of different radii using the power diagram. See [131] for the details.

[2]Polytope stands here for the union of the closure of the domain of the simplices, rather then the convex hull of a set of points in $\mathbb{R}^d$.

vertices, edges, triangles and tetrahedra, the boundary of the α-shape consists only of vertices, edges and triangles. In surface reconstruction where one is concerned with triangles contributing to the reconstructed surface, the focus has mainly been on the boundary of the α-shape.

It is actually possible to assign to each simplex of the Delaunay triangulation an interval specifying whether it is present in the α-complex for a given value of α, and similarly for the simplices in the boundary of the α-shape. The intervals for the boundary are contained in the intervals for the α-complex.

May be a more intuitive characterization of the points of appearance and disappearance of simplices in the boundary of the α-shape is as follows: let balls grow at the sample points with uniform speed. A simplex appears in the boundary of the α-shape, when the balls corresponding to the vertices of the simplex intersect for the first time. Note that this intersection takes place on the dual Voronoi face of this simplex. It disappears when the common intersection of the balls corresponding to the vertices of the simplex completely contains the dual Voronoi face of the simplex. This growing process is illustrated in Fig. 6.8. In terms of growing process, the differences between the α-complex and the α-shape are twofold: first, once a simplex appears in the α-complex, it stays forever; second, the α-complex also contains Delaunay tetrahedra.

Note that α can be interpreted as a spatial scale parameter. If P is a uniform sample of the surface S then there exist α-values such that the boundaries of the corresponding α-shapes of P provide a reasonable reconstruction of S.

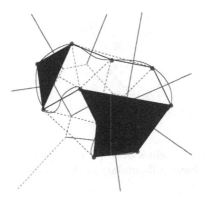

Fig. 6.6. Diagrams restricted to a curve

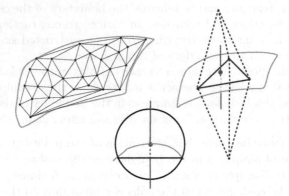

Fig. 6.7. Triangulation restricted to a surface

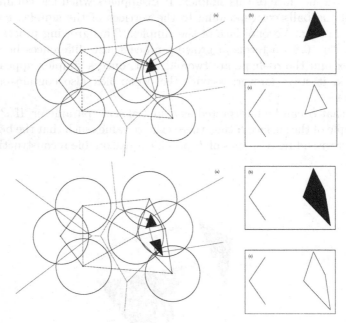

Fig. 6.8. At two different values of α: (a)α-complex with solid triangles scaled to avoid cluttering (b)α-shape (c)boundary of the α-shape

Topological Ball Property.

The restricted Voronoi diagram $V_S(P)$ of a sample P of a surface S has the *topological ball property* if the intersection of S with every Voronoi face in $V(P)$ is homeomorphic to a closed ball whose dimension one smaller then that of the Voronoi face. (Notice that the transverse intersection of a Voronoi cell of dimension k with a manifold of dimension $d - 1$ has dimension equal to $k + (d - 1) - d = k - 1$.) Edelsbrunner and Shah [138] were able to relate the topology of the restricted Delaunay triangulation $D_S(P)$ to the topology of S.

Theorem 1. *Let S be a surface and P be a sample of S such that $V_S(P)$ has the closed all property. Then $D_S(P)$ and S are homeomorphic.*

Power Diagram and Regular Triangulation.

The concepts of Voronoi- and Delaunay diagrams are easily generalized to sets of weighted points. A *weighted point p* in $\mathbb{R}^3$ is a tuple (z, w) where $z \in \mathbb{R}^3$ denotes the point itself and $w \in \mathbb{R}$ its weight. Every weighted point gives rise to a distance function, namely the *power distance function*,

$$\pi_p : \mathbb{R}^3 \to \mathbb{R}, x \mapsto \|x - z\|^2 - w.$$

Let P now be a set of weighted point in $\mathbb{R}^3$. The power diagram of P is a decomposition of $\mathbb{R}^3$ into the *power cells* of the points in P. The power cell of $p \in P$ is given as

$$V_p = \{x \in \mathbb{R}^3 : \forall q \in P, \pi_p(x) \leq \pi_q(x)\}.$$

The points that have the same power distance from two weighted points in P form a hyperplane. Thus V_p is either a convex polyhedron or empty. Closed facets shared by two power cells are called *power facets*, closed edges shared by three power cells are called *power edges* and the points shared by four power cells are called *power vertices*. The term *power face* can denote either a power cell, facet, edge or vertex. The *power diagram* of P is the collection of all power faces.

The dual of the power diagram of P is called the regular triangulation of P. The duality is defined in exactly the same way as for Voronoi diagrams and Delaunay triangulations. That is the reason why regular triangulations are also referred to as weighted Delaunay triangulations.

Natural Neighbors.

Given a Delaunay triangulation, it is natural to define the neighborhood of a vertex as the set of vertices this vertex is connected to. This information is of combinatorial nature and can be made quantitative using the so-called *natural coordinates* which were introduced by Sibson [318].

Given a point $x \in \mathbb{R}^3$ which is not a sample point, define $V^+(P) = V(P \cup \{x\})$, $D^+(P) = D(P \cup \{x\})$, and denote by V_x^+ the Voronoi cell of x in $V^+(P)$. In addition, for any sample point $p \in P$ define $V_{(x,p)} = V_x^+ \cap V_p$ and denote by $w_p(x)$ the volume of $V_{(x,p)}$. The *natural neighbors* of a point x are the sample points in P that are connected to x in $D^+(P)$. Equivalently, these are the points $p \in P$ for which $V_{(x,p)} \neq \emptyset$. The *natural coordinate* associated with a natural neighbor is the quantity

$$\lambda_p(x) = \frac{w_p(x)}{w(x)}, \text{ with } w(x) = \sum_{p \in P} w_p(x). \tag{6.1}$$

For an illustration of these definitions see Fig. 6.9.

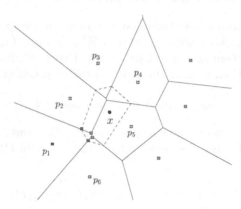

Fig. 6.9. Point x has six natural neighbors

The term *coordinate* is clearly evocative of barycentric coordinates. Recall that in any three-dimensional affine space, a set of four affinely independent points $p_i, i = 1, \ldots, 4$ define a basis of the affine space. Moreover, every point x decomposes uniquely as $x = \sum_{i=1,\ldots,4} \lambda_{p_i}(x) p_i$, with $\lambda_{p_i}(x)$ the barycentric coordinate of x with respect to p_i. Natural coordinates provide an elegant extension of barycentric coordinates to the case where one has more than four points. The following results have been proven in a number of ways [318, 36, 72, 210].

Theorem 2. *The natural coordinates satisfy the requirements of a coordinate system, namely,*

(1) for any $p, q \in P$, $\lambda_p(q) = \delta_{pq}$ where δ_{pq} is the Kronecker symbol and

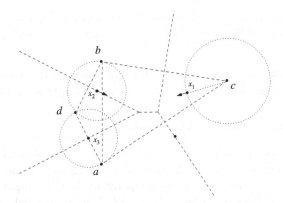

Fig. 6.10. Critical points of the distance function. Points x_1, x_2 are regular, but point x_3 is critical

(2) the point x is the weighted center of mass of its neighbors. That is,

$$x = \sum_{p \in P} \lambda_p(x)\, p, \quad with \quad \sum_{p \in P} \lambda_p(x) = 1. \tag{6.2}$$

Induced Distance Function.

Voronoi diagrams of a sample P are closely related to the distance function

$$h : \mathbb{R}^3 \to \mathbb{R}, x \mapsto \min_{p \in P} \|x - p\| \tag{6.3}$$

induced by the set of sample points. This distance function is smooth everywhere besides at the points in P and on the lower dimensional Voronoi faces, i.e., on the facets, edges and vertices.

At every point x inside a Voronoi region, the gradient of h is the unit vector pointed away from the center of the region. Interestingly, for points x on lower dimensional Voronoi faces, one can define a generalized gradient, as depicted in Fig. 6.10. Let x be a point and denote by $C(x)$ the set of closest points in P to x, i.e., $C(x)$ consists of the vertices dual to the (lowest dimensional) Voronoi face containing x. If x does not belong to the convex hull of $C(x)$, then the generalized gradient at x is the unit vector that points from the closest neighbor of x in the convex hull of $C(x)$ to x. Otherwise, i.e., x is contained int eh convex hull of $C(x)$, the point x is called a critical point.

It was was observed by Edelsbrunner [132] and later proved by Giesen and John [184] that the critical points of the distance function, i.e., the local extrema and the saddle points, can be characterized in terms of Delaunay simplices and Voronoi faces.

Theorem 3. *The critical points of h are the intersection points of Voronoi faces and their dual Delaunay simplices. The local maxima are Voronoi vertices contained in their dual Delaunay cell. The saddle points are intersection* *points of Voronoi facets and their dual Delaunay edges and intersection points of Voronoi edges and their dual Delaunay triangles. All sample points are minima.*

The *index* of a critical point is the dimension of the Delaunay simplex involved in its definition. See Fig. 6.2.1 for an example in two dimensions.

Induced Flow and Stable Manifolds.

As in the case of smooth functions there is a unique direction of steepest ascent of h at every non-critical point of h. Assigning to the critical points of h the zero vector and to every other point in $\mathbb{R}^3$ the unique unit vector of steepest ascent defines a *vector field* on $\mathbb{R}^3$. This vector field is not smooth but nevertheless gives rise to a *flow* on $\mathbb{R}^3$, i.e., a mapping

$$\phi : [0, \infty) \times \mathbb{R}^3 \to \mathbb{R}^3,$$

such that at every point $(t, x) \in [0, \infty) \times \mathbb{R}^3$ the right derivative

$$\lim_{t \leftarrow t'} \frac{\phi(t, x) - \phi(t', x)}{t - t'}$$

exists and is equal to the unique unit vector of steepest ascent at x. The flow basically tells how a point moved if it would always follow the steepest ascent of the distance function h. The curve that a point x follows is given by $\phi_x : \mathbb{R} \to \mathbb{R}^3, t \mapsto \phi(t, x)$ and called the *orbit* of x. See Fig. 6.2.1 for some example orbits in two dimensions.

Given a critical point x of h the set of all points whose orbit ends in x, i.e., the set of all points that flow into x, is called the *stable manifold* of x. The collection of all stable manifolds forms a cell complex which is called *flow complex*. See Fig. 6.2.1 for examples of stable manifolds in two dimensions.

6.2.2 Medial Axis and Derived Concepts

Medial Axis.

The medial axis $M(S)$ of a closed subset $S \subset \mathbb{R}^3$ consists of all points in $\mathbb{R}^3 \backslash S$ having two or more nearest points on S. In a way the medial axis generalizes the concept of the Voronoi diagram of a point set. We have seen when discussing the empty ball property that the Voronoi faces of dimension k with $k = 0, \ldots, 2$ consist of all points equidistant from $3 - k + 1$ sample points.

Smooth surfaces S play a special role in reconstruction since for their reconstruction several guarantees can be provided under a certain sampling condition. This sampling condition is based on the medial axis of S. That is the reason why we here provide some more details on the structure of the medial axis of a smooth surface S.

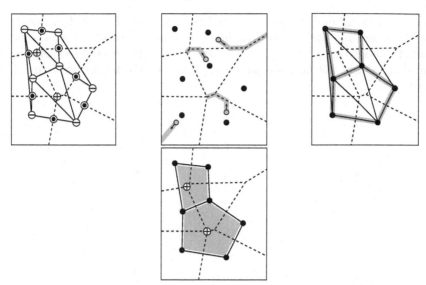

Fig. 6.11. From the left: 1) The local minima $\ominus$, saddle points $\odot$ and local maxima $\oplus$ of the distance function induced by the sample points (local minima). 2) Some orbits of the flow induced by the sample points. 3) The stable manifolds of the saddle points. 4) The stable manifolds of the local maxima

Structure of the Medial Axis of a Smooth Surface.

The medial axis of a smooth surface S shares another structural property with the Voronoi diagram of a finite point set, namely, it has a stratified structure. For the Voronoi diagram this structure means that a Voronoi facet is the common intersection of two Voronoi regions, a Voronoi edge is the common intersection of three Voronoi facets and a Voronoi vertex is the common intersection of four Voronoi edges. To precisely describe the stratified structure of $M(S)$ one needs the notion of contact between a sphere and the surface. Informally, the contact of a sphere at a point p of S tells how much the sphere and the surface agree at p. More precisely, an A_1 contact means that the tangent plane to the sphere and to S agree at p; an A_2 contact point has the property of an A_1 point with the additional property that the radius of the sphere is the inverse of a principal curvature of S at p; at last, an A_3 contact is like an A_2 contact with the additional property that the curvature involved is an extreme along the corresponding line of curvature. Focusing on the centers of the contact spheres rather than the contact points themselves, and denoting A_1^k a set of $k \geq 1$ simultaneous A_1 contacts between a sphere and the surface, the structure of the $M(S)$ is described by the following theorem [347, 181] which is illustrated by an example in Fig. 6.12.

Theorem 4. *The medial axis of a smooth surface S in $\mathbb{R}^3$ is a stratified variety containing sheets, curves and points. The sheets correspond to A_1^2 contacts,* the curves to A_1^3 and A_3 contacts, and the points to A_1^4 and A_3A_1 contacts. Moreover, one has the following incidences. At an A_1^4 point, six A_1^2 sheets and four A_1^3 curves meet. Along an A_1^3 curve, three A_1^2 sheets meet. A_3 curves bound A_1^2 sheets. At last, the point where an A_1^2 sheet vanishes is an A_3A_1 point.

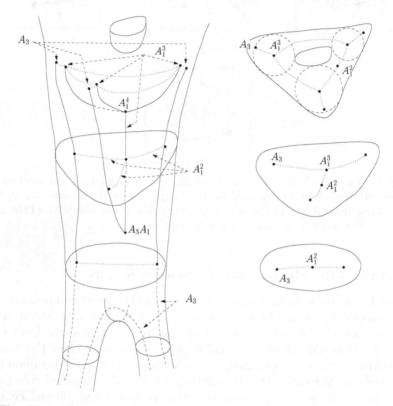

Fig. 6.12. The stratified structure of the medial axis of a smooth surface

Medial Axis Transform.

A concept closely related to the medial axis of a closed subset $S \subset \mathbb{R}^3$ is the *skeleton* of $\mathbb{R}^3 \backslash S$, which consists of the centers of maximal spheres included in $\mathbb{R}^3 \backslash S$. Here maximal is meant with respect to inclusion among spheres. For a smooth surface S the closure of the medial axis is actually equal to the skeleton of $\mathbb{R}^3 \backslash S$. The *medial axis transform* builds on the close relationship of the skeleton and the medial axis, namely, the medial axis transform is the collection of maximal empty balls centered at the medial axis of S. It can be shown that a smooth surface S can be recovered as the envelope of its medial axis transform.

Tubular Neighborhoods.

A natural tool involved in the analysis of several reconstruction algorithms is that of *tubular neighborhood* or *tube* of a surface S. As indicated by the name, a tube of a surface is a thickening of the surface such that within the volume of the thickening, the projection of a point x to the nearest point $\pi(x)$ on S remains well defined. Following our discussion of the medial axis, a surface can always be thickened provided the thickening avoids the medial axis. Moreover, it is easily checked that the projection onto S proceeds along the normal at the projection point. This property provides a way to *retract* the neighborhood onto the surface.

Local Feature Size.

The *local feature size* is a function lfs: $S \to \mathbb{R}$ on the surface S that assigns to each point in S its least distance to the medial axis of S. An immediate consequence of the triangle inequality is that the local feature size of a smooth surface is Lipschitz continuous with Lipschitz constant 1, see Fig. 6.13 for an illustration. The local feature size can be used to establish another quantitative connection of a surface and its medial axis [59] by using the following theorem.

Theorem 5. *Let B be a ball centered at $x \in \mathbb{R}^3$ with radius r that intersects the surface S. If this intersection is not a topological ball then B contains a point of the medial axis of S.*

From this theorem we can conclude that any ball centered at any point $p \in S$ whose radius is smaller then the local feature size lfs(p) at p intersects S in a topological disk.

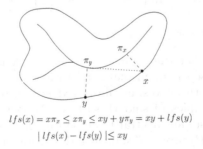

$$lfs(x) = x\pi_x \leq x\pi_y \leq xy + y\pi_y = xy + lfs(y)$$
$$|\,lfs(x) - lfs(y)\,| \leq xy$$

Fig. 6.13. The local feature size is 1-Lipschitz

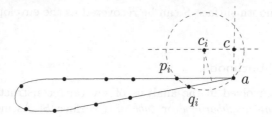

Fig. 6.14. For a non-smooth curve, some Voronoi centers may not converge to the medial axis

ε-sample.

Amenta and Bern [23, 22] introduced a non-uniform measure of sampling density using the local feature size. For $\varepsilon > 0$ a sample P of a surface S is called an ε-sample of S if every point x on S has a point of P in distance at most $\varepsilon\,\mathrm{lfs}(x)$.

We next provide three theorems that involve ε-samples. The first theorem is concerned with the topological equivalence of the restricted Delaunay triangulation $D_S(P)$ and a surface S for an ε-sample P. The second theorem is concerned with the convergence of Voronoi vertices of the Voronoi diagram of an ε-sample of a smooth surface S towards the medial axis $M(S)$ of S. The last theorem provides a good approximation of the normal of S at some sample point in an ε-sample P.

Amenta and Bern [22] stated the following theorem, which provides a topological guarantee for a value of ε less than ~ 0.3. The theorem is rigorously proven in [88]. In the context of surface reconstruction, this theorem should be put in perspective with respect to Theorem 1:

Theorem 6. *If P is an ε-sample of S such that ε satisfies*

$$\cos\left(\arcsin\left(\frac{2\varepsilon}{1-\varepsilon}\right) + \frac{\varepsilon}{1-3\varepsilon}\right) > \frac{\varepsilon}{1-\varepsilon}$$

then $V_S(P)$ has the topological ball property.

It can be shown that the Voronoi vertices of a dense sample of a planar smooth curve lie close to the medial axis of the curve. This result is false in general for non smooth curves, as illustrated in Fig. 6.14. It is also false in general for dense samples of smooth surfaces. In fact for almost any point $x \in \mathbb{R}^3 \backslash S$, there exists an arbitrarily dense sample P of S such that x is a Voronoi vertex of $V(P)$ provided some non-degeneracy holds. To see this grow a ball around x until it touches S. Now grow it a little bit further and put four sample points on the intersection of S with the boundary of the ball. Then x is shared by the Voronoi cells of the four points, i.e., it is a Voronoi vertex if the four points are in general position.

Fortunately, it was observed by Amenta and Bern [22] that the poles of the Voronoi diagram of a sample of a smooth surface converge to the medial axis.

Theorem 7. *Let P be an ε-sample of a smooth surface S. The poles of the Voronoi diagram $V(P)$ converge to the medial axis $M(S)$ of S as ε goes to zero.*

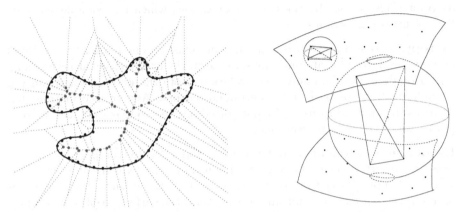

Fig. 6.15. In $2D$, all Voronoi vertices converge to the medial axis. In $3D$, some Voronoi vertices may be far from the medial axis but poles are guaranteed to converge to the medial axis

Finally, also the following theorem is due to Amenta and Bern [22]. It follows from Theorem 7.

Theorem 8. *Let P be an ε-sample of a smooth surface S. For any sample point $p \in P$ let p^+ be the positive pole of the Voronoi cell V_p. The angle between the normal of S at p and the vector $p - p^+$ if oriented properly can be bounded by $2\arcsin\left(\frac{\varepsilon}{1-\varepsilon}\right)$.*

6.2.3 Topological and Geometric Equivalences

To assess the quality of a reconstruction we need topological and geometric concepts. Our presentation of these concepts is informal, and the reader is referred to [208] for an exposition involving the apparatus of differential topology.

Topological Concepts.

Homeomorphy. Two surfaces are called *homeomorphic* if there is a *homeomorphism* between them. A *homeomorphism* is a continuous bijection of one surface onto the other, such that the inverse is also continuous. Two home-omorphic surfaces have the same properties regarding open and closed sets, and also neighborhoods. Note that homeomorphy is an equivalence relation. Surfaces that are embedded in $\mathbb{R}^3$ can be completely classified with respect to homeomorphy by their genus, i.e., the number of holes. For example the torus of revolution, i.e., a doughnut, and a "knotted" torus are homeomorphic since both have genus 1. This example shows that homeomorphy is a weak concept in the sense that it does not take the ambient space (here $\mathbb{R}^3$) into account. This is done by the concept of *isotopy* which for example accounts for the knottedness of a torus.

Isotopy. Two surfaces are *isotopic* if there exists a one-parameter family of embeddings into $\mathbb{R}^3$ that continuously deform the first surface into the second one. Isotopy is also an equivalence relation. Note the knotted torus can not be deformed continuously into the unknotted one. Any transformation that deforms the knotted torus in the unknotted one has to tear the torus at some point. Thus it cannot be continuous.

Homotopy equivalence. If we want to topologically compare the medial axes of two surfaces even the concept of homeomorphy (which can be extended to more complex faces than surfaces) seems too strong since the medial axis of a surface is a more complicated object than the surface itself. For comparing medial axes the concept of *homotopy equivalence* seems to be appropriate. Intuitively, the *homotopy type* of a space encodes its system of internal closed paths, regardless of size, shape and dimension. For example, an annulus has the homotopy type of a circle. Two topological spaces are homotopy equivalent if they have the same homotopy type. Homotopy equivalence is another equivalence relation on topological spaces.

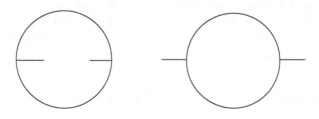

Fig. 6.16. Two homeomorphic topological spaces

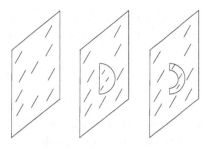

Fig. 6.17. The first two figures have the same homotopy type, but are not homeomorphic. The third one has a different homotopy type

Geometric Concepts.

Hausdorff distance. The *Hausdorff distance* is a measure for the distance of two subsets of some metric space. We are interested in the case where these subsets are surfaces or medial axes of surfaces in $\mathbb{R}^3$.

Given two closed subsets X, Y of $\mathbb{R}^3$, the one-sided Hausdorff distance $h(X, Y)$ is defined as $h(X, Y) = \max_{x \in X} \min_{y \in Y} \|x - y\|$. The one-sided Hausdorff distance is not a distance measure since in general it is not symmetric. Symmetrizing h yields the Hausdorff distance as $H(X, Y) = \max\{h(X, Y), h(Y, X)\}$. See also Fig. 6.18.

Intuitively, the Hausdorff distance is the smallest thickening such that the tubular neighborhood of X contains Y and the tubular neighborhood of Y contains X.

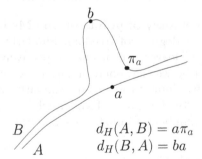

Fig. 6.18. The one-sided Hausdorff distance is not symmetric

Normals and tangent planes. Given two surfaces, their Hausdorff distance just takes into account their relative positions. In the context of surface reconstruction, we are also be interested in differential properties of the reconstructed surface with respect to the sampled surface. At the first order, such a measure is provided by the tangent planes (or the normals) to the surfaces, a quantity known to play a key role in the definition of metric properties of surfaces [259].

6.2.4 Exercises

The following exercises are meant to make sure the important notions have been understood. We also provide selected references to further investigate the problems addressed.

Exercise 1 (Sampling conditions and *reasonable* reconstructions). Consider the one-parameter family of curves that are indicated in Fig. 6.19. Let C_d be the curve corresponding to a value $d \in [0, \infty)$. For $d = 0$, C_d is a single curve without boundary, and for $d > 0$, C_d consists of two connected components with boundaries.

Plot the medial axis of $\mathbb{R}^2 \setminus C_d$ for $d = 0$ and $d > 0$. Assume we are given sample points equally spaced along C_d. Discuss what a *reasonable* reconstruction would be depending on d.

Exercise 2 (Sorting Gabriel edges). Consider a curve bounding a *stadium*, i.e., two line-segments joined by two half-circles. Also consider a dense sampling of the curve, that is the distance between to samples along the curve is much smaller than the radius of the circles and the distance between the two line-segments.

Plot the Delaunay triangulation of the samples, and report the Gabriel edges. Explain which of the Gabriel edges are relevant for the reconstruction of the curve, and which are not. Now, perturb *slightly* the boundary of the stadium, pick samples on the new curve, and answer the same questions.

Exercise 3 (Local geometry of points on the Medial Axis). Specifying a sphere in $\mathbb{R}^3$ leaves 4 degrees of freedom. Similarly, choosing a point on a surface or curve in $\mathbb{R}^3$ leaves two or one, respectively, degrees of freedom. Consider an A_1 contact between a sphere and a smooth surface. Such a contact fixes three degrees of freedom of the sphere. Similarly, an A_3 contact fixes all four degrees of freedom of the sphere. Explain why.

By matching the number of constraints and the number of degrees of freedom, show that

(i) A_1^2 points of the medial axis of a surface are expected to be on sheets of the medial axis, and

(ii) A_1^3 and A_3 points are expected on curves of the medial axis, and

(iii) A_1^4 and $A_3 A_1$ points are expected at isolated points of the medial axis.

Exercise 4 (Using poles). Let S be a closed surface which is in C^0 but not in C^1, that is, there are curves on S along which the normal to the surface is not continuous. Describe the geometry of the Voronoi cell of a sample point on such a curve. Which problem may arise in using the poles associated to such a sample point? Answer the same question assuming the surface S is smooth but has boundaries.

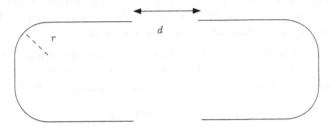

Fig. 6.19. A one parameter family of shapes

Further reading.

- Exercise 1. For curve reconstruction, see [24, 125, 127].
- Exercise 2. The separation of critical points of the distance functions to an ε-sample a smooth surface is studied in [122].
- Exercise 3. An intuitive presentation of the local geometry of the medial axis is provided in [181].
- Exercise 4. To learn more on the geometry of Voronoi cells, refer to [21] and [120].

6.3 Overview of the Algorithms

6.3.1 Tangent Plane Based Methods

We assume the sampled surface S is smooth, i.e., there exists a well defined tangent plane at each point of the surface. Since we only deal with surfaces of co-dimension 1, i.e., surfaces embedded in $\mathbb{R}^3$, approximating the tangent plane at some point of the surface is equivalent to approximating the normal at this point. Thus here tangent plane based methods include normal based methods. The first algorithm based on the tangent planes at the sample points is Boissonnat's [54] algorithm. Boissonnat's paper probably is the first reference to the surface reconstruction problem at all. Simply put, his algorithm reduces the reconstruction problem to the computation of local reconstructions in the tangent planes at the sample points. These local reconstructions have to be pasted together in the end.

Lower Dimensional Localized Delaunay Triangulation.

Gopi, Krishan and Silva [188] designed an algorithm that is very similar in nature to Boissonnat's early algorithm.

▷ **Bottom-line.** This algorithm has three major steps. First, normal and tangent plane approximation at the sample points. Second, selection of a neighborhood of sample points for each sample point. Third, projection of the neighborhood of a sample point on its tangent plane and computation of the Delaunay neighborhood of the sample point in its projected neighborhood. Sample points $p, q, r \in P$ form a triangle in the reconstruction if they all are mutually contained in their Delaunay neighborhoods.

▷ **Algorithm.** The normal and tangent plane approximation at the sample points is done using the eigenvectors of the covariance matrix of the k nearest neighbors of the sample point p. The covariance matrix is the 3×3 matrix

$$C = \sum_i (q_i - \hat{p})(q_i - \hat{p})^T$$

where the sum is taken over the k nearest neighbors of p in P and $\hat{p}$ is the centroid of these neighbors. The eigenvector corresponding to the smallest eigenvalue of the positive definite, symmetric matrix C is taken as the approximate normal at p. The remaining two eigenvectors span an approximate tangent plane at p. The approximate normals at the sample points are consistently oriented by propagating the orientation at some seed sample point along the edges of the Euclidean minimum spanning tree of P.

From the approximated normals at the sample points the directional normal variations and even the principal curvatures at the sample points can be estimated using again the k nearest neighbors of the sample points. The approximated principal curvatures $k_{\min}(p)$ and $k_{\max}(p)$ at a sample point p are used to locally approximate the unknown surface S by a height function

$$h(x, y) = \frac{1}{2}(k_{\min}(p)x^2\theta + k_{\max}(p)y^2\theta)$$

parameterized by Cartesian coordinates x and y over the approximated tangent plane at p. The neighborhood of a sample point p contains all sample points in P whose distance to p and whose height value are bounded by functions of $2k_{\max}(p)$ and $k_{\min}(p)$.

The neighbors of a sample point p are projected onto the approximated tangent plane at p by rotating the all the vectors from p to its neighbors into the tangent plane. In the tangent plane the Delaunay neighbors of p are determined by computing a two dimensional Delaunay triangulation of p and its projected neighbors. The output of the algorithm consists of all triangles with vertices in P whose vertices are mutual Delaunay neighbors.

▷ **Complexity.** The complexity of the algorithm was not theoretically analyzed. But is seems reasonable to assume that the local operations at each sample point can be done in constant time each, which would amount to a linear time complexity in total. But there are also the global operations of determining the neighborhoods of the sample points and of consistently orienting the normals. Although the latter operation is not really needed for the algorithm to work.

▷ **Guarantees.** The triangles output by the algorithm form surface homeo-morphic to S provided a curvature based, locally uniform sampling condition holds. This sampling condition also takes care of different parts of S coming close together.

▷ **Extensions.** Some heuristics are given to deal with samplings that do not fulfill the sampling condition. Especially the case of under-sampling is dealt with, though even the extensions do not make sure that the output is a topological surface in practice. Of course also oversampling can cause problems since at some points of the algorithm the k nearest neighbors of a sample point are used. This k-neighborhood can be spatially biased in the case of oversampling. This bias can invalidate the geometric approximations of normals, tangent planes and curvatures.

Greedy Algorithm.

The Greedy algorithm was introduced by Cohen-Steiner and Da in [94]. It incrementally grows a surface from a seed triangle guided by the intuition that the normals vary smoothly over the surface S.

▷ **Bottom-line.** The greedy algorithm incrementally reconstructs an oriented surface $\hat{S}$ by selecting triangles from the Delaunay triangulation $D(P)$ of P and stitching them to $\hat{S}$. The guideline for the selection is straightforward: the incremental construction should make easy decisions first by stitching triangles, which do not yield ambiguities.

▷ **Algorithm.** When extending the surface, a *valid* triangle is a triangle whose stitching does not create a *topological singularity*, and admissible glue opera-tions are of four types *extension, gluing, hole filing, ear filling* —See Fig. 6.20.
Let e be a boundary edge of $\hat{S}$. Out of all the valid triangles t incident to e, one of them is chosen as candidate for the surface extension. To define the candidate, denote r_t the radius of the smallest empty ball circumscribing a triangle t. Among all the triangles whose dihedral angle β_t across e is larger than some threshold α_s (an angle near π), the candidate is the triangle with least r_t. Since a greedy approach is used, one needs to rate the different candidates. To do so, each triangle is assigned a score, which is $1/r_t$ if β_t is larger than a threshold β, and $-\beta_t$ otherwise.
The threshold α_s prevents from considering facets whose stitching would cause a *fold-over*, i.e., a large dihedral angle. Notice also the scoring strategy favors small triangles provided the dihedral angle is less than the threshold β.

Equipped with these notions, the algorithm consists of the initialization and extension stages. First, the triangle with least circumradius is chosen as a seed, and its edges are pushed into a priority queue Q. Next, the algorithm iterates over Q and processes triangles in order of decreasing confidence. Once a candidate triangle has been popped, a check is performed to see whether a possible extension is possible. This might not be the case anymore due to potential changes in the environment of the triangle. In any case, the priority queue and the surface are updated.

By construction, the output of the Greedy algorithm is a triangulated and oriented surface, which may not interpolate all the samples since the used thresholds might leave some sample points without incident triangle.

▷ **Complexity.** The algorithm uses the Delaunay triangulation of the samples together with a priority queue. Both data structures determine the complexity of the algorithm.

▷ **Guarantees.** No guarantee can be provided on the quality of the reconstruction due to the difficulty of handling clusters of flat tetrahedra. As the surface extension is incremental, such clusters can be approached in various manners from different directions, thus making it impossible to close the surface.

▷ **Extensions.** Two heuristics are used to accommodate boundaries as well as sharp features. For boundaries, a candidate triangle is discarded as soon as the radius of its empty ball is significantly larger than that of the triangle it would be stitched to. Sharp edges are detected and removed through the removal of samples, which are not part of the output surface.

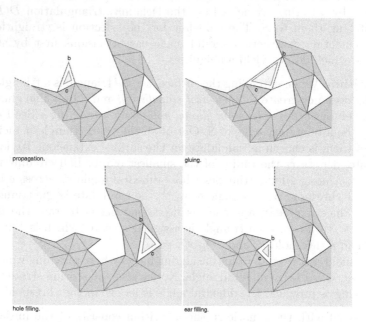

Fig. 6.20. Glue operations of the Greegy algorithm [94]

6.3.2 Restricted Delaunay Based Methods

Since all Delaunay based surface reconstruction algorithms filter out a subset of the Delaunay triangulation $D(P)$ of the sampling P it seems natural and very appealing to choose just these simplices from $D(P)$ that are restricted to some subset of $\mathbb{R}^3$ that is a good approximation of the unknown surface S and can be computed efficiently from P. This paradigm is motivated by the fact that if we could directly compute the Delaunay triangulation $D_S(P)$ of P restricted to S we would be done since due to the theorems of Edelsbrunner and Shah (Theorem 1) and Amenta and Bern (Theorem 6), respectively, for sufficiently dense ε-samples $D_S(P)$ is homeomorphic to S.

Crust.

The Crust algorithm was designed by Bern and Amenta [22] who also were the first to provide detailed guarantees for the reconstruction provided some ε-sampling condition is fulfilled.

▷ **Bottom-line.** The Crust is based on the Delaunay triangulation $D(P \cup Q)$ of P and the set Q of poles of the Voronoi diagram $V(P)$. Let V be the union of the Voronoi cells of the points in P in the Voronoi diagram $V(P \cup Q)$ of $P \cup Q$. In a nutshell the Crust is the Delaunay triangulation of P restricted to V. The rationale behind this approach is that $\mathbb{R}^3 \setminus V$ should cover the medial axis $M(S)$ of the surface S. Thus restricting the Delaunay triangulation of P to V should remove all simplices from $D(P)$ that cross the medial axis $M(S)$ of S. On the other hand V should provide a thickened version of S and thus the restricted Delaunay triangulation $D_V(P)$ should contain all simplices from $D_S(P)$. See Fig. 6.21 for an illustration in two dimensions.

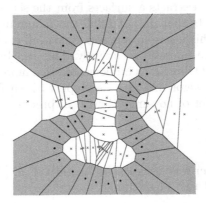

Fig. 6.21. The set V is a thickening of the surface S that avoids the medial axis of S

▷ **Algorithm.** The Crust algorithm proceeds as follows: After the set Q of poles is computed from $V(P)$ the Delaunay triangulation $D(P \cup Q)$ of the union of P and Q is constructed. From this Delaunay triangulation all triangles that have all three vertices in P are retained. The retained triangles are called *candidate triangles*. The candidate triangles not necessarily form a surface, but they contain at least one if P is sufficiently dense. One of these surfaces is extracted from the candidate triangles in the final step of the Crust algorithm, which walks along the inside or outside of the candidate triangle set and reports all triangles visited. While walking care has to be taken of dangling triangles. These are triangles that have an edge that is incident only to this triangle in the set of candidate triangles. The dangling triangles are recursively removed from the set of candidate triangles before the actual walk starts.

▷ **Guarantees.** It can be shown that for a closed smooth surface S and a sufficiently dense ε-sampling P of S the set of candidate triangles contains all the triangles of the Delaunay triangulation $D_S(P)$ of P restricted to S. This implies that the set of candidate triangles contains at least one surface, namely the restricted Delaunay triangulation $D_S(P)$. Thus the manifold extraction step of the Crust algorithm, i.e., the removal of the dangling triangles from the set of candidate triangles and the actual walking are safe in the sense that the reported triangles actually form a surface.

▷ **Complexity.** The worst case running time and memory consumption of the Crust algorithm are $\Theta(m^2)$, where m is the size of $P \cup Q$. Note, that two Delaunay triangulations have to be computed, one from n and the other from m points. The latter computation determines the asymptotic complexity of the algorithm.

▷ **Extensions.** As described above the Crust algorithm only works for smooth closed surfaces and sufficiently dense ε-samples. Main problem is the final step of the algorithm that extracts a surfaces from the set of candidate triangles. For practical data sets that do not fulfill the requirements of the algorithm it can happen that this last step removes almost all candidate triangles since dangling triangles are removed recursively. This can be prevented if the removal of the dangling triangles is implemented in a more conservative fashion. This done the Crust algorithm can also cope with surfaces with boundaries and a "certain amount of non-smoothness" in practice.

Cocone.

The Cocone algorithm was designed by Amenta et al. [25] as a successor and improvement of the Crust algorithm.

▷ **Bottom-line.** The Cocone algorithm builds as the Crust algorithm on the idea of approximating the Delaunay triangulation $D_S(P)$ of P restricted to S by computing a subset $C \subset \mathbb{R}^3$ from P, which is a thickened version of S such that the Delaunay triangulation $D_C(P)$ of P restricted to C can computed.

The subset C is defined as follows: for every sample point $p \in P$ approximate the normal of S at p using the pole of the Voronoi cell V_p in $V(P)$, see Theorem 8. The co-cone at p is now defined as the intersection of V_p with the complement of a double cone with apex p and fixed opening angle around the approximate normal at p, see Fig. 6.22 for a two dimensional example. The set C is the union of all such co-cones. Note, that C can be computed just from P. Theorem 7 implies that the local thickening of S using co-cones is small compared to the local feature size and thus C is a reasonable approximation of S.

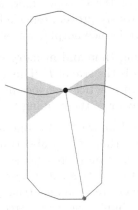

Fig. 6.22. The co-cone of a sample on a curve together with the Voronoi cell of the sample point and its pole

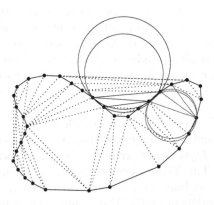

Fig. 6.23. Balls of opposite (same) color intersect shallowly (deeply)

▷ **Algorithm.** As the Crust algorithm the Cocone algorithm first computes a subset of candidate triangles from the triangles in $D(P)$. A triangle t in $D(P)$

is a candidate triangle if its dual Voronoi edge e intersects any of the co-cones. This intersection test boils down to go through the vertices of t and check if e intersects the co-cone of one of the vertices, which is checking the angles the of vectors from a vertex v incident to t to the endpoints of e with the approximate normal at v. As for the Crust algorithm the candidate triangles form not necessarily a surface, but they contain at least one if P is sufficiently dense. Finally, the last step of the Crust algorithm is used to extract one of these surfaces.

▷ **Guarantees.** The same guarantees as for the Crust algorithm hold under the same conditions. But for the Cocone algorithm it is the first time that these guarantees were rigorously proven, especially the fact that the surface S and its reconstruction are homeomorphic for dense enough sampling.

▷ **Complexity.** The running time and memory consumption of the Cocone algorithm is $\Theta(n^2)$ where n is the size of P. This complexity is determined by the computation of $D(P)$. In practice one does not observe the quadratic but a slightly super linear behavior of the running time.

▷ **Extensions.** As described above the Cocone algorithm has the same restrictions as the Crust algorithm, which also can be mitigated in the same way. But for the Cocone algorithm there exist several extensions.

In the vein of large datasets, two extensions were designed. In order to avoid the calculation of the entire Delaunay triangulation, Dey et al. propose to split the point cloud into chunks using an octree [121]. Cocone is then called on each chunk, and the matching of surface pieces from adjacent octree cells is obtained by duplicating selected sample points at the cells' boundaries.

Alternatively, the complexity of the algorithm was reduced by Funke and Ramos [174] to $\Theta(n \log n)$ by avoiding the computation of $D(P)$. They use a data-structure called *well separated pair decomposition* that allows to compute efficiently nearest neighbors of any sample point $p \in P$ in all spatial directions. These neighbors approximate the Voronoi neighbors of p, i.e., the sample points connected to p with an edge in $D(P)$. From these neighbors the normal of S at p can be approximated, and candidate triangles incident to p and two of the approximate Voronoi neighbors can be computed as in the Cocone algorithm.

The output of the Cocone algorithm with robust manifold extraction step is a surface with boundary. This surface might contain small unpleasant holes. An extension called *Tight Cocone* removes these unpleasant holes provided the surface S is closed. The Tight Cocone algorithm falls in the class of inside / outside labeling algorithms, i.e., it removes tetrahedra from the outside of the Delaunay triangulation $D(P)$. The stopping criterion for the tetrahedron removal is based on the triangles computed by the Cocone algorithm. The latter triangles have to be contained in the reconstruction. The Tight Cocone algorithm was designed by Dey and Goswami [123] and got its name from the fact that its output is a watertight surface, i.e., a surface that bounds a solid, which might be pinched together at some points.

The Tight Cocone algorithm was even further extended to deal with a noisy sampling P. This extension called *Robust Cocone* was also developed by Dey and Goswami [124]. The Robust Cocone algorithm employs a fact that was first used in the Power Crust algorithm of Amenta and Choi [27, 26], namely, the balls circumscribing adjacent Delaunay tetrahedra intersect deeply if both tetrahedra belong to the same component, i.e., either outside or inside. The tetrahedra only have a shallow intersection if they belong to different components. Dey and Goswami observed that this might not be true for tetrahedra in the noisy regions around the surface S. But these tetrahedra have comparatively small circumscribing balls and thus can be detected. In the Robust Cocone algorithm only the sample points on the boundary of the noise layer either facing the outside or the inside are retained. Finally the Tight Cocone algorithm is run on the retained subset of the sample points.

6.3.3 Inside / Outside Labeling

The common ancestor of all algorithms that are based on an approximate inside / outside labeling of the Delaunay tetrahedra with respect to the unknown closed surface S is the algorithm by Boissonnat [54], from the same paper that introduces the first tangent plane based surface reconstruction algorithm. In this seminal paper Boissonnat describes a sculpturing technique, i.e., removing tetrahedron from the Delaunay triangulation from the outside in order to sculpture a solid whose boundary is the reconstruction. Boissonnat uses a priority queue and weights on the tetrahedra still present in the shape to decide which tetrahedron to remove next. The tetrahedron removal is controlled by topological constraints, i.e., by prescribing the genus of the surface of the resulting solid.

As depicted on 6.24, surfaces with boundary do not define an inside and an outside, so that the methods described in this section may not work for such surfaces.

Power Crust.

The Power Crust algorithm was introduced by Amenta et al. in [27, 26].

▷ **Bottom-line.** To put it briefly, the Power Crust algorithm is an approximate medial axis transform built from empty balls rather than maximal balls, which is asymptotically licit by Theorem 7.

▷ **Algorithm.** Consider a solid T bounded by a closed surface S. By the medial axis transform, the solid T can be expressed as an infinite union of spheres centered on the inner medial axis, i.e., the medial axis of $\mathbb{R}^3 \backslash S$ restricted to T. But if one has a dense point set on S and since the inner poles, i.e., the poles contained in T, converge to the inner medial axis, the surface S can certainly be approximated by the boundary of the union of balls centered

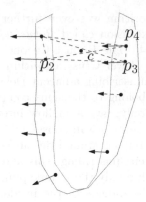

Fig. 6.24. Pole c lies near the medial axis, but cannot be tagged as inside or outside
.

on the inner poles, their radii being the distance to the corresponding sample points. Similarly, the complement of the solid can be approximated by a union of balls centered on the outer poles, i.e., the complement of the set of inner poles. Therefore and since we have two collections of balls (inner and outer), the surface S should also be approximated by the interface between the inner and outer cells of the power diagram defined by the union of inner and outer polar balls. This interface that consists of facets in a power diagram is called the power crust. At last, the dual of the complement of the power crust in the power diagram is called the power shape. The power shape is a subset of the regular triangulation and is expected to approximate the medial axis of $\mathbb{R}^3 \backslash S$.

From an algorithmic standpoint, the critical part consists of tagging the poles as inner or outer. Assuming the samples are enclosed within a bounding box, and starting from the poles of the sample points located on the bounding box, a greedy algorithm is developed. The strategy consists of iterating on the poles using a priority queue, which depends on the angle defined by the two poles of a sample, as well as the intersection angle between polar balls. The algorithm is provably good for dense enough samples, yet may fail in case of under-sampling, at sharp edges or in the presence of noise.

The surface reported is a watertight, i.e., a closed, piecewise linear surface consisting of power facets, yet possibly pinched at some points. All the sample points are interpolated, yet the result contains additional vertices.

▷ **Complexity.** Three data structures are used: the Delaunay triangulation of the original samples, the power diagram of the poles, and the priority queue of the tagging algorithm.

▷ **Guarantees.** It is first proved that an inner and an outer ball intersect shallowly, see Fig. 6.23 for an illustration. This result was already present in [30] for the two-dimensional case.

Denoting U_I (U_O) the boundary of the union of inner (outer) balls, the following properties are proved. The one-sided Hausdorff distance between U_I and S is small, and so is the distance between U_O and S, as well as between the power-crust and S. The angle between the normal at a point of U_I or U_O (where it is defined, i.e., on the interior of the spherical caps of U_I and U_O) and the normal at the nearest point on S is also small. This geometric property can be used to show that the projection of U_I (or U_O) to the nearest point on S defines a homeomorphism. Using a similar construction, it is also proved that the power crust and the surface are homeomorphic. At last, the power shape is homotopy equivalent to the complement of the surface. Notice that providing more accurate guarantees on the power shape is significantly more difficult due to the intricate structure of the medial axis, and also due to the fact that the power shape may contain flat tetrahedra.

▷ **Extensions.** Noise and under-sampling are detected by analyzing the roundness of Voronoi cells. Badly shaped cells and the corresponding poles are discarded. Discarding both poles of a sample that fails the *skinniness* test allows to accommodate sharp edges as the intersection of two facets of the power diagram. At last, large facets of the power crust witnessed by an inner and an outer ball intersecting deeply can be removed, thus leaving a surface with boundary.

Natural Neighbors.

Natural neighbors were first used for reconstruction in [56, 58].

▷ **Bottom-line.** It is well known by a theorem of Whitney that any smooth surface occurs as the solution set $f^{-1}(0)$ for some smooth function $f : \mathbb{R}^3 \mapsto \mathbb{R}$. The Natural Neighbors reconstruction method is based on the definition of such a function based on two ingredients: an estimate of the tangent plane based on the poles, and the natural coordinates defined with respect to the Voronoi diagram of the samples.

▷ **Algorithm.** For the sake of clarity, first assume that each sample point is given with its normal vector n_i. Denoting $NNs(x)$ the natural neighbors of a point x and $\lambda_i(x)$ the natural coordinate of x with respect to p_i, the method is based upon the following implicit function $f : \mathbb{R}^3 \mapsto \mathbb{R}$:

$$f(x) = \sum_{p_i \in NNs(x)} \lambda_i(x) \langle p_i x, n_i \rangle.$$

The inner product $\langle p_i x, n_i \rangle$ measures the signed distance from x to the tangent plane at p_i, and the function f therefore averages the signed distances to the tangent planes of the natural neighbors of the point x. A direct consequence of the properties of natural coordinates is that the function f interpolates the point cloud, so that the reconstructed surface $\hat{S}$ is naturally defined as

$f^{-1}(0)$. It has been conjectured that $\hat{S}$ is a smooth surface, yet it remains to show that 0 is a regular value of f.

Since the natural coordinates have an involved expression, a triangulated approximation of $\hat{S}$ can be obtained as a subset of the Delaunay triangulation of the samples, namely as the restricted Delaunay triangulation $D_{f^{-1}(0)}(P)$. This triangulation is easily computed as follows. Denote by $c_1 c_2$ the dual Voronoi edge of a triangle t. If $f(c_1)f(c_2) < 0$, then triangle t belongs to the restricted Delaunay triangulation. Such triangles are also called bipolar in this case.

If the normals are unknown, they can be estimated using the poles. Orienting the normals is also possible using a greedy algorithm similar to the one used in [27] for the Power-Crust algorithm.

▷ **Complexity.** Apart from the Delaunay triangulation, a priority queue is required to sign the poles and orient the normals if the normals are not provided.

▷ **Guarantees.** It can be shown that the Hausdorff distance between $f^{-1}(0)$ and S tends to zero when the sampling density goes to infinity. There is no guarantee on the topological coherence between the bipolar facets that make up the reconstruction since non manifold edges may be encountered in case of boundaries, noise, or under-sampling. However, if a (strictly positive) lower bound on the local feature size is known, the mesh obtained can be refined using Delaunay refinement –see Chapter 5, until a mesh ambient isotopic to the original surface is obtained.

Topologically guided methods for the inside / outside labeling make use of the distance to the closest sample point. In Sect. 6.2 we have already summarized some properties of this function. In topological guided methods one wants to exploit the critical points of the distance function and their stable manifolds for reconstruction.

▷ **Extensions.** An interesting extension of the above scheme, geared towards large datasets, is developed in [55]. Starting with a subset of the point cloud to be processed, and observing that points located near the level set $f^{-1}(0)$ are redundant, the algorithm iteratively inserts points located far away from this zero level set. Upon termination, a mesh featuring a subset of the whole point cloud is returned, the points discarded being within a user specified tolerance from this mesh.

Wrap.

The Wrap algorithm was designed by Edelsbrunner [132].

▷ **Bottom-line.** The Wrap algorithm is based on the concepts of flow and stable manifolds. But instead of building directly on the flow induced by the sample points a *flow relation* is defined on the set of simplices of the Delaunay triangulation $D(P)$ of the sample points P. The critical points are defined exactly as for the distance function induced by P, i.e., as the intersection points of Delaunay- and their dual Voronoi objects. But their stable manifolds are now approximated by sub-complexes of $D(P)$. The reconstruction produced by the Wrap algorithm is the boundary of the union of stable manifolds of a subset of the maxima of the flow relation. As the boundary of a solid it is a surface.

▷ **More details.** The flow relation $\triangleleft \subset D \times D$ on the set D of Delaunay simplices is defined as follows: $\tau \triangleleft \nu \triangleleft \sigma$ if ν is a face of τ and σ and there exists a point x in the interior of ν such that there is an orbit ϕ_y that is passing from the interior of τ through x to σ. τ is called a *predecessor* and σ is called a *successor* of ν. The relation $\triangleleft$ is acyclic. A sink is a Delaunay tetrahedron that contains a maximum of the flow, i.e., its dual Voronoi vertex. The set of sinks is augmented by an artificial sink at infinity. The flow relation can be used to define the *ancestor* and *conservative ancestor* sets of a set B of sinks. These sets consist of Delaunay simplices that are linked to a tetrahedron in B by a chain in the flow relation. The wrapping surface of the point set P is the boundary of the union of the ancestor sets of all finite sinks or equivalently it consists of the complement in the Delaunay triangulation $D(P)$ of the conservative ancestor set of the sink at infinity. The wrapping surface is unique. It can be computed by collapsing certain simplices. The collapse operation removes simplices that are the unique co-face of one of their faces. A collapse does not change the homotopy type of the complex since it can be seen as a deformation retraction, which always keeps the homotopy type. Thus the complex bounded by the wrapping surface is homotopy equivalent to a point, i.e., the wrapping surface cannot be a torus for example. The latter disadvantage is bypassed by allowing a simplex removing operation that changes the homotopy type. The deletion is similar to the original definition of the wrapping surface. Instead of removing from the Delaunay triangulation $D(P)$ only the conservative ancestor set of the sink at infinity, the conservative ancestor sets of a set of sinks is removed. Consequently the wrapping surface is now the boundary of the union of the ancestor sets of the remaining sinks. The latter union need not be homotopy equivalent to a point, i.e., the wrapping surface can be topologically more complicated.

▷ **Guarantees.** No reconstruction guarantees are given besides the fact that the wrapping surface is always the boundary of a solid.

▷ **Complexity.** The running time of the Wrap algorithm is dominated by the time needed to compute the Delaunay triangulation $D(P)$, i.e., it is $\Theta(n^2)$ where n is the size of P.

▷ **Extensions.** No extensions to the Wrap algorithm are known.

Flow Complex.

The flow complex is very much related to the Wrap algorithm.

▷ **Bottom-line.** It was observed by Giesen and John [183] that a reconstruction similar to the one obtained by the Wrap algorithm can be derived from the flow complex. The flow complex has a recursive structure, i.e., the stable manifolds of a critical point is bounded by stable manifolds of critical points of lower index. The reconstruction is the boundary of the union of all stable manifolds of the local maxima of the induced distance function. The stable manifold of an index 2 saddle point can be in the boundary of either one or two stable manifolds of local maxima. As in the Wrap algorithm one can recursively use stable manifolds of index 2 saddles, which are in the boundary of only one stable manifold of a local maximum to push the reconstruction further to interior of the complex. The pushing is guided by considering the difference in value of the height function at the local maximum and the index 2 saddle point.

▷ **Algorithm.** The flow complex is not a subcomplex of the Delaunay triangulation $D(P)$ though $D(P)$ can be used to compute the flow complex. This computation is quite involved and makes use of the recursive structure of the stable manifolds. Here we want to refer the reader to [184] for a detailed description.

▷ **Guarantees.** No reconstruction guarantees are given besides the fact that the wrapping surface always is the boundary of a solid.

▷ **Complexity.** The combinatorial and algorithmic complexities of the flow complex are not known yet. The reconstruction has roughly three times as many triangles as other Delaunay based reconstruction algorithms.

▷ **Extensions.** The reconstruction algorithm of [183] is guided by the stable manifolds of index 2 saddles. An extension of this strategy is developed in [122], where it is first observed that critical points of the distance function to the samples are either located near the surface or near the medial axis. This distinction is algorithmic since one can actually classify critical points as medial axis or surface critical points using an angle-based criterion involving the poles. Based on this classification, it is sufficient to union all the stable manifolds of medial axis critical points, and report the boundary of this union.

Convection Algorithm.

The convection algorithm was designed by Chaine [82].

▷ **Bottom-line.** The Convection algorithm is the geometric implementation of the convection model introduced by Zhao, Osher and Fedkiw [350]. In this model it is proposed to use the surface that minimizes the following energy functional as the reconstruction of S from P:

$$E(S') = \left(\int_{x \in S'} h^p(x)\, \mathrm{d}x \right)^{1/p}, 1 \le p \le \infty,$$

where the integral is taken over a closed surface S' and h is the distance function induced by the sampling P of the surface S. Zhao et al. propose an evolution equation to construct the surface that minimizes the energy functional by deforming a good initial enclosing approximation of the surface. The evolution follows the gradient descent of the energy functional. Every point x of the surface S' evolves towards the interior of the surface along the normal direction $\mathbf{n}$ of S' at x with speed proportional to $-\nabla h(x) \cdot \mathbf{n} + t(x)$, where $t(x)$ is the surface tension of S' at x. In order to compute an initial approximation of S Zhao et al. change the evolution in the sense the velocity field at each point $x \in S'$ is replaced by $-\nabla h(x)$. Chaine discretizes the convection approach by introducing the concept of *pseudo-surfaces*. The facets of a pseudo-surface are Delaunay triangles that have the oriented Gabriel property, where the triangles are oriented such that their normal points inside the bounded component enclosed by the pseudo-surface. A pseudo surface can be pinched together along some of its subsets. To define the oriented Gabriel property let t be a Delaunay triangle with oriented normal $\mathbf{n}$. Let s be the half-sphere of the minimum enclosing sphere of t that is contained in the half-space bounded by the affine hull of t and pointed into by $\mathbf{n}$. The triangle t has the oriented Gabriel property if the half sphere s does not contain any point from P in its interior. Chaine proves the following theorem.

Theorem 9. *Given a closed surface S' enclosing the point set P then S' evolves under the convection $-\nabla h(x)$ to a set of closed, piecewise linear pseudo-surfaces.*

Furthermore, she observed that Theorem 9 can be turned into an algorithm based on the Delaunay triangulation $D(P)$ of the sample points P. In this algorithm the evolving pseudo surface is initialized with the boundary of the convex hull of P and all the triangles on this boundary are oriented to point inside the convex hull. The pseudo surface evolves by pushing into Delaunay tetrahedra. The pushing operations are determined by the vector field $-\nabla h(x)$ and topological constraints.

▷ **Algorithm.** The algorithm basically works as follows: as long as there is a facet f in the evolving, oriented, pseudo surface S' that does not have the oriented Gabriel property do the following. If the facet f with the inverse orientation also belongs to S' then remove f from S'. Otherwise replace f by the three Delaunay facets incident to the Delaunay tetrahedron, which is incident to f and on the positive side of f with respect to the orientation of f. Orient the three new facets in S' properly such that their normals point into the interior of the evolving pseudo surface.

▷ **Guarantees.** No reconstruction guarantees are given.

▷ **Complexity.** The running time of the Convection algorithm is dominated by the time needed to compute the Delaunay triangulation $D(P)$, i.e., it is $\Theta(n^2)$ where n is the size of P.

▷ **Extensions.** One modification of the Convection algorithm is to keep an oriented facet f in S' if the same facet with the inverse orientation is also in S'. In doing so the convection algorithm can also reconstruct surfaces with boundaries.

Sometimes the Convection algorithm stops too early, i.e., one would like to push the evolving surface even further. A heuristic to do so is provided.

Another extension, geared towards large datasets, is presented in [18]. Borrowing the *coarse-to-fine* strategy from [57], the method first extracts a triangulated surface corresponding to a subset of the point cloud. This surface can be further refined by locally updating the Delaunay triangulation, and updating the reconstruction accordingly —a local process which does not require running the convection algorithm from scratch.

▷ **Comments.** The Convection algorithm is dual to the Wrap algorithm (and the Flow complex) in the sense that the direction of "flow" is reversed. The Wrap algorithm retains the part of the Delaunay triangulation that does not "flow" to infinity whereas the Convection algorithm lets the convex hull of P "flow" towards the shape.

6.3.4 Empty Balls Methods

A triangle reported in a reconstructed surface should be local in some sense. One way to specify locality is to use the empty ball property.

Ball Pivoting Algorithm.

Bernardini et al. [52]designed the ball pivoting algorithm to compute a surface subset of an α-shape of a sampling P in linear time and space.

▷ **Bottom-line and algorithm.** Like in the definition of α-shapes a triangle pqr with vertices $p, q, r \in P$ forms a triangle in the reconstruction if there is a ball of radius α that contains p, q and r in its boundary and no point from P in its interior. Starting with a seed triangle in the α-shape the ball pivoting algorithm pivots a ball around an edge of this seed triangle, i.e., it revolves around an edge while keeping the edge's endpoints on its boundary, until it touches another point from P, forming another triangle in the α-shape. This process continues until all reachable edges have been processed.

▷ **Guarantees.** No guarantees are given.

▷ **Complexity.** Time and space complexity of the ball pivoting algorithm are linear under some assumptions, i.e., it is asymptotically faster than computing the Delaunay triangulation $D(P)$ of P.

▷ **Extensions.** If not all points of P have been processed by the algorithm then one can restart it with a new seed triangle until all points in P have been considered.

To accommodate non-uniform sampling the pivoting process can be repeated with a larger value for α.

Conformal α-shapes.

Conformal α-shapes were introduced in [81] to circumvent the uniformity limitations inherent to α-shapes.

▷ **Bottom-line.** In the context of surface reconstruction, the size of the smallest empty ball associated with a simplex does not have an absolute meaning: a ball of a given radius may be associated to neighbors on the surface at one location, but may connect points across the surface elsewhere. To get around this difficulty, conformal α-shapes re-scale the size of balls by taking into account the information provided by the poles.

▷ **Algorithm.** Consider the Delaunay triangulation of the sample points, and the associated α-complex. The α values associated to the simplices incident to a sample p span the range $[0, \alpha_p^+]$, with α_p^+ the distance from p to its pole. In contrast to α-shapes, where α spans the real numbers, consider now a parameter $\hat{\alpha} \in [0, 1]$. Placing around each sample point p a ball of radius $\alpha_p(\hat{\alpha}) = \alpha_p^+ \hat{\alpha}$, denote by $C_p^{\hat{\alpha}}$ the intersection of that ball with the Voronoi cell of p. The *conformal alpha shape* is the Delaunay triangulation of the sample points P restricted to $\cup_{p \in P} C_p^{\hat{\alpha}}$. Notice that the radii of the balls get scaled by a factor equal to the distance to the poles instead of using the same radius α for all balls.

▷ **Guarantees.** For an ε-sample of a surface, it can be shown that the conformal alpha shape contains the restricted Delaunay triangulation as soon as $\hat{\alpha} \geq \eta$, with $\eta = \varepsilon/(1 - \varepsilon)$. It can also be shown that the conformal alpha shape does not contain *large* simplices for *small* values of $\hat{\alpha}$. Note that such guarantees cannot be provided for ordinary α-shapes and are not known to hold for any method based on weighted α-shapes (an extension of α-shapes to weighted points, especially balls).

▷ **Complexity.** Commuting the conformal α-shapes requires to compute the moment of appearance of the simplices as a function of $\hat{\alpha}$. This is straightforward from the α-values. Thus, the time complexity is the same as for ordinary α-shapes.

Regular Interpolant.

The regular interpolant was introduced in [281] by Petitjean and Boyer. Their work stresses the importance of Gabriel triangles for surface reconstruction, an observation also raised in [34].

▷ **Bottom-line.** The framework of ε-samples might not be the definitive set-up for solving practical problems. To bypass this difficulty, Petitjean and Boyer address the issue of finding an interpolant encoding the properties of the sampling P rather than those of an hypothetical smooth surface S. To see how, we first introduce the relevant notions.

An interpolant O in $\mathbb{R}^3$ is a 2-simplicial complex having P as vertex set. The interpolant is closed if each simplex bounds two distinct connected components of the ambient space.

Given a sample point $p \in P$, its *granularity* $g(p)$ is defined as the radius of the largest ball circumscribing a triangle incident to p.

Now, given an interpolant, its associated *discrete medial axis* is the Voronoi diagram from which one removes the Voronoi cells dual to simplices of the interpolant. Notice that the process leaves Voronoi cells of dimensions from two to zero, and in particular all the Voronoi vertices.

The *discrete local feature size* or local thickness $t(p)$ at a sample point p is its least distance to the discrete medial axis with the convention $t(p) = 0$ if p is on the boundary of a connected component of $\mathbb{R}^3 \backslash O$, which does not contain any piece of the discrete medial axis.

Equipped with these notions, an interpolant is called *regular* is $g(p) < t(p)$ for all sample points p. Getting back to the point cloud, P is said to be regular if it admits at least one regular interpolant. These notions are depicted in Fig. 6.25.

Regular interpolants do not exist in general due to the presence of slivers, see Fig. 6.3. When such a tetrahedron is located near its equatorial plane, the granularity is indeed larger than the distance to the discrete medial axis, which contains at least the circumcenter of the tetrahedron.

▷ **Algorithm.** For a regular interpolant, the triangles contributing to the interpolant are the Gabriel triangles minimizing the granularity at the vertices. They can be retrieved in an incremental fashion.

▷ **Guarantees.** No guarantees are given.

▷ **Complexity.** The Gabriel property must be checked for triangles incident to an edge. Using the Delaunay triangulation this can be done in time $O(n^2)$.

▷ **Extensions.** For non-regular point-sets, triangles are first selected so as to minimize the granularity, and are further decimated if they are not Gabriel. The interpolant built in this way is called a *minimal interpolant*. It is not manifold in general. A manifold extraction step can be applied, which consists of reporting groups of simplices that are simply connected, i.e., contractible to a point.

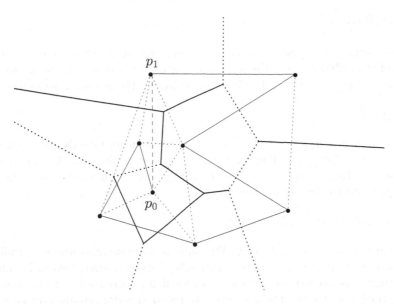

Fig. 6.25. A discrete version of the medial axis. Solid segments: interpolant; dotted segments: belonging to Delaunay but not the interpolant. The medial axis consists of the Voronoi segments dual to Delaunay edges, which do not contribute to the interpolant. The Voronoi edges not belonging to the medial axis are dotted too

6.4 Evaluating Surface Reconstruction Algorithms

Evaluating surface reconstruction algorithms is a difficult task. Some of the algorithms presented in this chapter come with theoretical guarantees under certain conditions. But if these conditions are not met, their behavior is not specified. Thus it is an interesting question how reconstruction algorithms perform on "real data". In order to assess the performance of different algorithms on real data two surface reconstruction challenges have been organized. One challenge was organized within the *Effective Computational Geometry* project, a project funded by the European Union. The other challenge was organized within a DIMACS Workshop. For both challenges several data sets featuring the following difficulties were selected: undersampling, sharp features, thin parts, boundaries, high genus, noise.

The reader is referred to `www-sop.inria.fr/prisme/manifestations/` `ECG02/SurfReconsTestbed.html`   and   `www.cse.ohio-state.edu/` `dimacs-sr-challenge` where the data sets used in the challenges are available. Some of these models are presented on Fig. 6.28, 6.29, 6.30, 6.27.

6.5 Software

In this section, we provide information on the availability of implementations of the different algorithms, and on the projects they have been used for. Whenever the information has been provided by the authors, we indicate so.

Greedy [94].

Information provided by D. Cohen-Steiner. Algorithm Greedy has been marketed by the Geometry Factory, the company selling the Computational Geometry Algorithms Library library, and is also available through the web site cgal.inria.fr/Reconstruction.

Cocone and variants [25, 123, 124].

Information provided by T. Dey. The suite of Cocone algorithms is available from www.cse.ohio-state.edu/~tamaldey/cocone.html. Depending on the constraints, users can choose from Cocone which reconstructs with boundaries, tight tcocone which returns a water-tight reconstruction, robust cocone which handles noise. Current implementations are based upon version 2.3 of the *Computational Geometry Algorithms Library*, www.cgal.org.

Power crust algorithm [27, 26].

Information provided by N. Amenta. The power crust software was released in 2002 at www.cs.utexas.edu/users/amenta/powercrust/welcome.html. The software was ported into the Visual Toolkit VTK by Tim Hutton —see www.sq3.org.uk/powercrust. Unfortunately since powercrust was released under the GPL licence, it cannot be officially included in the VTK distribution.

Natural Neighbors [56, 58].

Information provided by F. Cazals and A. Lieutier. The surface reconstruction algorithm based on Natural Neighbors was purchased by Dassault Sytèmes, the editor of CAGD system CATIA, and has been integrated into the *Digital Shape Editor* of CATIA V5R6 since spring 2001.

Wrap [132].

As pointed out in [132], Algorithm Wrap has been implemented in 1996 at Raindrop Geomagic, and successfully commercialized as geomagic Wrap. It is also protected by the U.S patent No. 6,3777,865.

Ball pivoting algorithm [52].

Information provided by F. Bernardini. The algorithm is patented, US6968299: Method and apparatus for reconstructing a surface using a ball-pivoting algorithm. The code is copyright of IBM and not commercially available. The Ball Pivoting Algorithm has been used in two projects sponsored by IBM Corporate Community Relations: Michelangelo's Florence Pieta `www.research.ibm.com/pieta`, Eternal Egypt `www.eternalegypt.org`. The algorithm is part of a scanning system that IBM has made available to the Egyptian Center for Documentation of Cultural and Natural History (CULTNAT)

6.6 Research Problems

Exercise 5 (Independence from the Delaunay triangulation). It has been shown in [33] that the complexity of the Delaunay triangulation for *reasonable* point sets sampled from a smooth generic surface is $O(n \log n)$, which is better than the $\Theta(n^2)$ worst-case bound on the complexity of the Delaunay triangulation. Therefore, one challenge is to design a surface reconstruction algorithm whose running time is always independent of the size of the Delaunay triangulation $D(P)$. Even better, the running time could be output sensitive in the size of the reconstructed surface. An example such an algorithm running in time $O(n \log n)$ is the modification of the Cocone algorithm by Funke and Ramos [174].

Exercise 6 (Boundaries). Define a meaningful sampling theory for a smooth surface with boundaries. Design an algorithm that comes with guarantees in terms of your sampling theory not only for the sampled surface, but also for its bounding curves.

Exercise 7 (ε-samples). The major drawback of the ε-sample framework is that the sufficient conditions of algorithms developed under its auspices cannot be checked as a pre-condition. Propose a more constructive framework.

Fig. 6.26. The evolution and progress in the Cocone family of algorithms, ilustrated on the Stanford bunny, 36k points. From left to right: Cocone, Tight Cocone and Robust Cocone. Triangles featuring non manifold edges and vertices are colored

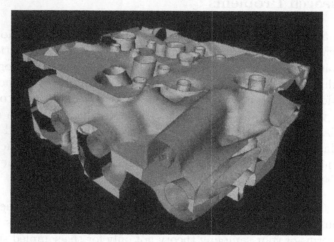

Fig. 6.27. Pump, 47k points. Reconstructed with [94]

Fig. 6.28. Mechanical part, 12k points. Reconstructed with [56]

Fig. 6.29. Vase, 2.7k points. Reconstructed with [56]

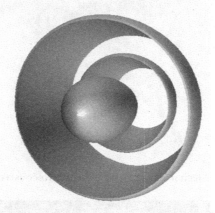

Fig. 6.30. Plane engine, 11k points. Reconstructed with [82]

7

Computational Topology: An Introduction

Günter Rote and Gert Vegter*

7.1 Introduction

Topology studies point sets and their invariants under continuous deformations, invariants such as the number of connected components, holes, tunnels, or cavities. Metric properties such as the position of a point, the distance between points, or the curvature of a surface, are irrelevant to topology. Computational topology deals with the complexity of topological problems, and with the design of efficient algorithms for their solution, in case these problems are tractable. These algorithms can deal only with spaces and maps that have a finite representation. To this end we restrict ourselves to simplicial complexes and maps. In particular we study algebraic invariants of topological spaces like Euler characteristics and Betti numbers, which are in general easier to compute than topological invariants.

Many computational problems in topology are algorithmically undecidable. The mathematical literature of the 20th century contains many (beautiful) topological algorithms, usually reducing to decision procedures, in many cases with exponential-time complexity. The quest for efficient algorithms for topological problems has started rather recently. The overviews by Dey, Edelsbrunner and Guha [119], Edelsbrunner [133], Vegter [329], and the book by Zomorodian [351] provide further background on this fascinating area.

This chapter provides a tutorial introduction to computational aspects of algebraic topology. It introduces the language of combinatorial topology, relevant for a rigorous mathematical description of geometric objects like meshes, arrangements and subdivisions appearing in other chapters of this book, and in the computational geometry literature in general.

Computational methods are emphasized, so the main topological objects are simplicial complexes, combinatorial surfaces and submanifolds of some Euclidean space. These objects are introduced in Sect. 7.2. Here we also introduce the notions of homotopy and isotopy, which also feature in other

---

* Chapter coordinator

parts of this book, like Chapter 5. Most of the computational techniques are introduced in Sect. 7.3. Topological invariants, like Betti numbers and Euler characteristic, are introduced and methods for computing such invariants are presented. Morse theory plays an important role in many recent advances in computational geometry and topology. See, e.g., Sect. 5.5.2. This theory is introduced in Sect. 7.4.

Given our focus on computational aspects, topological invariants like Betti numbers are defined using simplicial homology, even though a more advanced study of deeper mathematical aspects of algebraic topology could better be based on singular homology, introduced in most modern textbooks on algebraic topology. Other topological invariants, like homotopy groups, are harder to compute in general; These are not discussed in this chapter.

The chapter is far from a complete overview of computational algebraic topology, and it does not discuss recent advances in this field. However, reading this chapter paves the way for studying recent books and papers on computational topology. Topological algorithms are currently being used in applied fields, like image processing and scattered data interpolation. Most of these applications use some of the tools presented in this chapter.

7.2 Simplicial complexes

Topological spaces.

In this chapter a *topological space* X (or space, for short) is a subset of some Euclidean space $\mathbb{R}^d$, endowed with the induced topology of $\mathbb{R}^d$. In particular, an ε-neighborhood ($\varepsilon > 0$) of a point x in X is the set of all points in X within Euclidean distance ε from x. A subset O of X is *open* if every point of O contains an ε-neighborhood contained in O, for some $\varepsilon > 0$. A subset of X is *closed* if its complement in X is open. The *interior* of a set X is the set of all points having an ε-neighborhood contained in X, for some $\varepsilon > 0$. The *closure* of a subset X of $\mathbb{R}^d$ is the set of points x in $\mathbb{R}^d$ every ε-neigborhood of which has non-empty intersection with X. The *boundary* of a subset X is the set of points in the closure of X that are not interior points of X. In particular, every ε-neighborhood of a point in the boundary of X has non-empty intersection with both X and the complement of X. See [28, Sect. 2.1] for a more complete introduction of the basic concepts and properties of point set topology.

The space $\mathbb{R}^d$ is called the *ambient space* of X. Examples of topological spaces are:

1. The interval $[0, 1]$ in $\mathbb{R}$;
2. The open unit d-ball: $\mathbb{B}^d = \{(x_1, \dots, x_d) \in \mathbb{R}^d \mid x_1^2 + \dots + x_d^2 < 1\}$;
3. The closed unit d-ball: $\overline{\mathbb{B}}^d = \{(x_1, \dots, x_d) \in \mathbb{R}^d \mid x_1^2 + \dots + x_d^2 \leq 1\}$ (the closure of $\mathbb{B}^d$);

4. The unit d-sphere $\mathbb{S}^d = \{(x_1, \ldots, x_{d+1}) \in \mathbb{R}^{d+1} \mid x_1^2 + \cdots + x_{d+1}^2 = 1\}$ (the boundary of the $(d+1)$-ball);

5. A *d-simplex*, i.e., the convex hull of $d + 1$ affinely independent points in some Euclidean space (obviously, the dimension of the Euclidean space cannot be smaller than d). The number d is called the dimension of the simplex. Fig. 7.1 shows simplices of dimensions up to and including three.

Fig. 7.1. Simplices of dimension zero, one, two and three

Homeomorphisms.

A *homeomorphism* is a 1–1 map $h: X \to Y$ from a space X to a space Y with a continuous inverse. (In this chapter a *map* is always continuous by definition.) In this case we say that X is *homeomorphic to* Y, or, simply, that X and Y are *homeomorphic*.

1. The unit d-sphere is homeomorphic to the subset Σ of $\mathbb{R}^m$ defined by $\Sigma = \{(x_1, \ldots, x_{d+1}, 0, \ldots, 0) \in \mathbb{R}^m \mid x_1^2 + \cdots + x_{d+1}^2 = 1\}$ $(m > d)$. Indeed, the map $h: \mathbb{S}^d \to \Sigma$, defined by $h(x_1, \ldots, x_{d+1}) = (x_1, \ldots, x_{d+1}, 0, \ldots, 0)$, is a homeomorphism. Loosely speaking, the ambient space does not matter from a topological point of view.

2. The map $h: \mathbb{R}^k \to \mathbb{R}^m$, $m > k$, defined by

$$h(x_1, \ldots, x_k) = (x_1, \ldots, x_k, 0, \ldots, 0),$$

   is *not* a homeomorphism.

3. Any invertible affine map between two Euclidean spaces (of necessarily equal dimension) is a homeomorphism.

4. Any two d-simplices are homeomorphic. (If the simplices lie in the same ambient space of dimension $d - 1$, there is a unique invertible affine map sending the vertices of the first simplex to the vertices of the second simplex. For other, possibly unequal dimensions of the ambient space one can construct an invertible affine map between the affine hulls of the simplices.)

5. The boundary of a d-simplex is homeomorphic to the unit d-sphere. (Consider a d-simplex in $\mathbb{R}^{d+1}$. The projection of its boundary from a fixed point in its interior onto its circumscribed d-sphere is a homeomorphism. See Fig. 7.2. The circumscribed d-sphere is homeomorphic to the unit d-sphere.)

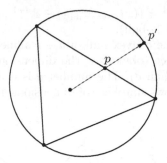

Fig. 7.2. The point p on the boundary of a 3-simplex is mapped onto the point p' on the 2-sphere. This mapping defines a homeomorphism between the 2-simplex and the 2-sphere

Simplices.

Consider a k-simplex σ, which is the convex hull of a set A of $k+1$ independent points $a_0, \ldots, a_k$ in some Euclidean space $\mathbb{R}^d$ (so $d \geq k$). A is said to *span* the simplex σ. A simplex spanned by a subset A' of A is called a *face* of σ. If τ is a face of σ we write $\tau \preceq \sigma$. The face is *proper* if $\emptyset \neq A' \neq A$. The *dimension* of the face is $|A'| - 1$. A 0-dimensional face is called a *vertex*, a 1-dimensional face is called an *edge*. An *orientation* of σ is induced by an ordering of its vertices, denoted by $\langle a_0 \cdots a_k \rangle$, as follows: For any permutation π of $0, \ldots, k$, the orientation $\langle a_{\pi(0)} \cdots a_{\pi(k)} \rangle$ is equal to $(-1)^{\text{sign}(\pi)} \langle a_0 \cdots a_k \rangle$, where $\text{sign}(\pi)$ is the number of transpositions of π (so each simplex has two distinct orientations). A simplex together with a specific choice of orientation is called an *oriented simplex*. If τ is a $(k-1)$-dimensional face of σ, obtained by omitting the vertex a_i, then the *induced orientation* on τ is $(-1)^i \langle a_0 \cdots \hat{a}_i \cdots a_k \rangle$, where the hat indicates omission of a_i.

Simplicial complexes.

A *simplicial complex* K is a finite set of simplices in some Euclidean space $\mathbb{R}^m$, such that (i) if σ is a simplex of K and τ is a face of σ, then τ is a simplex of K, and (ii) if σ and τ are simplices of K, then $\sigma \cap \tau$ is either empty or a common face of σ and τ. The *dimension* of K is the maximum of the dimensions of its simplices. The *underlying space* of K, denoted by $|K|$, is the union of all simplices of K, endowed with the subspace topology of $\mathbb{R}^m$. The *i-skeleton* of K, denoted by K^i, is the union of all simplices of K of dimension at most i. A *subcomplex* L of K is a subset of K that is a simplicial complex. A *triangulation* of a topological space X is a pair (K, h), where K is a simplicial complex and h is a homeomorphism from the underlying space $|K|$ to X. The *Euler characteristic* of a simplicial d-complex K, denoted by

$\chi(K)$, is the number $\sum_{i=0}^{d}(-1)^i \alpha_i$, where α_i is the number of i-simplices of K. Examples of simplicial complexes are:

1. A *graph* is a 1-dimensional simplicial complex (think of a graph as being embedded in $\mathbb{R}^3$). The complete graph with n vertices is the 1-skeleton of an $(n-1)$-simplex.
2. The Delaunay triangulation of a set of points in general position in $\mathbb{R}^d$ is a simplicial complex.

Combinatorial surfaces.

A *Combinatorial closed surface* is a finite two-dimensional simplicial complex in which each edge (1-simplex) is incident with two triangles (2-simplices), and the set of triangles incident to a vertex can be cyclically ordered $t_0, t_1, \ldots, t_{k-1}$ so that t_i has exactly one edge in common with $t_{i+1 \bmod k}$, and these are the only common edges. Stillwell [325, page 69 ff] contains historical background and the basic theorem on the classification combinatorial surfaces.

Homotopy and Isotopy: Continuous Deformations.

Homotopy is a fundamental topological concept that describes equivalence between curves, surfaces, or more general topological subspaces within a given topological space, up to "continuous deformations".

Technically, homotopy is defined between two maps $g, h\colon X \to Y$ from a space X into a space Y. The maps g and h are *homotopic* if there is a continuous map

$$f\colon X \times [0,1] \to Y$$

such that $f(x,0) = g(x)$ and $f(x,1) = h(x)$ for all $x \in X$. The map f is then called a homotopy between g and h. It is easy to see that homotopy is an equivalence relation, since a homotopy can be "inverted" and two homotopies can be "concatenated".

When g and h are two curves in $Y = \mathbb{R}^n$ defined over the same interval $X = [a,b]$, the homotopy f defines, for each "time" t, $0 \leq t \leq 1$, a curve $f(\cdot, t)\colon [a,b] \to \mathbb{R}^n$ that interpolates smoothly between $f(\cdot, 0) = g$ and $f(\cdot, 1) = h$.[1]

To define homotopy for two surfaces or more general spaces S and T, we start with the identity map on S and deform it into a homeomorphism from S to T. Two topological subspaces $S, T \subseteq X$ are called *homotopic* if there is a continuous mapping

$$\gamma\colon S \times [0,1] \to X$$

such that $\gamma(\cdot, 0)$ is the identity map on S and $\gamma(\cdot, 1)$ is a homeomorphism from S to T.

---

[1]In the case of curves with the same endpoints $g(a) = h(a)$ and $g(b) = h(b)$, one usually requires also that these endpoints remain fixed during the deformation: $f(a,t) = g(a)$ and $f(b,t) = g(b)$ for all t.

By the requirement that we have a homeomorphism at time $t = 1$, one can see that this definition is symmetric in S and T. Note that we do not require $\gamma(\cdot, t)$ to be a homeomorphism at all times t. Thus, a clockwise cycle and a counterclockwise cycle in the plane are homotopic. In fact, all closed curves in the plane are homotopic: every cycle can be contracted into a point (which is a special case of a closed curve). A connected topological space with this property is called *simply connected*.

Examples of spaces which are not simply connected are a plane with a point removed, or a (solid or hollow) torus. For example, on the hollow torus in Fig. 7.3, the closed curve in the figure is not homotopic to its inverse.

If we require that $\gamma(\cdot, t)$ is a homeomorphism at all times during the deformation we arrive the stronger concept of *isotopy*. For example, the smooth closed curves without self-intersections in the plane fall into two isotopy classes, according to their orientation (clockwise or counterclockwise). Isotopy is usually what is meant when speaking about a "topologically correct" approximation of a given surface, as discussed in Sect. 5.1, where the stronger concept of *ambient isotopy* is also defined (Definition 1, p. 183).

A map $f \colon X \to Y$ is a *homotopy equivalence* if there is a map $g \colon Y \to X$ such that the composed maps gf and fg are homotopy equivalent to the identity map (on X and Y, respectively). The map g is a homotopy inverse of f. The spaces X and Y are called *homotopy equivalent*. A space is *contractible* if it is homotopy equivalent to a point.

1. The unit ball in a Euclidean space is contractible. Let $f \colon \{0\} \to \mathbb{B}^d$ be the inclusion map. The constant map $g \colon \mathbb{B}^d \to \{0\}$ is a homotopy inverse of f. To see this, observe that the map fg is the identity, and gf is homotopic to the identity map on $\mathbb{B}^d$, the homotopy being the map $F \colon \mathbb{B}^d \times [0, 1] \to \mathbb{B}^d$ defined by $F(x, t) = tx$.
2. The solid torus is homotopy equivalent to the circle. More generally, the cartesian product of a topological space X and a contractible space is homotopy equivalent to X.
3. A punctured d-dimensional Euclidean space $\mathbb{R}^d \setminus \{0\}$ is homotopy equivalent to a $(d-1)$-sphere.

Note that homotopy equivalent spaces need not be homeomorphic. However, such spaces share important topological properties, like having the same Betti numbers (to be introduced in the next section). Section 6.2.3 (p. 250) describes how this concept is applied in surface reconstruction.

7.3 Simplicial homology

A calculus of closed loops.

Intuitively, it is clear that the sphere and the torus have different shapes in the sense that these surfaces are not homeomorphic. A formal proof of this observation could be based on the Jordan curve theorem: take a simple closed curve on the torus that does not disconnect the torus. Such curves, the complement of which is connected, do exist, as can be seen from Fig. 7.3. If there exists a homeomorphism from the torus to the sphere, the image of the curve on the torus would be a simple closed curve on the sphere. By the Jordan curve theorem, the complement of this curve is disconnected. Since connectedness is preserved by homeomorphisms, the complements of the curves on the torus and the sphere are not homeomorphic. This contradiction proves that the torus and the sphere are not homeomorphic.

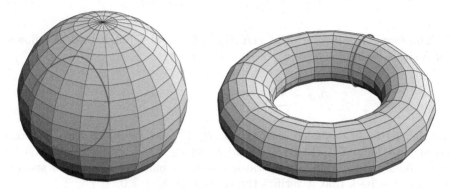

Fig. 7.3. Every simple closed curve on the sphere disconnects. Not every closed curve on the torus disconnects

This proof seems rather ad hoc: it only proves that the sphere is not homeomorphic to a closed surface with holes, but it cannot be used to show that a surface with more than one hole is not homeomorphic to the torus. Homology theory provides a systematic way to generalize the argument above to more general spaces.

In this chapter we present basic concepts and properties of *simplicial homology theory*, closely related to simplicial complexes and suitable for computational purposes. An alternative, more abstract approach is followed in the context of *singular homology theory*. This theory is more powerful when proving general results like topological invariance of homology spaces. Since we focus on basic computational techniques we will not discuss this theory here, but refer the reader to standard textbooks on algebraic topology, like [203]. The equivalence of Simplicial and Singular Homology is proven in [203, Sect. 2.1].

Chain spaces and simplicial homology.

Let K be a finite simplicial complex. In this chapter, an *simplicial k-chain* is a formal sum of the form $\sum_j a_j \sigma_j$ over the oriented k-simplices σ_j in K, with coefficients a_j in the field $\mathbb{Q}$ of rational numbers. In other words, it can be regarded as a rational vector whose entries are indexed by the oriented

k-simplices of K. Furthermore, by definition, $-\sigma = (-1)\sigma$ is the simplex obtained from σ by reversing its orientation. With the obvious definition for addition and multiplication by scalars (i.e., rational numbers), the set of all simplicial k-chains forms a vector space $C_k(K, \mathbb{Q})$, called the *vector space of simplicial k-chains* of K. The dimension of this vector space is equal to the number of k-simplices of K. Therefore, the Euler characteristic of a d-dimensional simplicial complex K can be expressed as an alternating sum of dimensions of the spaces of k-chains:

$$\chi(K) = \sum_{i=0}^{d} (-1)^i \dim C_k(K, \mathbb{Q}). \tag{7.1}$$

The *boundary operator* $\partial_k : C_k(K, \mathbb{Q}) \to C_{k-1}(K, \mathbb{Q})$ is defined as follows. For a single k-simplex $\sigma = \langle v_{i_0} \cdots v_{i_k} \rangle$, $k > 0$, let

$$\partial_k \sigma = \sum_{h=0}^{k} (-1)^h \langle v_{i_0} \cdots \hat{v}_{i_h} \cdots v_{i_k} \rangle,$$

and then let ∂_k be extended linearly, viz., $\partial_k(\sum_j a_j \sigma_j) = \sum_j a_j \partial_k \sigma_j$. For consistency we define $C_{-1}(K, \mathbb{Q}) = 0$, and we let $\partial_0 : C_0(K, \mathbb{Q}) \to C_{-1}(K, \mathbb{Q})$ be the zero-map. The boundary operator is a linear map between vector spaces. It is easy to check that it verifies the relation $\partial_k \partial_{k+1} = 0$.

Example: One-homologous chains.

In the simplicial complex of Fig. 7.4 we consider the 2-chain $\gamma = \langle v_1 v_4 v_2 \rangle + \langle v_2 v_4 v_5 \rangle + \langle v_2 v_5 v_3 \rangle + \langle v_3 v_5 v_6 \rangle + \langle v_1 v_3 v_6 \rangle + \langle v_1 v_6 v_4 \rangle$. Then $\partial_2 \gamma = \alpha - \beta$, where $\alpha = \langle v_4 v_5 \rangle + \langle v_5 v_6 \rangle - \langle v_4 v_6 \rangle$ and $\beta = \langle v_1 v_2 \rangle + \langle v_2 v_3 \rangle - \langle v_1 v_3 \rangle$. Since $\partial_1 \alpha = 0$ and $\partial_1 \beta = 0$, it follows that $\partial_1 \partial_2 \gamma = 0$.

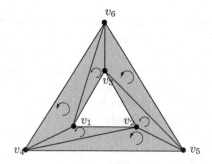

Fig. 7.4. One- and two-chains in an annulus

The vector space $Z_k(K, \mathbb{Q}) = \ker \partial_k$ is called the vector space of *simplicial k-cycles*. The vector space $B_k(K, \mathbb{Q}) = \operatorname{im} \partial_{k+1}$ is called vector space of simplicial k-*boundaries*. Since the boundary of a boundary is 0, $B_k(K, \mathbb{Q})$ is a subspace of $Z_k(K, \mathbb{Q})$. The quotient vector space $H_k(K, \mathbb{Q}) = Z_k(K, \mathbb{Q})/B_k(K, \mathbb{Q})$ is the k-*th homology vector space* of K. In particular, two k-cycles α and β are k-*homologous* if their difference is a k-boundary, i.e., if there is a $k+1$-chain γ such that $\alpha - \beta = \partial_{k+1}\gamma$. The homology class of $\alpha \in Z_k(K, \mathbb{Q})$ is denoted by $[\alpha]$. The k-*th Betti number* of the simplicial complex K, denoted by $\beta_k(K, \mathbb{Q})$, is the dimension of $H_k(K, \mathbb{Q})$. In particular:

$$\beta_k(K, \mathbb{Q}) = \dim Z_k(K, \mathbb{Q}) - \dim B_k(K, \mathbb{Q}). \tag{7.2}$$

Remark. In this chapter, the coefficients of simplicial chains are rational numbers. One usually takes these coefficients in a ring, like the set of integers. In that case one obtains homology *groups* in stead of homology vector spaces. Then, the Betti numbers are the ranks of these groups.

Example: Zero-homology of a connected simplicial complex.

Consider the connected simplicial complex K of Fig. 7.5. The 0-chains $\alpha = \langle v_6 \rangle$ and $\beta = \langle v_2 \rangle$ are 0-homologous since their difference is the boundary of the 1-chain $\gamma = -\langle v_1 v_2 \rangle + \langle v_1 v_4 \rangle + \langle v_4 v_6 \rangle$, since $\partial_1 \gamma = -(\langle v_2 \rangle - \langle v_1 \rangle) + (\langle v_4 \rangle - \langle v_1 \rangle) + (\langle v_6 \rangle - \langle v_4 \rangle) = \alpha - \beta$. In the same way one shows that every 0-chain of

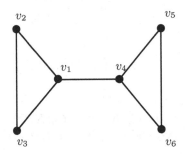

Fig. 7.5. Zero-homology of a graph

the form $\langle v_i \rangle$, $1 \le i \le 6$, is homologous to α. This implies that every 0-chain of K is of the form $c\langle \alpha \rangle$, for some $c \in \mathbb{Q}$. Hence: $H_0(K, \mathbb{Q}) = \mathbb{Q}$. It is not hard to generalize this property to all *connected* simplicial complexes: if K is a finite connected simplicial complex, then $H_0(K, \mathbb{Q}) = \mathbb{Q}$.

Example: One-homologous chains.

The boundary chains of the annulus in Fig. 7.4 are one-homologous. Indeed, the difference of the boundary chains $\alpha = \langle v_4 v_5 \rangle + \langle v_5 v_6 \rangle - \langle v_4 v_6 \rangle$ and $\beta = \langle v_1 v_2 \rangle + \langle v_2 v_3 \rangle - \langle v_1 v_3 \rangle$ is the boundary of the 2-chain $\gamma = \langle v_1 v_4 v_2 \rangle + \langle v_2 v_4 v_5 \rangle + \langle v_2 v_5 v_3 \rangle + \langle v_3 v_5 v_6 \rangle + \langle v_1 v_3 v_6 \rangle + \langle v_1 v_6 v_4 \rangle$.

Betti numbers.

We present a few examples, demonstrating the computation of Betti numbers directly from the definition.

1. Connected simplicial complex. If K is a connected simplicial complex, then $\beta_0(K, \mathbb{Q}) = 1$. In fact, we already did the example on 0-homologous chains of a connected simplicial complex K, proving that $H_0(K, \mathbb{Q}) = \mathbb{Q}$.

2. Betti numbers of a tree. The tree of Fig. 7.6 is a simplicial complex K with edges oriented according to the direction of the arrows, i.e., $e_1 = \langle v_1 v_2 \rangle$ and so on. Since it is connected, we have $\beta_0(K, \mathbb{Q}) = 1$. Furthermore, the matrix

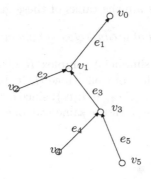

Fig. 7.6. A tree

of the boundary operator $\partial_1 : C_1(K, \mathbb{Q}) \to C_0(K, \mathbb{Q})$ with respect to the basis e_1, e_2, e_3, e_4, e_5 of $C_1(K, \mathbb{Q})$ and $\langle v_0 \rangle, \langle v_1 \rangle, \langle v_2 \rangle, \langle v_3 \rangle, \langle v_4 \rangle, \langle v_5 \rangle$ of $C_0(K, \mathbb{Q})$ is

∂_1	e_1	e_2	e_3	e_4	e_5
$\langle v_0 \rangle$	1	0	0	0	0
$\langle v_1 \rangle$	−1	1	1	0	0
$\langle v_2 \rangle$	0	−1	0	0	0
$\langle v_3 \rangle$	0	0	−1	1	1
$\langle v_4 \rangle$	0	0	0	−1	0
$\langle v_5 \rangle$	0	0	0	0	−1

(E.g., $\partial_1(e_1) = \langle v_0 \rangle - \langle v_1 \rangle = 1 \cdot \langle v_0 \rangle + (-1) \cdot \langle v_1 \rangle + 0 \cdot \langle v_2 \rangle + 0 \cdot \langle v_3 \rangle + 0 \cdot \langle v_4 \rangle + 0 \cdot \langle v_5 \rangle$.) Since the columns of this matrix are independent (why?), the image of ∂_1 has dimension 5. Therefore, $\beta_1(K, \mathbb{Q}) = \dim \ker \partial_1 = \dim C_1(K, \mathbb{Q}) - \dim \operatorname{im} \partial_1 = 0$.

3. Betti numbers of the 2-sphere. The simplicial complex K of Fig. 7.7 is the boundary of a 3-simplex, consisting of four 2-simplices, six 1-simplices and four

0-simplices. For convenience it is shown flattened on the plane, after cutting the edges incident to 0-simplex v_4. The underlying space $|K|$ is homeomorphic to the 2-sphere. Vertices with the same label have to be identified, like edges between vertices with the same label. The matrix of the boundary operator

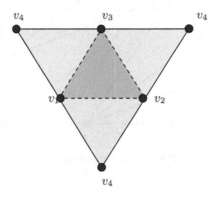

Fig. 7.7. A 2-sphere

∂_1 with respect to the canonical bases of $C_1(K, \mathbb{Q})$ and $C_0(K, \mathbb{Q})$ is

∂_1	$\langle v_1 v_2 \rangle$	$\langle v_1 v_3 \rangle$	$\langle v_1 v_4 \rangle$	$\langle v_2 v_3 \rangle$	$\langle v_2 v_4 \rangle$	$\langle v_3 v_4 \rangle$
$\langle v_1 \rangle$	-1	-1	-1	0	0	0
$\langle v_2 \rangle$	1	0	0	-1	-1	0
$\langle v_3 \rangle$	0	1	0	1	0	-1
$\langle v_4 \rangle$	0	0	1	0	1	1

It follows that $\dim C_0(K, \mathbb{Q}) = 4$, $\dim \operatorname{im} \partial_1 = 3$, and $\dim \ker \partial_1 = 3$. The matrix of the boundary operator ∂_2 with respect to the canonical bases of $C_2(K, \mathbb{Q})$ and $C_1(K, \mathbb{Q})$ is

∂_2	$\langle v_1 v_2 v_3 \rangle$	$\langle v_1 v_3 v_4 \rangle$	$\langle v_1 v_4 v_2 \rangle$	$\langle v_2 v_4 v_3 \rangle$
$\langle v_1 v_2 \rangle$	1	0	-1	0
$\langle v_1 v_3 \rangle$	-1	1	0	0
$\langle v_1 v_4 \rangle$	0	-1	1	0
$\langle v_2 v_3 \rangle$	1	0	0	-1
$\langle v_2 v_4 \rangle$	0	0	-1	1
$\langle v_3 v_4 \rangle$	0	1	0	-1

Therefore, $\dim \operatorname{im} \partial_2 = 3$ and $\dim \ker \partial_2 = 1$. Combining the previous results, we conclude that $\beta_0(K, \mathbb{Q}) = 1$, $\beta_1(K, \mathbb{Q}) = 0$ and $\beta_2(K, \mathbb{Q}) = 1$.

4. Betti numbers of the torus. Consider the simplicial complex of Fig. 7.8, which is a triangulation of the torus. It has 7 vertices, 21 oriented edges, and 14 oriented faces. The matrix of ∂_2 with respect to the canonical bases of

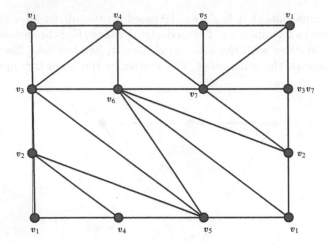

Fig. 7.8. A triangulation of the torus

$C_1(K, \mathbb{Q})$ and $C_2(K, \mathbb{Q})$ is

∂_2	142	245	253	356	165	126	276	237	173	157	475	467	134	364
$\overline{12}$	1	0	0	0	0	1	0	0	0	0	0	0	0	0
$\overline{13}$	0	0	0	0	0	0	0	0	-1	0	0	0	1	0
$\overline{14}$	1	0	0	0	0	0	0	0	0	0	0	0	-1	0
$\overline{15}$	0	0	0	0	-1	0	0	0	0	1	0	0	0	0
$\overline{16}$	0	0	0	0	1	-1	0	0	0	0	0	0	0	0
$\overline{17}$	0	0	0	0	0	0	0	0	1	-1	0	0	0	0
$\overline{23}$	0	0	-1	0	0	0	0	1	0	0	0	0	0	0
$\overline{24}$	-1	1	0	0	0	0	0	0	0	0	0	0	0	0
$\overline{25}$	0	-1	1	0	0	0	0	0	0	0	0	0	0	0
$\overline{26}$	0	0	0	0	0	1	-1	0	0	0	0	0	0	0
$\overline{27}$	0	0	0	0	0	0	1	-1	0	0	0	0	0	0
$\overline{34}$	0	0	0	0	0	0	0	0	0	0	0	0	1	-1
$\overline{35}$	0	0	-1	1	0	0	0	0	0	0	0	0	0	0
$\overline{36}$	0	0	0	-1	0	0	0	0	0	0	0	0	0	1
$\overline{37}$	0	0	0	0	0	0	0	1	-1	0	0	0	0	0
$\overline{45}$	0	1	0	0	0	0	0	0	0	0	-1	0	0	0
$\overline{46}$	0	0	0	0	0	0	0	0	0	0	0	1	0	-1
$\overline{47}$	0	0	0	0	0	0	0	0	0	0	1	-1	0	0
$\overline{56}$	0	0	0	1	-1	0	0	0	0	0	0	0	0	0
$\overline{57}$	0	0	0	0	0	0	0	0	0	1	-1	0	0	0
$\overline{67}$	0	0	0	0	0	0	-1	0	0	0	0	1	0	0

The matrix of ∂_1 with respect to the canonical bases of $C_0(K, \mathbb{Q})$ and $C_1(K, \mathbb{Q})$ is obtained similarly (preferably using a computer algebra system). Computing the dimensions of the kernel and image of these operators we finally get

$$\beta_0(K,\mathbb{Q}) = 1, \quad \beta_1(K,\mathbb{Q}) = 2, \quad \beta_2(K,\mathbb{Q}) = 1$$

Euler characteristic and Betti numbers.

One of the fundamental results of simplicial homology theory states that Betti numbers of the underlying space of finite simplicial complex does not depend on the triangulation.

Theorem 1. *Betti numbers are* **homotopy invariants***: if K and L are simplicial complexes with homotopy equivalent underlying spaces, then the i-th homology vector spaces of K and L are isomorphic. In particular,*

$$\beta_i(K,\mathbb{Q}) = \beta_i(L,\mathbb{Q}), \quad \text{for all } i.$$

The proof of this theorem is beyond the scope of these introductory notes. One usually introduces the more general *singular homology groups* for a topological space X, which are independent of any triangulation. Then one proves that these groups are isomorphic to the simplicial homology groups, obtained by taking simplicial chains with integer coefficients in stead of rational coefficients. In particular, the corresponding Betti numbers, being the ranks of these groups, are equal.

Theorem 2. *Let K be a d-dimensional simplicial complex. Then*

$$\chi(K) = \sum_{i=0}^{d} (-1)^i \beta_i(K,\mathbb{Q}).$$

Proof. Recall from (7.1) that $\chi(K) = \sum_{i=0}^{d} (-1)^i \dim C_k(K,\mathbb{Q})$. Since

$$H_i(K,\mathbb{Q}) = \left. \ker \partial_i \middle/ \operatorname{im} \partial_{i+1} \right. .$$

we see that

$$
\begin{aligned}
\beta_i(K,\mathbb{Q}) &= \dim H_i(K,\mathbb{Q}) \\
&= \dim \ker \partial_i - \dim \operatorname{im} \partial_{i+1} \\
&= \dim C_i(K,\mathbb{Q}) - \dim \operatorname{im} \partial_i - \dim \operatorname{im} \partial_{i+1}.
\end{aligned}
$$

Now:

$$\sum_{i=0}^{d} (-1)^i \left(\dim \operatorname{im} \partial_i + \dim \operatorname{im} \partial_{i+1} \right) = 0.$$

Hence:

$$\sum_{i=0}^{d} (-1)^i \beta_i(K,\mathbb{Q}) = \chi(K,\mathbb{Q}).$$

The claimed identities follow from the preceding derivation.

If X is a topological space with a simplicial complex K triangulating it, then we define $\chi(X) = \chi(K,\mathbb{Q})$. It follows from Theorem 1 and Theorem 2 that the Euler characteristic does not depend on the specific choice of the triangulation K.

Incremental algorithm for computation of Betti numbers.

As can be seen in the case of a simple space like the torus, the matrices of the boundary map become rather large, even for simple examples. Therefore alternative approaches have been developed for special cases. We start with an incremental approach, in which the simplicial complex is constructed by adding simplices one at a time, making sure that during the process all partial constructs are indeed simplicial complexes. The key idea is to maintain the Betti numbers of the partial complexes. The following result indicates how to do this.

Proposition 1. *Let K be a simplicial complex, and let K' be a simplicial complex such that $K' = K \cup \sigma$ for some k-simplex σ. Let ∂_i and ∂_i' be the boundary operators of the chain complexes associated with K and K', respectively. Furthermore, let $\gamma = \partial_k' \sigma$. If γ also bounds in K, i.e., $\partial_k' \sigma \in \operatorname{im} \partial_k$, then*

$$\beta_i(K', \mathbb{Q}) = \begin{cases} \beta_i(K, \mathbb{Q}) & \text{if } i \neq k \\ \beta_k(K, \mathbb{Q}) + 1 & \text{if } i = k \end{cases}$$

If γ does not bound in K, i.e., $\partial_k' \sigma \notin \operatorname{im} \partial_k$, then

$$\beta_i(K', \mathbb{Q}) = \begin{cases} \beta_i(K, \mathbb{Q}) & \text{if } i \neq k - 1 \\ \beta_{k-1}(K, \mathbb{Q}) - 1 & \text{if } i = k - 1 \end{cases}$$

Proof.

$$\cdots \xrightarrow{\partial_{k+1}'} \quad C_k(K', \mathbb{Q}) \quad \xrightarrow{\partial_k'} C_{k-1}(K', \mathbb{Q}) \xrightarrow{\partial_{k-1}'} \cdots$$
$$\|\qquad\qquad\qquad\qquad\|$$
$$C_k(K, \mathbb{Q}) \oplus \mathbb{Q}[\sigma] \qquad\quad C_{k-1}(K, \mathbb{Q})$$

Case 1: $\partial_k' \sigma \in \operatorname{im} \partial_k$. Then $\operatorname{im} \partial_i' = \operatorname{im} \partial_i$, for all i, so $\dim \operatorname{im} \partial_i' = \dim \operatorname{im} \partial_i$, for all i. Therefore:

$$\dim \ker \partial_k' = \dim C_k(K', \mathbb{Q}) - \dim \operatorname{im} \partial_k'$$
$$= 1 + \dim C_k(K, \mathbb{Q}) - \dim \operatorname{im} \partial_k$$
$$= 1 + \dim \ker \partial_k$$

Furthermore, for $i \neq k$ we have $\dim \ker \partial_i' = \dim \ker \partial_i$. Hence (recall $\dim H_i(K', \mathbb{Q}) = \dim \ker \partial_i' - \dim \operatorname{im} \partial_{i+1}'$):

$$\beta_i(K', \mathbb{Q}) = \dim H_i(K', \mathbb{Q}) = \begin{cases} \dim H_i(K, \mathbb{Q}) & \text{if } i \neq k \\ 1 + \dim H_k(K, \mathbb{Q}) & \text{if } i = k \end{cases}$$

Case 2: $\partial_k' \sigma \notin \operatorname{im} \partial_k$. Then

$$\dim \operatorname{im} \partial_i' = \begin{cases} \dim \operatorname{im} \partial_i & \text{if } i \neq k \\ \dim \operatorname{im} \partial_k + 1 & \text{if } i = k. \end{cases}$$

Hence:

$$\begin{aligned} \dim \ker \partial_i' &= \dim C_i(K', \mathbb{Q}) - \dim \operatorname{im} \partial_i' \\ &= \begin{cases} \dim C_i(K, \mathbb{Q}) - \dim \operatorname{im} \partial_i & \text{if } i \neq k \\ 1 + \dim C_k(K, \mathbb{Q}) - (1 + \dim \operatorname{im} \partial_k) & \text{if } i = k \end{cases} \\ &= \dim \ker \partial_i \end{aligned}$$

This result yields an incremental algorithm for the computation of Betti numbers. Whether this algorithm is efficient depends on the implementation of the test '$\partial_k' \sigma \notin \operatorname{im} \partial_k$'. The paper [113] presents an efficient implementation of this algorithm for subcomplexes of the three-sphere. This incremental method can be used to compute the Betti numbers of some familiar spaces. Before showing how to do this, we introduce some additional tools that are helpful in the computation of Betti numbers.

Chain maps and chain homotopy.

Just like maps between spaces provide information about the topology of these spaces, maps between homology spaces provide information about the homology of these spaces. The key stepping stone towards these maps are chain maps.

Let K and L be finite simplicial complexes. A chain map from K to L is a sequence of linear maps $f_k \colon C_k(K, \mathbb{Q}) \to C_k(L, \mathbb{Q})$ such that $\partial_{k+1} \circ f_{k+1} = f_k \circ \partial_{k+1}$. In other words, the sequence $\{f_k\}$ is a chain map if the following diagram is commutative:

$$\begin{array}{ccccccc} \cdots \xrightarrow{\partial_{k+2}} & C_{k+1}(K, \mathbb{Q}) & \xrightarrow{\partial_{k+1}} & C_k(K, \mathbb{Q}) & \xrightarrow{\partial_k} & \cdots \\ & \downarrow{f_{k+1}} & & \downarrow{f_k} & & \\ \cdots \xrightarrow{\partial_{k+2}} & C_{k+1}(L, \mathbb{Q}) & \xrightarrow{\partial_{k+1}} & C_k(L, \mathbb{Q}) & \xrightarrow{\partial_k} & \cdots \end{array}$$

This chain map is denoted by $f \colon C(K, \mathbb{Q}) \to C(L, \mathbb{Q})$. In fact, a chain map is a family of maps, containing one linear map for each dimension.

Proposition 2. *Let K, L and M be finite simplicial complexes.*

1. The sequence of identity maps $\operatorname{id}_k : C_k(K, \mathbb{Q}) \to C_k(K, \mathbb{Q})$ is a chain map.

2. The composition of a chain map from K to L and a chain map from L to M is a chain map from K to M.

The proof of this result is straightforward and left as an exercise (Exercise 4). Let $f \colon C(K, \mathbb{Q}) \to C(L, \mathbb{Q})$ be a chain map. The linear map $f_* \colon H(K, \mathbb{Q}) \to H(L, \mathbb{Q})$ is defined by

$$f_{*k}([\alpha]) = [f_k(\alpha)],$$

for $\alpha \in Z_k(K, \mathbb{Q})$. We say that f_* is the map induced by f at the level of homology. Using commutativity of the diagram above, it is easy to see that this map is well-defined, i.e., that $[f_k(\alpha)]$ is independent of the choice of the representative α of the homology class $[\alpha]$. This map has some natural properties, following in a straightforward way from the definition.

Proposition 3. *Let K, L and M be finite simplicial complexes.*

1. The identity chain map generates the identity map at the level of homology.

2. The map induced by a composition of chain maps is the composition of the maps induced by each chain map. In other words, for chain maps $f: C(K, \mathbb{Q}) \to C(L, \mathbb{Q})$ and $g: C(L, \mathbb{Q}) \to C(M, \mathbb{Q})$:

$$(g \circ f)_* = g_* \circ f_*.$$

A *chain homotopy* between two chain maps $f, g: C(K, \mathbb{Q}) \to C(L, \mathbb{Q})$ is a sequence $\{T_k\}$ of linear maps $T_k: C_k(K, \mathbb{Q}) \to C_{k+1}(L, \mathbb{Q})$ such that

$$T_{k-1} \circ \partial_k + \partial_{k+1} \circ T_k = f_k - g_k.$$

If such a chain homotopy exists, then f and g are called *chain-homotopic*. We shall frequently use the following result, the proof of which is a simple exercise in Linear Algebra (see Exercise 4).

Proposition 4. *Chain homotopic chain maps induce the same linear map at the level of homology.*

Simplical collapse.

We now consider *simplicial collapse*, a very simple transformation of simplicial complexes which does not alter homology in positive dimensions. This operation allows us to compute the Betti numbers of a simplicial complex K by simplifying K until we obtain another simplicial complex L for which the Betti numbers are known or easy to compute.

Let K be a finite simplicial complex, and let α and β be two simplices of K such that α is a face of β, and α is not a face of any other simplex of K. Let L be the subcomplex of K obtained by deleting the simplices α and β. The transformation from K to L is called an *elementary collapse*. See Fig. 7.9.

More generally, we say that K *collapses onto* a subcomplex L, denoted by $K \searrow L$, if there is a finite sequence of elementary collapses transforming K into L.

Proposition 5. *Let K and L be finite simplicial complexes such that K collapses onto L. Then $H_k(K, \mathbb{Q})$ and $H_k(L, \mathbb{Q})$ are isomorphic.*

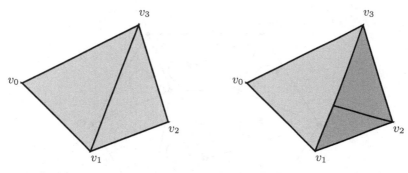

Fig. 7.9. An elementary collapse removes the simplices $v_0v_1v_2v_3$ and $v_1v_2v_3$ from the leftmost simplex

Proof. We give the proof for positive k, the case $k = 0$ being trivial. Our strategy consists of finding a chain homotopy inverse to the inclusion chain map $\iota \colon C(L, \mathbb{Q}) \to C(K, \mathbb{Q})$. To this end let α be a k-simplex, positively oriented in the boundary $\partial \beta$ of the $k + 1$-simplex β. Introduce the map $f \colon C(K, \mathbb{Q}) \to C(L, \mathbb{Q})$ by putting $f_k(\alpha) = \alpha - \partial\beta$, $f_{k+1}(\beta) = 0$, $f_i(\sigma) = \sigma$ for every i-simplex different from α and β, and extending linearly. It is not hard to prove that f is a chain map. Furthermore, $f \circ \iota$ is the identity chain map on $C(L, \mathbb{Q})$.

Let the sequence of linear maps $P_i \colon C_i(K, \mathbb{Q}) \to C_{i+1}(K, \mathbb{Q})$ be defined by $P_k(\alpha) = \beta$, and $P_i(\sigma) = 0$ for each i-simplex σ different from α. A straightforward computation shows that the sequence $\{P_i\}$ is a chain homotopy between the identity map on $C(K, \mathbb{Q})$ and the chain map $\iota \circ f$. From this we conclude that $\iota_i \colon H_i(L, \mathbb{Q}) \to H_i(K, \mathbb{Q})$ is an isomorphism, for $i > 0$. In particular, K and L have the same Betti numbers in positive dimension.

Example: Betti numbers of the projective plane.

The incremental algorithm, combined with the method of simplicial collapse, allows for rather painless computation of Betti numbers of familiar spaces. In this example we compute the Betti numbers of the projective plane $\mathbb{R}P_2$. The simplicial complex K of Fig. 7.10 is the unique triangulation of the projective plane with a minimal number of vertices. The vertices and edges on the boundary of the six-gon are identified in pairs, as indicated by the double occurrence of the vertex-labels v_1, v_2 and v_3. The arrows indicate the orientation of the simplices forming the basis of the chain space $C_2(K)$. We orient the edges of the simplex from the vertex with lower index to the vertex with higher index.

Let L be the simplicial complex obtained from K by deleting the oriented simplex $\tau = \langle v_4v_5v_6 \rangle$. The Betti numbers of L are easy to compute, since a sequence of simplicial collapses transforms L into the subcomplex L_0 with

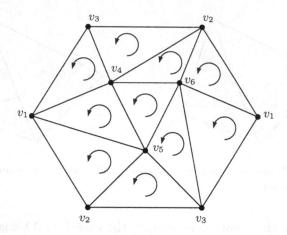

Fig. 7.10. A triangulation of the projective plane

vertices v_1, v_2 and v_3, and oriented edges $\langle v_1 v_2 \rangle$, $\langle v_2 v_3 \rangle$ and $\langle v_1 v_3 \rangle$. The simplicial complex L_0 is a 1-sphere, so $\beta_0(L) = \beta_0(L_0) = 1$, $\beta_1(L) = \beta_1(L_0) = 1$, and $\beta_i(L) = \beta_i(L_0) = 0$ for $i > 1$.

To relate the Betti numbers of K with those of L, we have to determine whether $\tau' = \partial_2 \tau$ is a boundary in L. Consider the special 2-chain α, which is the formal sum of all oriented 2-simplices in L. Taking the boundary of α, we see that all oriented 1-simplices not in $\partial_2 \tau$ occur twice, those in the interior of the six-gon in Fig. 7.10 with opposite coefficients and those in the boundary with the same coefficient. In other words, $\partial_2 \alpha = 2\gamma - \partial_2 \tau$, where γ is the 1-cycle $\langle v_1 v_2 \rangle + \langle v_2 v_3 \rangle - \langle v_1 v_3 \rangle$ of L. Therefore, $[\tau'] = 2[\gamma]$ in $H_1(L)$. Since $[\gamma]$ forms a basis for $H_1(L)$, we conclude that $[\tau'] \neq 0$ in $H_1(L)$. Hence τ' is *not* a boundary in L. Applying the incremental algorithm we see that $\beta_0(K) = \beta_0(L) = 1$, $\beta_1(K) = \beta_1(L) - 1 = 0$, and $\beta_2(K) = \beta_2(L) = 0$.

Example: Betti numbers depend on field of scalars.

Homology theory can be set up with coefficients in a general field. A priory, this leads to different Betti numbers. This is illustrated by revisiting the simplicial complex K of Fig. 7.10, and applying the same procedure to compute the Betti numbers over $\mathbb{Z}_2$. Using the same notation as in the preceding example, we see that $[\tau'] = 2[\gamma] = 0$ in $H_1(L, \mathbb{Z}_2)$, so τ' is a boundary in $C_2(L, \mathbb{Z}_2)$. Applying the incremental algorithm again we conclude that $\beta_i(K, \mathbb{Z}) = \beta_i(L, \mathbb{Z}) = 1$, for $i = 0, 1$, and $\beta_2(K, \mathbb{Z}) = \beta_2(L, \mathbb{Z}) + 1 = 1$. Note that the Euler characteristic is independent of the coefficient field.

7.4 Morse Theory

Finite dimensional Morse theory deals with the relation between the topology of a smooth manifold and the critical points of smooth real-valued functions on the manifold. It is the basic tool for the solution of fundamental problems in differential topology. Recently, basic notions from Morse theory have been used in the study of the geometry and topology of large molecules. We review some basic concepts from Morse theory, like in [329]. More elaborate treatments are [255] and [250].

7.4.1 Smooth functions and manifolds

Differential of a smooth map.

A function $f: \mathbb{R}^n \to \mathbb{R}$ is called *smooth* if all derivatives of any order exist. A map $\varphi: \mathbb{R}^n \to \mathbb{R}^m$ is called smooth if its component functions are smooth. The *differential* of φ at a point $q \in \mathbb{R}^n$ is the *linear map* $d\varphi_q: \mathbb{R}^n \to \mathbb{R}^m$ defined as follows. For $v \in \mathbb{R}^n$, let $\alpha: I \to \mathbb{R}^n$, with $I = (-\varepsilon, \varepsilon)$ for some positive ε, be defined by $\alpha(t) = \varphi(q + tv)$, then $d\varphi_q(v) = \alpha'(0)$. Let $\varphi(x_1, \ldots, x_n) = (\varphi_1(x_1, \ldots, x_n), \ldots, \varphi_m(x_1, \ldots, x_n))$. The differential $d\varphi_q$ is represented by the Jacobian matrix

$$
\begin{pmatrix}
\dfrac{\partial \varphi_1}{\partial x_1}(q) & \cdots & \dfrac{\partial \varphi_1}{\partial x_n}(q) \\
\vdots & & \vdots \\
\dfrac{\partial \varphi_m}{\partial x_1}(q) & \cdots & \dfrac{\partial \varphi_m}{\partial x_n}(q)
\end{pmatrix}.
$$

Regular surfaces in $\mathbb{R}^3$.

A subset S in $\mathbb{R}^3$ is a *smooth surface* if we can cover the surface with open coordinate neighborhoods. More precisely, a coordinate neighborhood of a point p on the surface is a subset of the form $V \cap S$, where V is an open subset of $\mathbb{R}^3$, for which there exists a smooth map $\varphi: U \to \mathbb{R}^3$ defined on an open subset U of $\mathbb{R}^2$, such that where V is an open subset of $\mathbb{R}^3$ containing p, for which there exists a smooth map $\varphi: U \to \mathbb{R}^3$ defined on an open subset U of $\mathbb{R}^2$, such that

(i) The map φ is a homeomorphism from U onto $V \cap S$;
(ii) If $\varphi(u, v) = (x(u, v), y(u, v), z(u, v))$, then the two tangent vectors

$$
\begin{pmatrix} \dfrac{\partial x}{\partial u} \\[4pt] \dfrac{\partial y}{\partial u} \\[4pt] \dfrac{\partial z}{\partial u} \end{pmatrix}, \quad
\begin{pmatrix} \dfrac{\partial x}{\partial v} \\[4pt] \dfrac{\partial y}{\partial v} \\[4pt] \dfrac{\partial z}{\partial v} \end{pmatrix}
$$

are non-zero and not parallel.

The map φ is called a parametrization or a system of local coordinates in p. The set S is a smooth surface if each point of S has a coordinate neighborhood. Note that condition (ii) is equivalent to the fact that the differential of φ at (u, v) is an injective map.

Example: spherical coordinates. Let S be a 2-sphere in $\mathbb{R}^3$ with radius R and center $(0, 0, 0) \in \mathbb{R}^3$. Consider the set $U = \{ (u, v) \mid 0 < u < 2\pi, -\pi/2 < v < \pi/2 \}$. The map $\varphi \colon U \to S$, given by

$$\varphi(u, v) = (R \cos u \cos v, R \sin u \cos v, R \sin v).$$

corresponds to the well-known spherical coordinates. Note that $\varphi(U)$ is the 2-sphere minus a meridian. Each point of $\varphi(U)$ has a system of local coordinates given by φ.

Example: coordinates on the upper and lower hemisphere. Again, let S be the sphere with radius R and center at the origin of $\mathbb{R}^3$, and let $U = \{ (x, y) \mid x^2 + y^2 < R^2 \}$. The (open) upper and lower hemispheres of the torus are the graph of a smooth function. More precisely, each point of the upper hemisphere has local coordinates given by the map

$$\varphi(x, y) = (x, y, \sqrt{R^2 - x^2 - y^2}).$$

A similar expression defines local coordinates at each point of the lower hemisphere. Covering the sphere by six hemispheres yields a system (at least one) of local coordinate system for each point of the sphere. Therefore, the sphere is a regular surface.

Example: coordinates on the torus of revolution. Let S be the torus obtained by rotating the circle in x, y-plane with center $(0, R, 0)$ and radius r around the x-axis, where $R > r$. We show that S is a smooth surface by introducing a system of local coordinates for all points of the torus. To this end, let $U = \{(u, v) \mid 0 < u, v < 2\pi\}$ and let $\varphi \colon U \to \mathbb{R}^3$ be the map defined by

$$\varphi(u, v) = (r \sin u, (R - r \cos u) \sin v, (R - r \cos u) \cos v).$$

It is not hard to check that $\varphi(U) \subset S$. In fact, the map φ covers the torus except for one meridian and one parallel circle. It is easy to find local coordinates in points of these two circles by translating the parameter domain U a little bit. Therefore, the torus is a regular surface.

Example: Local form of torus of revolution near $(0, 0, \pm(R - r))$. As in the example of hemispheres, parts of the torus are graphs of a smooth function. In particular, the points $(0, 0, \pm(R - r))$ have local coordinates of the form $\varphi(x, y) = (x, y, f_\pm(x, y))$, where

$$f_\pm(x, y) = \pm\sqrt{R^2 + r^2 - x^2 - y^2 - 2R\sqrt{r^2 - x^2}}.$$

Submanifolds of $\mathbb{R}^n$.

More generally, a subset M of $\mathbb{R}^n$ is an m-dimensional smooth *submanifold* of $\mathbb{R}^n$, $m \leq n$, if for each $p \in M$, there is an open set V in $\mathbb{R}^n$, containing p, and a map $\varphi \colon U \to M \cap V$ from an open subset U in $\mathbb{R}^m$ onto $V \cap M$ such that (i) φ is a smooth homeomorphism, (ii) the differential $d\varphi_q \colon \mathbb{R}^m \to \mathbb{R}^n$ is injective for each $q \in U$. Again, the map φ is called a parametrization or a system of local coordinates on M in p. In particular, the space $\mathbb{R}^n$ is a submanifold of $\mathbb{R}^n$. A subset N of a submanifold M of $\mathbb{R}^n$ is a submanifold of M if it is a submanifold of $\mathbb{R}^n$. The difference of the dimensions of M and N is called the *codimension* of N (in M).

Example: linear subspaces are submanifolds. The Euclidean space $\mathbb{R}^m$ is a smooth submanifold of $\mathbb{R}^n$, for $m \leq n$. For $m < n$, we identify $\mathbb{R}^m$ with the subset $\{(x_1, \ldots, x_n) \in \mathbb{R}^n \mid x_{m+1} = \cdots = x_n = 0\}$ of $\mathbb{R}^n$.

Example: $\mathbb{S}^{n-1}$ is a smooth submanifold of $\mathbb{R}^n$. A smooth parametrization of $\mathbb{S}^{n-1}$ at $(0, \ldots, 0, 1) \in \mathbb{S}^{n-1}$ is given by $\varphi \colon U \to \mathbb{R}^n$, with

$$U = \{(x_1, \ldots, x_{n-1}) \in \mathbb{R}^{n-1} \mid x_1^2 + \cdots + x_{n-1}^2 < 1\},$$

and

$$\varphi(x_1, \ldots, x_{n-1}) = (x_1, \ldots, x_{n-1}, \sqrt{1 - x_1^2 - \cdots - x_{n-1}^2}).$$

In fact, φ is a parametrization in every point of the upper hemisphere, i.e., the intersection of $\mathbb{S}^{n-1}$ and the upper half space $\{(y_1, \ldots, y_n) \mid y_n > 0\}$.

Example: codimension one submanifolds. The equator $\mathbb{S}^1 = \{(x_1, x_2, 0) \mid x_1^2 + x_2^2 = 1\}$ is a codimension one submanifold of $\mathbb{S}^2 = \{(x_1, x_2, x_3) \mid x_1^2 + x_2^2 + x_3^2 = 1\}$. More generally, every intersection of the 2-sphere with a plane at distance less than one from the origin is a codimension one submanifold.

Tangent space of a manifold.

The tangent vectors at a point p of a manifold form a vector space, called the tangent space of the manifold at p. More formally, a *tangent vector* of M at p is the tangent vector $\alpha'(0)$ of some smooth curve $\alpha \colon I \to M$ through p. Here a smooth curve through a point p on a smooth submanifold M of $\mathbb{R}^n$ is a smooth map $\alpha \colon I \to \mathbb{R}^n$, with $I = (-\varepsilon, \varepsilon)$ for some positive ε, satisfying $\alpha(t) \in M$, for $t \in I$, and $\alpha(0) = p$. The set T_pM of all tangent vectors of M at p is the *tangent space* of M at p.

If $\varphi \colon U \to M$ is a smooth parametrization of M at p, with $0 \in U$ and $\varphi(0) = p$, then T_pM is the m-dimensional subspace $d\varphi_0(\mathbb{R}^m)$ of $\mathbb{R}^n$, which passes through $\varphi(0) = p$. Let $\{e_1, \ldots, e_m\}$ be the standard basis of $\mathbb{R}^m$; define the tangent vector $\bar{e}_i \in T_pM$ by $\bar{e}_i = d\varphi_0(e_i)$. Then $\{\bar{e}_1, \ldots, \bar{e}_m\}$ is a basis of T_pM.

Example: tangent space of the sphere. The tangent space of the unit sphere $\mathbb{S}^{n-1} = \{(x_1, \ldots, x_n) \mid x_1^2 + \cdots + x_n^2 = 1\}$ at a point p is the hyperplane through p, perpendicular to the normal vector of the sphere at p.

Smooth function on a submanifold.

A function $f \colon M \to \mathbb{R}$ on an m-dimensional smooth submanifold M of $\mathbb{R}^n$ is smooth at $p \in M$ if there is a smooth parametrization $\varphi \colon U \to M \cap V$, with U an open set in $\mathbb{R}^m$ and V an open set in $\mathbb{R}^n$ containing p, such that the function $f \circ \varphi \colon U \to \mathbb{R}$ is smooth. A function on a manifold is called smooth if it is smooth at every point of the manifold.

Example: height function on a surface. The height function $h \colon S \to \mathbb{R}$ on a surface S in $\mathbb{R}^3$ is defined by $h(x, y, z) = z$, for $(x, y, z) \in S$. Let $\varphi(u, v) = (x(u, v), y(u, v), z(u, v))$ be a system of local coordinates in a point of the surface, then $h \circ \varphi(u, v) = z(u, v)$ is smooth. Therefore, the height function is a smooth function on S.

Regular and critical points.

A point $p \in M$ is a *critical point* of a smooth function $f \colon M \to \mathbb{R}$ if there is a local parametrization $\varphi \colon U \to \mathbb{R}^n$ of M at p, with $\varphi(0) = p$, such that 0 is a critical point of $f \circ \varphi \colon U \to \mathbb{R}$ (i.e., the differential of $f \circ \varphi$ at q is the zero function on $\mathbb{R}^n$). This condition does not depend on the particular parametrization.

A real number $c \in \mathbb{R}$ is a *regular value* of f if $f(p) \neq c$ for all critical points p of f, and a *critical value* otherwise.

Example: critical points of height function on the sphere. Consider the height function on the unit sphere in $\mathbb{R}^3$. Spherical coordinates define a parametrization $\varphi(u, v)$ in every point, except for the poles $(0, 0, \pm 1)$. With respect to this parametrization the height function h has the expression $\tilde{h}(u, v) = h(\varphi(u, v)) = \sin v$, so none of these points is singular (since $-\pi/2 < v < \pi/2$ away from the poles). Near the poles $(0, 0, \pm 1)$ we consider the sphere as the graph of a function, corresponding to the parametrization $\psi(x, y) = (x, y, \sqrt{1 - x^2 - y^2})$. The height function is expressed in these local coordinates as $\tilde{h}(x, y) = h(\psi(x, y)) = \pm\sqrt{1 - x^2 - y^2}$, so the singular points of h are $(0, 0, -1)$ (minimum), and $(0, 0, 1)$ (maximum).

Example: critical points of height function on the torus. The torus M in $\mathbb{R}^3$, obtained by rotating a circle in the x, y-plane with center $(0, R, 0)$ and radius r around the x-axis, is a smooth 2-manifold. Let $U = \{(u, v) \mid -\pi/2 < u, v < 3\pi/2\} \subset \mathbb{R}^2$, and let the map $\varphi \colon U \to \mathbb{R}^3$ be defined by

$$\varphi(u, v) = (r \sin u, (R - r \cos u) \sin v, (R - r \cos u) \cos v).$$

Then φ is a parametrization at all points of M, except for points on one latitudinal and one longitudinal circle. The height function on M is the function $h \colon M \to \mathbb{R}$ defined by $\tilde{h}(u, v) = h(\varphi(u, v)) = (R - r \cos u) \cos v$, so the singular points of h are:

(u,v)	$\varphi(u,v)$	type of singularity
$(0,0)$	$(0,0,R-r)$	saddle point
$(0,\pi)$	$(0,0,-R+r)$	saddle point
$(\pi,0)$	$(0,0,R+r)$	maximum
(π,π)	$(0,0,-R-r)$	minimum

The type of a singular point will be introduced in Sect. 7.4.2.

Implicit surfaces and manifolds.

In many cases a set is given as the zero set of a smooth function (or a system of functions). If this zero set contains no singular point of the function, then it is a manifold:

Proposition 6. (IMPLICIT FUNCTION THEOREM). *Let $f\colon M \to \mathbb{R}$ be a smooth function on the smooth submanifold M of $\mathbb{R}^n$. If c is a regular value of f, then the level set $f^{-1}(c)$ is a smooth submanifold of M of codimension one.*

A proof can be found in any book on analysis on manifolds, like [323].

Example: implicit surfaces in three-space. The unit sphere in three space is a regular surface, since 0 is a regular value of the function $f(x,y,z) = x^2 + y^2 + z^2 - 1$. The torus of revolution is a regular surface, since 0 is a regular value of the function $g(x,y,z) = (x^2 + y^2 + z^2 - R^2 - r^2)^2 - 4R^2(r^2 - x^2)$.

Hessian at a critical point.

Let M be a smooth submanifold of $\mathbb{R}^n$, and let $f\colon M \to \mathbb{R}$ be a smooth function. The *Hessian* of f at a critical point p is the quadratic form $H_p f$ on $T_p M$ defined as follows. For $v \in T_p M$, let $\alpha\colon (-\varepsilon,\varepsilon) \to M$ be a curve with $\alpha(0) = p$, and $\alpha'(0) = v$. Then

$$H_p f(v) = \left.\frac{d^2}{dt^2}\right|_{t=0} f(\alpha(t)).$$

The right hand side does not depend on the choice of α. To see this, let $\varphi\colon U \to M$ be a smooth parametrization of M at p, with $0 \in U$ and $\varphi(0) = p$, and let $v = v_1 \bar{e}_1 + \cdots + v_m \bar{e}_m \in T_p M$, where $\bar{e}_i = d\varphi_0(e_i)$. Then

$$H_p f(v) = \sum_{i,j=1}^{m} \frac{\partial^2(f \circ \varphi)}{\partial x_i \partial x_j}(0) v_i v_j.$$

In particular, the matrix of $H_f(p)$ with respect to this basis is

$$\begin{pmatrix} \dfrac{\partial^2(f \circ \varphi)}{\partial x_1^2}(0) & \cdots & \dfrac{\partial^2(f \circ \varphi)}{\partial x_1 \partial x_m}(0) \\ \vdots & & \vdots \\ \dfrac{\partial^2(f \circ \varphi)}{\partial x_1 \partial x_m}(0) & \cdots & \dfrac{\partial^2(f \circ \varphi)}{\partial x_m^2}(0) \end{pmatrix}. \tag{7.3}$$

It is not hard to check that the numbers of positive and negative eigenvalues of the Hessian do not depend on the choice of φ, since p is a critical point of f.

Non-degenerate critical point.

The critical point p of $f \colon M \to \mathbb{R}$ is *non-degenerate* if the Hessian $H_p f$ is non-degenerate. The *index* of the non-degenerate critical point p is the number of negative eigenvalues of the Hessian at p. If M is 2-dimensional, then a critical point of index 0, 1, or 2, is called a *minimum*, *saddle point*, or *maximum*, respectively.

7.4.2 Basic Results from Morse Theory

Morse function.

A smooth function on a manifold is a *Morse function* if all critical points are non-degerate. The *k-th Morse number* of a Morse function f, denoted by $\mu_k(f)$, is the number of critical points of f of index k.

Example: quadratic function on $\mathbb{R}^m$. The function $f \colon \mathbb{R}^m \to \mathbb{R}$, defined by $f(x_1, \ldots, x_m) = -x_1^2 - \ldots - x_k^2 + x_{k+1}^2 + \ldots + x_m^2$, is a Morse function, with a single critical point $(0, \ldots, 0)$. This point is a non-degenerate critical point, since the Hessian matrix at this point is $\mathrm{diag}(-2, \ldots, -2, 2, \ldots, 2)$, with k entries on the diagonal equal to -2. In particular, the index of the critical point is k.

Example: singularities of the height function on S^{m-1}. The height function on the standard unit sphere $\mathbb{S}^{m-1}$ in $\mathbb{R}^m$ is a Morse function. This function is defined by $h(x_1, \ldots, x_m) = x_m$ for $(x_1, \ldots, x_m) \in \mathbb{S}^{m-1}$, With respect to the parametrization $\varphi(x_1, \ldots, x_{m-1}) = (x_1, \ldots, x_{m-1}, \sqrt{1 - x_1^2 - \cdots - x_{m-1}^2})$, the expression of the height function is

$$h \circ \varphi(x_1, \ldots, x_{m-1}) = \sqrt{1 - x_1^2 - \cdots - x_{m-1}^2}.$$

Therefore, the only critical point of h on the upper hemisphere is $(0, \ldots, 0, 1)$. The Hessian matrix (7.3) is the diagonal matrix $\mathrm{diag}(-1, -1, \ldots, -1)$, so this critical point has index $m-1$. Similarly, $(0, \ldots, 0, -1)$ is the only critical point on the lower hemisphere. It is a critical point of index 0.

Example: singularities of the height function on the torus. The singular points of the height function on the torus of revolution with radii R and r are $(0, 0, -R-r)$, $(0, 0, -R+r)$, $(0, 0, R-r)$, and $(0, 0, R+r)$. See also Sect. 7.4.1. A parametrization of this torus near the singular points $\pm(R-r)$ is $\varphi(x, y) = (x, y, f_\pm(x, y))$, where $f_\pm(x, y) = \pm\sqrt{R^2 + r^2 - x^2 - y^2 - 2R\sqrt{r^2 - x^2}}$. The expression $h(x, y) = f_\pm(x, y)$ of the height function with respect to these local coordinates at $(x, y) = (0, 0)$ is

$$h(x,y) = \pm\left(R - r - \frac{1}{2r}x^2 + \frac{1}{2(R-r)}y^2\right) + \text{Higher Order Terms.}$$

Hence the singular points corresponding to $(x, y) = (0, 0)$, i.e., $(0, 0, \pm(R-r))$, are saddle points, i.e., singular points of index one. Similarly, the singular point $(0, 0, R+r)$ is a maximum (index two), and the singular point $(0, 0, -R-r)$ is a minimum (index zero), and the

Regular level sets.

Let M be an m-dimensional submanifold of $\mathbb{R}^n$, and let $f\colon M \to \mathbb{R}$ be a smooth function. The set $f^{-1}(h) := \{q \in M | f(q) = h\}$ of points where f has a fixed value h is called a *level set* (at level h). If $h \in \mathbb{R}$ is a regular value of f, then $f^{-1}(h)$ is a smooth $(m-1)$-dimensional submanifold of $\mathbb{R}^n$.

Similarly, we define the *lower level set* (also called *excursion set*) at some level $h \in \mathbb{R}$ as $M_h = \{q \in M \mid f(q) \le h\}$. If f has no critical values in $[a, b]$, for $a < b$, then the subsets M_a and M_b of M are homeomorphic (and even isotopic).

The Morse Lemma.

Let $f\colon M \to \mathbb{R}$ be a smooth function on a smooth m-dimensional submanifold M of $\mathbb{R}^n$, and let p be a non-degenerate critical point of index k. Then there is a smooth parametrization $\varphi\colon U \to M$ of M at p, with U an open neighborhood of $0 \in \mathbb{R}^m$ and $\varphi(0) = p$, such that

$$f \circ \varphi(x_1, \ldots, x_m) = f(p) - x_1^2 - \cdots - x_k^2 + x_{k+1}^2 + \cdots + x_m^2.$$

In particular, a critical point of index 0 is a local minimum of f, whereas a critical point of index m is a local maximum of f. See Fig. 7.11.

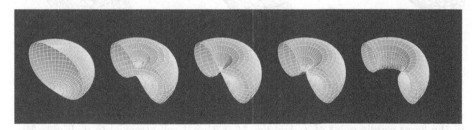

Fig. 7.11. Passing a critical level set of a Morse function in three-space. The critical point has index 1. A local model of the function near the critical point is $f(x_1, x_2, x_3) = -x_1^2 + x_2^2 + x_3^2$, with the x_1-axis running vertically

Abundance of Morse functions.

(i) *Morse functions are generic.* Every smooth compact submanifold of $\mathbb{R}^n$ has a Morse function. (In fact, if we endow the set $C^\infty(M)$ of smooth functions on M with the so-called Whitney topology, then the set of Morse functions on M is an open and dense subset of $C^\infty(M)$. In particular, there are Morse functions arbitrarily close to any smooth function on M.)

(ii) *Generic height functions are Morse functions.* Let M be an m-dimensional submanifold of $\mathbb{R}^{m+1}$ (e.g., a smooth surface in $\mathbb{R}^3$). For $v \in \mathbb{S}^m$, the height-function $h_v \colon M \to \mathbb{R}$ with respect to the direction v is defined by $h_v(p) = \langle v, p \rangle$. The set of v for which h_v is not a Morse function has measure zero in $\mathbb{S}^m$.

Passing critical levels.

One can build complicated spaces from simple ones by attaching a number of cells. Let X and Y be topological spaces, such that $X \subset Y$. We say that Y is obtained by *attaching a k-cell* to X if $Y \setminus X$ is homeomorphic to an open k-ball. More precisely, there is a map $f \colon \overline{\mathbb{B}}^k \to \overline{Y \setminus X}$, such that $f(\mathbb{S}^{k-1}) \subset X$ and the restriction $f \mid \mathbb{B}^k$ is a homeomorphism $\mathbb{B}^k \to Y \setminus X$. Let $f \colon M \to \mathbb{R}$ be a smooth Morse function with exactly one critical level in (a, b), and a and b are regular values of f. Then M_b is homotopy equivalent to M_a with a cell of dimension k attached, where k is the index of the critical point in $f^{-1}([a, b])$. See Fig. 7.12.

Fig. 7.12. Passing a critical level of index 1 corresponds to attaching a 1-cell. Here M is the 2-torus embedded in $\mathbb{R}^3$, in standard vertical position, and f is the height function with respect to the vertical direction. Left: M_a, for a below the critical level of the lower saddle point of f. Middle: M_a with a 1-cell attached to it. Right: M_b, for b above the critical level of the lower saddle point of f. This set is homotopy equivalent to the set in the middle part of the figure

Morse inequalities.

Let f be a Morse function on a compact m-dimensional smooth submanifold of $\mathbb{R}^n$. For each k, $0 \le k \le m$, the k-th Morse number of f dominates the

k-th Betti number of M:

$$\mu_k(f) \geq \beta_k(M, \mathbb{Q}).$$

An intuitive explanation is based on the observation that passing a critical level of a critical point of index k is equivalent corresponds to the attachment of a k-cell at the level of homotopy equivalence. Therefore, either the k-th Betti number increases by one, or the $k-1$-st Betti number decreases by one, cf the incremental algorithm for computing Betti numbers in Sect. 7.3, while none of the other Betti numbers changes. Since only the k-th Morse number changes, more precisely, increases by one, the Morse inequalities are invariant upon passage of a critical level.

In the same spirit one can show that the Morse numbers of f are related to the Betti numbers and the Euler characteristic of M by the following identity:

$$\sum_{k=1}^{m}(-1)^k \mu_k(f) = \sum_{k=1}^{m}(-1)^k \beta_k(M, \mathbb{Q}) = \chi(M).$$

Gradient vector fields.

Consider a smooth function $f: M \to \mathbb{R}$, where M is a smooth m-dimensional submanifold of $\mathbb{R}^n$. The *gradient of f* is a smooth map $\operatorname{grad} f: M \to \mathbb{R}^n$, which assigns to each point $p \in M$ a vector $\operatorname{grad} f(p) \in T_pM \subset \mathbb{R}^n$, such that

$$\langle \operatorname{grad} f(p), v \rangle = df_p(v), \quad \text{for all } v \in T_pM.$$

Since $df_p(v)$ is a linear form in v, the vector $\operatorname{grad} f(p)$ is well defined by the preceding identity. This definition has a few straightforward implications. The gradient of f vanishes at a point p if and only if p is a singular point of f. If p is not a singular point of f, then $df_p(v)$ is maximal for a unit vector $v \in T_pM$ iff $v = \operatorname{grad} f(p)/\|\operatorname{grad} f(p)\|$. In other words, $\operatorname{grad} f(p)$ is the direction of steepest ascent of f at p. Furthermore, if $c \in \mathbb{R}$ is a regular value of the function f, then $\operatorname{grad} f$ is perpendicular to the level set $f^{-1}(c)$ at every point.

To express $\operatorname{grad} f$ in local coordinates, let $\varphi: U \to M$ be a system of local coordinates at $p \in M$. Let $\overline{e_1}, \dots, \overline{e_m}$ be the basis of T_pM corresponding to the standard basis $e_1, \dots, e_m$ of $\mathbb{R}^m$. In other words: $\overline{e_i} = d\varphi_q(e_i)$, where $q \in U$ is the pre-image of p under φ. We denote the standard coordinates on $\mathbb{R}^m$ by $x_1, \dots, x_m$. Then

$$\operatorname{grad} f(p) = \sum_{i=1}^{m} a_i(q)\overline{e_i},$$

where $a_i: U \to \mathbb{R}$ is the smooth function defined by the set of linear equations

$$\sum_{j=1}^{m} g_{ij}(q)a_j(q) = \frac{\partial(f \circ \varphi)}{\partial x_i}(q), \quad (1 \leq i \leq m),$$

with $g_{ij}(q) = \langle \overline{e_i}, \overline{e_j} \rangle$. Since the coefficients are the entries $\langle \overline{e_i}, \overline{e_j} \rangle$ of a Gram matrix, the system is non-singular. Note that $a_i = \dfrac{\partial(f \circ \varphi)}{\partial x_i}$ if the system of coordinates is orthonormal at p, that is $g_{ij}(q) = 1$, if $i = j$ and $g_{ij}(q) = 0$, if $i \neq j$. This holds in particular if $U = M = \mathbb{R}^n$ and φ is the identity map on U, so the definition agrees with the usual definition in a Euclidean space.

Integral lines, and their local structure near singular points.

In the sequel M is a *compact* submanifold of $\mathbb{R}^n$. The gradient of a smooth function f on M is a smooth vector field on M. For every point p of M, there is a unique curve $x: \mathbb{R} \to M$, such that $x(0) = p$ and $x'(t) = \operatorname{grad} f(x(t))$, for all $t \in \mathbb{R}$. The image $x(\mathbb{R})$ is called the *integral curve* of the gradient vector field through p.

Lemma 1. *Let $f: M \to \mathbb{R}$ be a smooth function on a submanifold M of $\mathbb{R}^n$.*

1. *The integral curves of a gradient vector field of f form a partition of M.*
2. *The integral curve $x(t)$ through a singular point p of f is the constant curve $x(t) = p$.*
3. *The integral curve $x(t)$ through a regular point p of f is injective, and both $\lim_{t \to \infty} x(t)$ and $\lim_{t \to -\infty} x(t)$ exist. These limits are singular points of f.*
4. *The function f is strictly increasing along the integral curve of a regular point of f.*
5. *Integral curves are perpendicular to regular level sets of f.*

The proof is a bit technical, so we skip it. See [194] for details. The first property implies that the integral curves through two points of M are disjoint or coincide. The third property implies that a gradient vector field does not have closed integral curves. The limit $\lim_{t \to \infty} x(t)$ is called the ω-limit of p, and is denoted by $\omega(p)$. Similarly, $\lim_{t \to -\infty} x(t)$ is the α-limit of p, denoted by $\alpha(p)$. Note that all points on an integral curve have the same α-limit and the same ω-limit. Therefore, it makes sense to refer to these points as the α-limit and ω-limit of the integral curve. It follows from Lemma 1.2 that $\omega(p) = p$ and $\alpha(p) = p$ for a singular point p.

Stable and unstable manifolds.

The structure of integral lines of a gradient vector field $\operatorname{grad} f$ near a singular point can be quite complicated. However, for Morse functions, the situation is simple. To gain some intuition, let us consider the simple example of the function $f(x_1, x_2) = x_1^2 - x_2^2$ on a neighborhood of the non-degenerate singular point $0 \in \mathbb{R}^2$. The gradient vector field is $2x_1 e_1 - 2x_2 e_2$, where e_1, e_2 is the standard basis of $\mathbb{R}^2$. The integral line $(x_1(t), x_2(t))$ through a point $p = (p_1, p_2)$ is determined by $x_1(0) = p_1$, $x_2(0) = p_2$, and

$$\begin{cases} x_1'(t) = 2x_1(t) \\ x_2'(t) = -2x_2(t) \end{cases}$$

Therefore, the integral curve through p is $(x_1(t), x_2(t)) = (p_1 e^{2t}, p_2 e^{-2t})$, which is of the form $x_1 x_2 = c$. See Fig. 7.13 (Left). The singular point

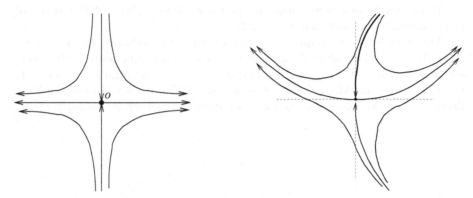

Fig. 7.13. Left: Integral curves of the gradient of $f(x_1, x_2) = x_1^2 - x_2^2$ on a neighborhood of the singular point $(0,0) \in \mathbb{R}^2$. Right: Integral curves of the gradient vector field near a general saddle point of a function on $\mathbb{R}^2$

$o = (0,0)$ is the α-limit of all points on the horizontal axis, and the ω-limit of all points on the vertical axis. The general structure of integral curves near a saddle point is similar, as indicated by Fig. 7.13 (Right). The stable curve of p consists of all points with ω-limit equal to p. The unstable curve is defined similarly. These curves intersect each other at p, and are perpendicular there.

More generally, the stable manifold of a singular point p is the set $W^s(p) = \{q \in M \mid \omega(q) = p\}$. Similarly, the unstable manifold of p is the set $W^u(p) = \{q \in M \mid \alpha(q) = p\}$. Note that both $W^s(p)$ and $W^u(p)$ contain the singular point p itself. Furthermore, the intersection of the stable and unstable manifolds of a singular point consists just of the singular point: $W^s(p) \cap W^u(p) = \{p\}$. Stable and unstable manifolds of gradient systems are submanifolds [214, Chapter 6]. The dimension of $W^s(p)$ is equal to the number of negative eigenvalues of the Hessian of f at p, whereas the dimension of $W^u(p)$ is equal to the number of positive eigenvalues of this Hessian. Stable and unstable manifolds of gradient systems are submanifolds [214, Chapter 6].

The Morse-Smale complex.

A Morse function on M is called a *Morse-Smale function* if its stable and unstable manifolds intersect transversally, i.e., at a point of intersection the tangent spaces of the stable and unstable manifolds together span the tangent space of M. If p and q are distinct singular points, the intersection $W^s(p) \cap$

$W^u(q)$ consists of all regular integral curves with ω-limit equal to p and α-limit equal to q. In particular, a Morse-Smale function on a two-dimensional manifold has no integral curves connecting two saddle points, since the stable manifold of one of the saddle points and the unstable manifold of the second saddle point would intersect non-transversally along this connecting integral curve.

Morse-Smale functions form an open and dense subset of the space of functions on a compact manifold [320].

The *Morse-Smale complex* associated with a Morse-Smale function f on M is the subdivision of M formed by the connected components of the intersections $W^s(p) \cap W^u(q)$, where p and q range over all singular points of f, see Fig. 7.14. The Morse-Smale complex is a CW-complex. In geographical literature, the Morse-Smale complex is known as the *surface network*.

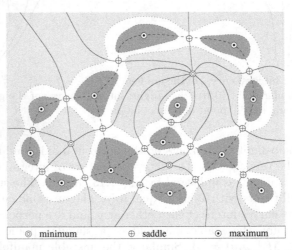

| | ◎ minimum | ⊕ saddle | ⊙ maximum |

Fig. 7.14. The Morse-Smale complex of a function on the plane. The stable one-manifolds are solid, the unstable one-manifolds are dashed (Courtesy Herbert Edelsbrunner.)

The Morse-Smale complex on a two-manifold consists of cells of dimension 0, 1 and 2, called vertices, edges and regions. According to the *Quadrangle Lemma* [135], each region of the Morse-Smale complex is a quadrangle with vertices of index 0, 1, 2, 1, in this order around the region. Hence the complex is not necessarily a regular CW-complex, since the boundary of a cell is possibly glued to itself along vertices and arcs.

Using a paradigm called *simulation of differentiability*, in [135] the concept of Morse-Smale complex is also defined for piecewise linear functions, and an algorithm for its construction is applied to geographic terrain data. In [134] this work is extend to piecewise linear 3-manifolds.

Reeb graphs and contour trees.

The level sets $f^{-1}(h)$ of a Morse function f on a two-dimensional domain change as h varies. At certain values of h, components of the level set may disappear, new components may appear, or a component may split into two components, or two components may merge. A component of a level set is called a *contour*. The Reeb graph (after the American journalist John Reeb, 1887–1920 [293]) encodes the changes of contours. It is obtained by contracting every contour to a single point. When f is defined on a simply connected domain (for example, a box), the Reeb graph is a tree, and it is also referred to as the *contour tree*. Fig. 7.15b shows an example of a contour tree of a bivariate function $h = f(x, y)$ defined on a square domain. The vertical axis

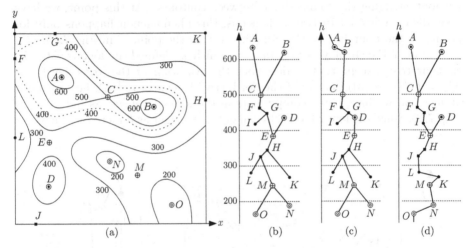

Fig. 7.15. (a) a contour map of level sets (isolines), (b) the corresponding contour tree, (c) the join tree, and (d) the split tree. As in Fig. 7.14, minima and maxima are indicated by empty and full circles, and crosses denote saddle points. The points where a contour touches the boundary play also a role in the contour tree (for example, they may be local minima or maxima) but they are not critical points in the sense of having derivative 0. The level sets in (a) are labeled with the height values, and these values are indicated in the trees of (b), (c), and (d). The critical point F changes only the topology of a contour and not the number of contours; when the contour tree is viewed as a discrete structure, F is not a vertex of the tree

of the contour tree represents the value h of the function. The intersection of a horizontal line at a given value h with the contour tree yields all contours at that level, and the merging or splitting, appearance or disappearance of contours is reflected in vertices of degree 3 and 1 in the contour tree, respectively. Saddle points become vertices of degree 3, and minima and maxima become vertices of degree 1. A contour tree is therefore a good tool to visualize the behavior of a function on a global scale, in particular when it is a function of more than two variables, see [224, 223]. In these applications, f is usually a continuous piecewise linear function interpolating data at given sample points. These functions are not smooth and therefore not Morse functions, but the notion of level sets and Reeb graphs extends without difficulty to this class of functions. It is not uncommon to have *multiple saddle points*, where more than two contours meet at the same time. The Reeb graph has then vertices of degree higher than three. More examples of contour trees are shown in Fig. 5.23 of Sect. 5.5.2.

Note that the Reeb graph only regards the number of components (the 0-homology) of the level sets, it does not reflect every change of topology. For example, in three dimensions, a contour might start as a ball, and as h increases, it might extrude two arms that meet each other, forming a torus, without changing the connectivity between contours. (At this point, we have a saddle of index 1.) In two dimensions, this phenomenon happens only for points on the boundary of the domain, such as the point F in Fig. 7.15.

Figure 7.15c displays the *join tree*, which is defined analogously to the contour tree, except that it describes the evolution of the lower level sets $M_h = f^{-1}([-\infty, h])$ instead of the "ordinary" level sets $f^{-1}(h)$. For example, at $h = 300$ we have three components in the lower level set, as indicated in Fig. 7.16. Since the lower level sets can only get bigger as h increases, they can

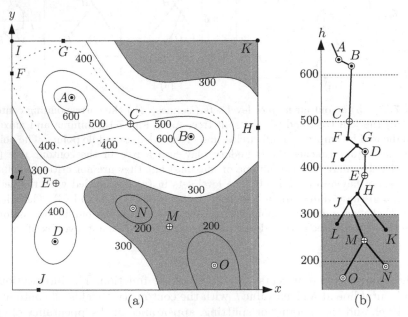

Fig. 7.16. (a) the lower level set at level 300 and (b) the corresponding part in the join tree

only join and never split (hence the name join tree): the tree is a directed tree with the root at the highest vertex. The *split tree* (Fig. 7.15d) can be defined analogously for upper level sets. The join and the split tree are important because it is easier to construct these trees first instead of constructing the contour tree directly. As shown by Carr, Snoeyink, and Axen [78] the contour tree can then be built from the join tree and the split tree in linear time.

The simplest and fastest way to construct the join (and split) tree of a piecewise linear function is the method of *monotone paths*, as described in [92]. We sketch the main idea. This method requires an initial identification of all "critical" vertices: vertices where the topology of the level set changes locally as the level set passes through them. This condition can be checked by scanning the neighboring faces of each vertex independently. These vertices are candidates for becoming vertices of the join tree. They are sorted by function values and processed in increasing order. At each critical vertex v which is not a minimum, we start a monotone decreasing path into each different "local component" of the lower level set in the neighborhood of v. For example, if we increase h in Fig. 7.16, the next critical point that is processed is J, see Fig. 7.17. Into each of the two shaded regions, we start a descending

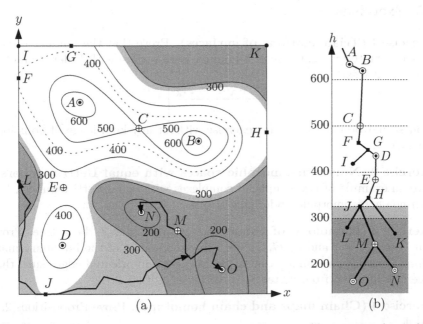

Fig. 7.17. Identifying the components that are to be merged by growing descending paths

path. Each path is continued until it reaches a local minimum (such as the point L) or a previously constructed path (such as the descending path from

M that ends in O). If we have stored the appropriate information with each path, we can identify the components of the lower level sets that need to be merged (namely, the component L and the component MNO; the component K remains separate). Since each path can only descend, it is guaranteed that it cannot leave the lower level set into which it belongs, and therefore it identifies the correct component. It can happen that two descending paths reach the same component. In this case we only have a change of topology of the contour, without changing the number of contours.

This algorithm works in any dimension. If the piecewise linear function f is defined on a triangulated mesh with t cells and there are n_c critical points, the algorithm $O(t + n_c \log n_c)$ time and $O(t)$ space.

Note that the descending paths do not have to follow the steepest direction; thus, unlike the integral curves of the gradient vector field, they can cross the boundaries of the Morse-Smale complex.

With few exeptions [99], the efficient computation of Reeb graphs has been studied mostly for functions on simply connected domains, and hence under the heading of contour *trees*.

7.5 Exercises

Exercise 1 (Triangulations of surfaces). Prove that the number of vertices in a finite triangulation of a boundaryless surface with Euler characteristic χ is at least

$$\left\lceil \frac{7 + \sqrt{49 - 24\chi}}{2} \right\rceil .$$

(You should be able to do this exercise without any knowledge of homology theory.)

Exercise 2 (Non-homeomorphic spaces with equal Betti numbers). Give an example of two simplicial complexes with equal Betti numbers, but with non-homeomorphic underlying spaces.

Exercise 3 (Homology of connected graphs). Let G be a tree. Prove that $\beta_0(G, \mathbb{Q}) = 1$ and $\beta_1(G, \mathbb{Q}) = 0$ using the matrix of the boundary map. (*Hint:* Consider an enumeration of the vertices and oriented edges such that edge e_i is directed from vertex v_j to vertex v_i, with $j > i$.)

Exercise 4 (Chain maps and chain homotopy). Prove Propositions 2, 3 and 4.

Exercise 5 (Cone construction and Betti numbers of spheres). Let L be a finite simplicial complex in $\mathbb{R}^n$, and regard $\mathbb{R}^n$ as the subspace of $\mathbb{R}^{n+1}$ with final coordinate zero. Let v be a point in $\mathbb{R}^{n+1} \setminus \mathbb{R}^n$. If σ is a k-simplex of L with vertices $v_0, \ldots, v_k$, then the $(k+1)$-simplex with vertices $v, v_0, \ldots, v_k$ is called the *join* of σ and v. The *cone of L with apex v* is the simplicial complex consisting of the simplices of L, the join of each of these simplices and v, and the 0-simplex $\langle v \rangle$ itself. (One can check that these simplices form a simplicial complex.) Let K be the cone of L.

1. Let the map $T_k \colon C_k(K, \mathbb{Q}) \to C_{k+1}(K, \mathbb{Q})$ be defined as follows: Let $\sigma = \langle v_0, \ldots, v_k \rangle$ be a k-simplex of K. If σ is also a k-simplex of L, then $T_k(\sigma) = \langle v, v_0, \ldots, v_k \rangle$, otherwise $T_k(\sigma) = 0$. Prove that the sequence $\{T_k\}$ is a chain homotopy between the identity map and the zero map on the chain complex $C(K, \mathbb{Q})$.
2. Conclude that $H_k(K, \mathbb{Q}) = 0$, for $k > 0$. What is $H_0(K, \mathbb{Q})$?
3. Determine the Betti numbers of the d-dimensional disk, i.e., the space $\mathbb{B}^d = \{(x_1, \ldots, x_d) \in \mathbb{R}^d \mid x_1^2 + \cdots + x_d^2 \leq 1\}$. (*Hint:* Note that a disk is homeomorphic to a d-simplex.)
4. Use the previous result, and the incremental homology algorithm to determine the Betti numbers of the d-sphere.

Exercise 6 (Homology of orientable surfaces).

1. Prove that $\beta_0(K) = 1$ for every triangulation K of an orientable surface of genus g (a sphere with g handles).
2. Let K be a simplicial complex whose underlying space is the torus, and let all simplices of K be oriented compatibly. Let $\alpha = \sum_\sigma \sigma$, where the sum ranges over all (oriented) simplices of K. Prove that $Z_2(K, \mathbb{Q}) = \mathbb{Q}\alpha$, and that $\beta_2(K, \mathbb{Q}) = 1$.
3. Use the same technique as in part 2 of this exercise to prove that $\beta_2(K, \mathbb{Q}) = 1$ for every triangulation K of an orientable surface of genus g.
4. Let L be the subcomplex of K obtained by deleting an arbitrary 2-simplex. Use the incremental algorithm to prove that $\beta_2(L, \mathbb{Q}) = \beta_2(K, \mathbb{Q}) - 1$, and $\beta_i(L, \mathbb{Q}) = \beta_i(K, \mathbb{Q})$, for $i = 0, 1$.
5. Now let K be the simplicial complex of Fig. 7.8. Prove that L simplicially collapses onto the subcomplex M, the subgraph of L consisting of the vertices $v_1, \ldots, v_5$ and the edges $v_1 v_2$, $v_2 v_3$, $v_3 v_1$, $v_1 v_4$, $v_4 v_5$, and $v_5 v_1$. Conclude that $\beta_1(K, \mathbb{Q}) = 2$, and $\beta_0(K, \mathbb{Q}) = 1$.
6. Try to generalize this exercise to an orientable surface of genus g.

Exercise 7 (Morse Theory yields Betti numbers).

1. Use Morse theory to compute the Betti numbers of the d-sphere $\mathbb{S}^d$.
2. Compute the Euler characteristic of a surface M with g handles by defining a suitable Morse function on it. Then compute the Betti numbers of this surface. (*Hint:* You may want to use the first and third result of Exercise 6).
3. For a Morse function f, let s be a critical point with Morse index i. Consider the intersection $L^-(s)$ of the lower level set $f^{-1}((-\infty, f(s)])$ with a small sphere around s. Prove that the Euler characteristic of $L^-(s)$ equals $1 - (-1)^i$.

Exercise 8 (The mountaineer's equation). For a smooth Morse function on the 2-sphere $\mathbb{S}^2$, the number of peaks and pits (maxima and minima) exceeds the number of passes (saddles) by 2.

Exercise 9 (Contour trees for bivariate Morse functions). Show that, for a smooth Morse function on the 2-sphere $\mathbb{S}^2$, a saddle point will always generate a vertex of degree three in the Reeb graph. Use this observation and the previous exercise to prove that the Reeb graph is in fact a tree in this case.

8

Appendix - Generic Programming and The CGAL Library

Efi Fogel and Monique Teillaud

8.1 The CGAL Open Source Project

Several research groups in Europe had started to develop small geometry libraries on their own in the early 1990s. A consortium of eight sites in Europe and Israel was founded to cultivate the labour of these groups and gather their produce in a common library called CGAL — the Computational Geometry Algorithms Library [2].

The goal of CGAL was to promote the research in Computational Geometry and translate the results into useful, reliable, and efficient programs for industrial and academic applications, the very same goal that governs CGAL developers to date. In fact, CGAL meets two recommendations of the Computational Geometry Impact Task Force Report [86, 87], which was published roughly when CGAL came to existence: production and distribution of usable (and useful) geometric software was a key recommendation, which came with the need for creating a reward structure for implementations in academia.

An INRIA startup, GEOMETRY FACTORY,[1] was founded on January 2003. The company sells CGAL commercial licenses, support for CGAL, and customized developments based on CGAL.

In November 2003, when Version 3.0 was released, CGAL became an Open Source Project, allowing new contributions from various resources. Common parts of CGAL (i.e., the so-called *kernel* and *support* libraries) are now distributed under the GNU Lesser General Public License (or GNU LGPL for short) and the remaining part (i.e., the *basic* library) is distributed under the terms of the Q Public License (QPL).

The implementations of the CGAL software modules described in this book are complete and robust, as they handle all degenerate cases. They rigorously adapt the *generic programming paradigm*, briefly reviewed in the next section to overcome problems encountered when effective computational geometry software is implemented. *Geometric programming* is discussed in the

---

[1] http://www.geometryfactory.com/.

succeeding section. Finally, a glimpse at the structure of CGAL is given in the concluding section.

8.2 Generic Programming

Several definitions of the term *generic programming* have been proposed since it was first coined about four decades ago along with the introduction of the LISP programming language. Since then several approaches have been put into trial through the introduction of new features in existing computer languages, or even new computer languages all together. Here we confine ourself to the classic notion first described by David Musser, Alexander Stepanov, Deepak Kapur, and collaborators, who considered generic programming as a discipline that consists of the gradual lifting of concrete algorithms abstracting over details, while retaining the algorithm semantics and efficiency [271].

One crucial abstraction supported by all contemporary computer languages is the subroutine (also known as procedure or function, depending on the programming language). Another abstraction supported by C++ is that of abstract data typing, where a new data type is defined together with its basic operations. C++ also supports object-oriented programming, which emphasizes on packaging data and functionality together into units within a running program, and is manifested in hierarchies of polymorphic data types related by inheritance. It allows referring to a value and manipulating it without needing to specify its exact type. As a consequence, one can write a single function that operates on a number of types within an inheritance hierarchy. Generic programming identifies a more powerful abstraction (perhaps less tangible than other abstractions), making extensive use of C++ class-templates and function-templates. It is a formal hierarchy of abstract requirements on data types referred to as *concepts*, and a set of classes that conform precisely to the specified requirements, referred to as *models*. Models that describe behaviours are referred to as *traits* classes [272]. Traits classes typically add a level of indirection in template instantiation to avoid accreting parameters to templates.

A generic algorithm has two parts: the actual instructions that describe the steps of the algorithm, and a set of requirements that specify which properties its argument types must satisfy. The following *swap* function is an example of the first part of a generic algorithm.

```
template <class T> void swap(T & a, T & b) {
  T tmp = a; a = b; b = tmp;
}
```

When the function call is compiled, it is instantiated with a data type that must have an assignment operator. A data type that fulfils this requirement is a model of a concept commonly called *Assignable* [38]. The instantiated data type must also model the concept *CopyConstructible*. The *int* data type, for example, is a model of these two concepts associated with the template parameter T. Thus, it can be used to instantiate the function template [38].[2]

There are many data types that model both concepts *Assignable* and *Copy-Constructible* in conjunction, as they consist of only a single requirement each. Thus, it is hard to refer to any of its models as a traits. On the other hand, consider an imaginary generic implementation of a data structure that handles geometric arrangements. Its prototype is listed below. The *Arrangement_2* class must be instantiated with a class that must in turn define a type that represents a certain family of curves, and some functions that operate on curves of this family.

```
template <class Traits> class Arrangement_2 {
  // the code
};
```

It is natural to refer to a model of this concept as a traits class. One important objective is to minimize the set of requirements the traits concept imposes. A tight traits concept may save tremendously in analysis and programming of classes that model the concept. Another important reason for reaching the minimal requirements is to avoid computing the same algebraic entity in different ways. The importance of this is amplified in the context of computational geometry, as a non tight model that consists of duplicate, but slightly different, implementations of the same algebraic entity, can lead to superficial degenerate conditions, which in turn can drastically increase running times.

An algorithm implemented according to the standard object-oriented paradigm alone may resort to use dynamic cast to achieve flexibility, is enforced to have tight coupling through the inheritance relationship, may require additional memory for each object to accommodate the virtual-function table-pointer, and adds for each call to a virtual member function an indirection through the virtual function table. An algorithm implemented according to the generic programming paradigm does not suffer from these disadvantages. The set of requirements on data types is not tied to a specific C++ language feature. Therefore it might be more difficult to grasp. In return, a generic implementation gains stronger type checking at compile time and a higher level of flexibility, without loss of efficiency. In fact, it may expedite the computation. Many articles and a few books have been written on the subject. We refer the reader to [38] for a complete introduction.

---

[2]See http://www.sgi.com/tech/stl/ for a complete specification of the SGI STL.

The prime example of generic programming was STL, the C++ Standard Template Library, that became part of the C++ standard library in 1994. Since then a few other generic-programming libraries emerged. The most notable in our context were LEDA (Library of Efficient Data Types and Algorithms), a library of combinatorial and geometric data types and algorithms [8, 251], and CGAL [2, 156], the Computational Geometry Algorithms Library. Early development of LEDA started in 1988, ten years before the first public release of CGAL became available. While LEDA is mostly a large collection of fundamental graph related and general purpose data structures and algorithms, CGAL is a collection of large and complex data structures and algorithms focusing on geometry.

A noticeable influence on generic programming is conducted by the Boost online community, which encourages the development of free C++ software gathered in the Boost library collection [1]. It is a large set of portable and high quality C++ libraries that work well with, and are in the same spirit as, the C++ Standard Library. The Boost Graph Library (BGL), which consists of generic graph algorithms, serves a particularly important role in our context. It can be used for example to implement the underlying topological data structure of an arrangement instance, that is, a model of the concept *ArrangementDcel*; see Sect. 1.4.1 for more details. Using some generic programming techniques, an arrangement instance can be adapted as a BGL graph, and passed as input to generic algorithms already implemented in the BGL, such as the Dijkstra shortest path algorithm.

8.3 Geometric Programming and CGAL

Implementing geometric algorithms and data structures is notoriously difficult, as transforming such algorithms and data structures into effective computer programs is a process full of pitfalls. However, the last decade has seen significant progress in the development of software for computational geometry. The mission of such a task, which Kettner and Näher [222] call *geometric programming*, is to develop software that is correct, efficient, flexible (namely adaptable and extensible[3]), and easy to use.

The use of the *generic programming paradigm* enables a convenient separation between the topology and the geometry of data structures. This is a key aspect, for example, of the design of CGAL polyhedra, CGAL triangulations, and CGAL arrangements (explored in Sect. 1.4.1) .

This way algorithms and data structures can be nicely abstracted in combinatorial and topological terms, regardless of the specific geometry and algebra of the objects at hand. This abstraction is realized through class and function

---

[3] *Adaptability* refers to the ability to incorporate existing user code, and *extendibility* refers to the ability to enhance the software with more code in the same style.

templates that represent specific data structures and algorithmic frameworks, respectively. The main class or function template that implements a data structure or an algorithm is typically instantiated with yet another class, referred to as a *traits* class, that defines the set of geometric objects and operations on them required to handle a concrete type of objects.

Generic programming is a key ingredient of flexibility. Changing only the traits class allows for instance to reuse the generic 2D class `Delaunay_triangulation_2` to compute a terrain in 3D: the traits class `Triangulation_euclidean_traits_xy_3` defines point sites as 3D points, and computes the elementary predicates on the first two coordinates only.

```
typedef CGAL::Exact_predicates_inexact_constructions_kernel Kernel;

typedef CGAL::Delaunay_triangulation_2<Kernel> Delaunay;
// the kernel Kernel defines the orientation and the in_circle
// tests on 2D points

typedef CGAL::Triangulation_euclidean_traits_xy_3<Kernel> Traits;
typedef CGAL::Delaunay_triangulation_2<Traits> Terrain;
// the traits class Traits defines the orientation and the in_circle
// tests on the 2D projections of the 3D points
```

An immediate advantage of the separation between the topology and the geometry of data structures is that users with limited expertise in computational geometry can employ the data structure with their own special type of objects. They must however supply the relevant traits class, which mainly involve algebraic computations. A traits class also encapsulates the number types used to represent coordinates of geometric objects and to carry out algebraic operations on them. It encapsulates the type of coordinate system used (e.g., Cartesian, Homogeneous), and the geometric or algebraic computation methods themselves. Naturally, a prospective user of the package that develops a traits class would like to face as few requirements as possible in terms of traits development.

Another advantage gained by the use of generic programming is the convenient handling of numerical issues to expedite exact geometric computation. In the classic computational-geometry literature two assumptions are usually made to simplify the design and analysis of geometric algorithms: First, inputs are in "general position". That is, degenerate input (e.g., three curves intersecting at a common point) is precluded. Secondly, operations on real numbers yield accurate results (the "real RAM" model [288], which also assumes that each basic operation takes constant time). Unfortunately, these assumptions do not hold in practice, as numerical errors are inevitable. Thus, an algorithm implemented without keeping this in mind may yield incorrect results (see [221, 305] for examples).

In a geometric algorithm each computational step is either a construction step or a conditional step based on the result of a predicate. The former produces a new geometric object such as the intersection point of two segments. The latter typically computes the sign of an expression used by the program control. Different computational paths lead to results with different combinatorial characteristics. Although numerical errors can sometimes be tolerated and interpreted as small perturbations in the input, they may lead to invalid combinatorial structures or inconsistent state during a program execution. Thus, it suffices to ensure that all predicates are evaluated correctly to eliminate inconsistencies and guarantee combinatorially correct results.

Exact Geometric Computation (EGC), as summarized by Yap [346], simply amounts to ensuring that we never err in predicate evaluations. EGC represents a significant relaxation from the naive concept of numerical exactness. We only need to compute to sufficient precision to make the correct predicate evaluation. This has led to the development of several techniques such as precision-driven computation, lazy evaluation, adaptive computation, and floating point filters, some of which are implemented in CGAL, such as numerical filtering. Here, computation is carried out using a number type that supports only inexact arithmetic (e.g., double floating point), while applying a filter that indicates whether the result is exact. If the filter fails, the computation is re-done using exact arithmetic.

Switching between number types and exact computation techniques, and choosing the appropriate components that best suit the application needs, is conveniently enabled through the generic programming paradigm, as it typically requires only a minor code change reflected in the instantiating of just a few data types.

8.4 CGAL Contents

CGAL is written in C++ according to the *generic programming* paradigm described above. It has a common programming style, which is very similar to that of the STL. Its Application Programming Interface (API) is homogeneous, and allows for a convenient and consistent interfacing with other software packages and applications. .

The library consists of about 500,000 lines of code divided among approximately 150 classes. CGAL also comes with numerous examples and demos. The manual has about 3,000 pages. There are roughly 50 chapters that are grouped in several parts for a rough description.

The first part is the kernels [155], which consist of constant size nonmodifiable geometric primitive objects and operations on these objects. The objects are represented both as stand-alone classes that are instantiated by a kernel class, and as members of the kernel classes. The latter option allows for more flexibility and adaptability of the kernel.

In addition, CGAL offers a collection of basic geometric data structures and algorithms such as convex hull, polygons and polyhedra and operations on them (Boolean operations, polygon offsetting), 2D arrangements, 2D and 3D triangulations, Voronoi diagrams, surface meshing and surface subdivision, search structures, geometric optimization, interpolation, and kinetic data structures. These data structures and algorithms are parameterized by traits classes, that define the interface between them and the primitives they use. In many cases, the kernel can be used as a traits class, or the kernel classes provided in CGAL can be used as components of traits classes for these data structures and algorithms.

The third part of the library consists of non-geometric support facilities, such as circulators, random generators, I/O support for debugging and for interfacing CGAL with various visualization tools. This part also provides the user with number type support.

CGAL kernel classes are parameterized by number types. Instantiating a kernel with a particular number type is a trade-off between efficiency and accuracy. The choice depends on the algorithm implementation and the expected input data to be handled. Number types must fulfil certain requirements, so that they can be successfully used by the kernel code. The list of requirements establishes a concept of a number type. A few number-type concepts have been introduced by CGAL, e.g., *RingNumberType* and *FieldNumberType*. Naturally, number types have evident semantic constraints. That is, they should be meaningful in the sense that they approximate some subfield of the real numbers. CGAL provides several models of its number-type concepts, some of them implement techniques to expedite exact computation mentioned in the previous paragraph. CGAL also provides a glue layer that adapts number-type classes implemented by external libraries as models of its number-type concepts.

The above describes the accessibility model of CGAL at the time this book was written. Constant and persistent improvement to the source code and the didactic manuals, review of packages by the Editorial board and exhaustive testing, through the years led to a state of excellent quality internationally recognized as an unrivalled tool in its field. At the time these lines are written, CGAL already has a foothold in many domains related to computational geometry and could be found in many academic and research institutes as well as commercial entities. Release 3.1 was downloaded more than 14.500 times, and the public discussion list counts more than 950 subscribed users.

Acknowledgements

The development of CGAL was supported by two European Projects CGAL and GALIA during three years in total (1996–1999). Several sites have kept on working on CGAL after the European support stopped.

The new European project ACS (Algorithms for Complex Shapes with certified topology and numerics)[4] provides again partial support for new research and developments in CGAL.

―――――――――

[4]http://acs.cs.rug.nl/.

References

The page numbers where each reference is cited are listed in brackets at the end of each item.

1. BOOST, C++ libraries.
   http://www.boost.org. [154, 316]
2. CGAL, Computational Geometry Algorithms Library.
   http://www.cgal.org. [VI, 3, 26, 29, 55, 114, 154, 232, 313, 316]
3. CORE number library.
   http://cs.nyu.edu/exact/core_pages. [11, 35, 38, 155]
4. EXACUS, Efficient and exact algorithms for curves and surfaces.
   http://www.mpi-inf.mpg.de/projects/EXACUS. [4, 13, 36]
5. FGB/RS.
   http://fgbrs.lip6.fr. [155]
6. GMP, GNU multiple precision arithmetic library.
   http://www.swox.com/gmp. [38, 119, 154]
7. IRIT modeling environment.
   http://www.cs.technion.ac.il/~irit/. [61]
8. LEDA, Library for efficient data types and algorithms.
   http://www.algorithmic-solutions.com/enleda.htm. [316]
9. MPFI, Multiple precision interval arithmetic library.
   http://perso.ens-lyon.fr/nathalie.revol/software.html. [154]
10. MPFR, Multiple-precision floating-point computations.
    http://www.mpfr.org. [154]
11. QI: Quadrics intersection.
    http://www.loria.fr/equipes/isa/qi. [50]
12. SINGULAR, Computer algebra system for polynomial computations.
    http://www.singular.uni-kl.de. [155]
13. SYNAPS, a library for symbolic and numeric computation.
    http://www-sop.inria.fr/galaad/logiciels/synaps. [155]
14. K. Abdel-Malek and H.-J. Yeh. On the determination of starting points for parametric surface intersections. *Computer-Aided Design*, 28:21–35, 1997. [117]
15. P. K. Agarwal and M. Sharir. Arrangements and their applications. In J.-R. Sack and J. Urrutia, editors, *Handbook of Computational Geometry*, pages 49–119. Elsevier Science Publishers B.V. North-Holland, Amsterdam, 2000. [2, 41, 66]

16. M. Alexa, J. Behr, D. Cohen-Or, S. Fleishman, D. Levin, and C. Silva. Point set surfaces. In IEEE, editor, *Visualization*, pages 21–28, 2001. [232]
17. B. S. Alin Bostan, Philippe Flajolet and É. Schost. Fast computation of special resultants. *Journal of Symbolic Computation*, 41(1):1–29, January 2006. [128]
18. R. Allegre, R. Chaine, and S. Akkouche. Convection-driven dynamic surface reconstruction. In *Proc. of Shape Modeling International*, pages 33–42, Cambridge, MA, USA, June 15–17 2005. [268]
19. P. Alliez, D. Cohen-Steiner, O. Devillers, B. Lévy, and M. Desbrun. Anisotropic polygonal remeshing. *ACM Trans. Graph.*, 22(3):485–493, 2003. [179]
20. H. Alt, O. Cheong, and A. Vigneron. The Voronoi diagram of curved objects. *Discrete and Computational Geometry*, 34(3):439–453, 2002. [104, 105]
21. N. Amenta and M. Bern. Surface reconsruction by Voronoi filtering. In *Proc. 14th Ann. Sympos. Comput. Geom.*, pages 39–48, 1998. [178, 253]
22. N. Amenta and M. Bern. Surface reconstruction by Voronoi filtering. *Discrete Comput. Geom.*, 22(4):481–504, 1999. [202, 203, 248, 249, 257]
23. N. Amenta, M. Bern, and D. Eppstein. The crust and the β-skeleton: Combinatorial curve reconstruction. *Graphical Models and Image Processing*, 60:125–135, 1998. [248]
24. N. Amenta, M. Bern, and D. Eppstein. The crust and the β-skeleton: Combinatorial curve reconstruction. *Graphical models and image processing: GMIP*, 60(2):125–135, 1998. [253]
25. N. Amenta, S. Choi, T. K. Dey, and N. Leekha. A simple algorithm for homeomorphic surface reconstruction. In *Proc. 16th Ann. Sympos. Comput. Geom.*, pages 213–222, 2000. [178, 258, 272]
26. N. Amenta, S. Choi, and R. Kolluri. The power crust. In *ACM Solid Modeling*, pages 249–260, 2001. [261, 272]
27. N. Amenta, S. Choi, and R. K. Kolluri. The power crust, unions of balls, and the medial axis transform. *Comput. Geom. Theory Appl.*, 19:127–153, 2001. [261, 264, 272]
28. M. Armstrong. *Basic Topology.* Undergraduate Texts in Mathematics. Springer-Verlag, New York, Berlin, Heidelberg, 1983. [278]
29. V. I. Arnol'd, S. M. Guseĭn-Zade, and A. N. Varchenko. *Singularities of differentiable maps. Vol. I: The classification of critical points, caustics and wave fronts*, volume 82 of *Monographs in Mathematics*. Birkhäuser Boston Inc., Boston, MA, 1985. [117]
30. D. Attali. r-regular shape reconstruction from unorganized points. *Comput. Geom. Theory Appl.*, 10:239–247, 1998. [262]
31. D. Attali and J.-D. Boissonnat. A linear bound on the complexity of the delaunay triangulation of points on polyhedral surfaces. *Discrete and Comp. Geometry*, 31:369–384, 2004. [77]
32. D. Attali, J.-D. Boissonnat, and H. Edelsbrunner. Stability and computation of medial axes: a state of the art report. In T. Möller, B. Hamann, and B. Russell, editors, *Mathematical Foundations of Scientific Visualization, Computer Graphics, and Massive Data Exploration*, Mathematics and Visualization. Springer-Verlag. [110, 113]
33. D. Attali, J.-D. Boissonnat, and A. Lieutier. Complexity of the Delaunay triangulation of points on surfaces the smooth case. In *Proc. 19th Ann. Symposium on Computational Geometry*, pages 201–210, San Diego, 2003. ACM Press. [77, 232, 273]

34. M. Attene and M. Spagnuolo. Automatic surface reconstruction from point sets in space. In *Eurographics*, pages 457–465. ACM Press, 2000. [269]

35. F. Aurenhammer. Power diagrams: properties, algorithms and applications. *SIAM J. Comput.*, 16:78–96, 1987. [83, 84, 90, 96]

36. F. Aurenhammer. Linear combinations from power domains. *Geom. Dedicata*, 28:45–52, 1988. [242]

37. F. Aurenhammer and R. Klein. Voronoi diagrams. In J.-R. Sack and J. Urrutia, editors, *Handbook of Computational Geometry*, pages 201–290. Elsevier Science Publishers B.V. North-Holland, Amsterdam, 2000. [69]

38. M. H. Austern. *Generic Programming and the STL*. Addison Wesley, 1999. [36, 315]

39. W. Auzinger and H. J. Stetter. An elimination algorithm for the computation of all zeros of a system of multivariate polynomial equations. In *Proc. Intern. Conf. on Numerical Math.*, volume 86 of *Int. Series of Numerical Math*, pages 12–30. Birkhäuser Verlag, 1988. [141]

40. C. Bajaj and M.-S. Kim. Generation of configuration space obstacles: the case of moving algebraic surfaces. *Algorithmica*, 4:155–172, 1989. [57]

41. T. F. Banchoff. Critical points and curvature for embedded polyhedra. *J. Diff. Geom.*, 1:245–256, 1967. [226]

42. T. F. Banchoff. Critical points and curvature for embedded polyhedral surfaces. *Amer. Math. Monthly*, 77:475–485, 1970. [165, 226]

43. G. Barequet and V. Rogol. Maximizing the area of an axially-symmetric polygon inscribed by a simple polygon. In *16th Canadian Conf. on Computational Geometry*, pages 128–131, 2004. [62]

44. S. Basu, R. Pollack, and M.-F. Roy. *Algorithms in Real Algebraic Geometry*. Springer-Verlag, Berlin, 2003. ISBN 3-540-00973-6. [110, 125, 130, 133, 138, 142]

45. J. L. Bentley. Multidimensional binary search trees used for associative searching. *Commun. ACM*, 18(9):509–517, Sept. 1975. [23]

46. J. L. Bentley and T. A. Ottmann. Algorithms for reporting and counting geometric intersections. *IEEE Trans. Comput.*, C-28(9):643–647, Sept. 1979. [4, 7]

47. E. Berberich. *Exact Arrangements of Quadric Intersection Curves*. Universität des Saarlandes, Saarbrücken, 2004. Master Thesis. [48]

48. E. Berberich, A. Eigenwillig, M. Hemmer, S. Hert, L. Kettner, K. Mehlhorn, J. Reichel, S. Schmitt, E. Schömer, and N. Wolpert. EXACUS: Efficient and exact algorithms for curves and surfaces. In *Proc. 13th Annu. European Sympos. Algorithms (ESA'05)*, volume 3669 of *Lecture Notes in Computer Science*, pages 155–166, Oct. 2005. [4, 36]

49. E. Berberich, A. Eigenwillig, M. Hemmer, S. Hert, K. Mehlhorn, and E. Schömer. A computational basis for conic arcs and Boolean operations on conic polygons. In *Proc. 10th European Symposium on Algorithms*, volume 2461 of *Lecture Notes in Computer Science*, pages 174–186. Springer-Verlag, 2002. [4, 9, 11, 36, 55]

50. E. Berberich, M. Hemmer, L. Kettner, E. Schömer, and N. Wolpert. An exact, complete and efficient implementation for computing planar maps of quadric intersection curves. In *Proc. 21th Annual Symposium on Computational Geometry*, pages 99–106, 2005. [4, 36]

51. M. Berger and B. Gostiaux. *Differential Geometry: Manifolds, Curves and Surfaces*. Graduate Texts in Mathematics No. 115. Springer-Verlag, New York, 1988. (translated by S. Levy). [158, 160, 161]

52. F. Bernardini, J. Mittleman, H. Rushmeir, C. Silva, and G. Taubin. The ball-pivoting algorithm for surface reconstruction. *IEEE Transactions on Visualization and Computer Graphics*, 5(4):349–359, 1999. [268, 273]

53. H. Bieri. Nef polyhedra: A brief introduction. *Computing Suppl.*, 10:43–60, 1995. [55]

54. J.-D. Boissonnat. Geometric structures for three-dimensional shape representation. *ACM Trans. Graph.*, 3(4):266–286, 1984. [232, 253, 261]

55. J.-D. Boissonnat and F. Cazals. Coarse-to-fine surface simplification with geometric guarantees. In *Eurographics*, 01. [264]

56. J.-D. Boissonnat and F. Cazals. Smooth surface reconstruction via natural neighbour interpolation of distance functions. In *Proc. 16th Ann. Sympos. Comput. Geom.*, pages 223–232, 2000. [263, 272, 275]

57. J.-D. Boissonnat and F. Cazals. Smooth surface reconstruction via natural neighbour interpolation of distance functions. In *Proc. 16th Ann. Symposium on Computational Geometry*, pages 223–232. ACM Press, 2000. [268]

58. J.-D. Boissonnat and F. Cazals. Coarse-to-fine surface simplification with geometric guarantees. In A. Chalmers and T.-M. Rhyne, editors, *Eurographics'01*, pages 490–499, Manchester, 2001. Blackwell. [263, 272]

59. J.-D. Boissonnat and F. Cazals. Natural coordinates of points on a surface. *Comput. Geom. Theory Appl.*, 19:155–173, 2001. [247]

60. J.-D. Boissonnat, A. Cérézo, O. Devillers, and M. Teillaud. Output-sensitive construction of the Delaunay triangulation of points lying in two planes. *Internat. J. Comput. Geom. Appl.*, 6(1):1–14, 1996. [69]

61. J.-D. Boissonnat, D. Cohen-Steiner, and G. Vegter. Isotopic implicit surface meshing. In *Proc. 36th Ann. ACM Symposium on Theory of Computing*, pages 301–309, New York, June 2004. ACM Press. [187, 224, 226]

62. J.-D. Boissonnat and C. Delage. Convex hulls and Voronoi diagrams of additively weighted points. In *Proc. 13th European Symposium on Algorithms*, Lecture Notes in Computer Science, pages 367–378. Springer, 2005. [91, 115]

63. J.-D. Boissonnat and M. Karavelas. On the combinatorial complexity of Euclidean Voronoi cells and convex hulls of *d*-dimensional spheres. In *Proc. 14th ACM-SIAM Sympos. Discrete Algorithms (SODA)*, pages 305–312, 2003. [81, 90]

64. J.-D. Boissonnat and S. Oudot. Provably good surface sampling and approximation. In *SGP '03: Proceedings of the 2003 Eurographics/ACM SIGGRAPH symposium on Geometry processing*, pages 9–18, Aire-la-Ville, Switzerland, Switzerland, 2003. Eurographics Association. [187, 203, 205, 207]

65. J.-D. Boissonnat and S. Oudot. Provably good sampling and meshing of surfaces. *Graphical Models*, 67:405–451, 2005. [114, 203, 208]

66. J.-D. Boissonnat and S. Oudot. Provably good sampling and meshing of lipschitz surfaces. In *Proc. 22nd Ann. Sympos. Comput. Geom.*, 2006. [208]

67. J.-D. Boissonnat and M. Yvinec. *Algorithmic Geometry*. Cambridge University Press, UK, 1998. [69, 100, 104, 105, 109]

68. V. Borrelli, F. Cazals, and J.-M. Morvan. On the angular defect of triangulations and the pointwise approximation of curvatures. *Comput. Aided Geom. Design*, 20:319–341, 2003. [160, 166]

69. H. Brönnimann, C. Burnikel, and S. Pion. Interval arithmetic yields efficient dynamic filters for computational geometry. *Discrete Applied Mathematics*, 109:25–47, 2001. [121, 154]

70. H. Brönnimann, L. Kettner, S. Schirra, and R. Veltkamp. Applications of the generic programming paradigm in the design of CGAL. In M. Jazayeri, R. Loos, and D. Musser, editors, *Generic Programming—Proceedings of a Dagstuhl Seminar*, volume 1766 of *Lecture Notes in Computer Science*, pages 206–217. Springer-Verlag, 2000. [36]

71. H. Brönnimann, G. Melquiond, and S. Pion. The Boost interval arithmetic library. In *Proc. 5th Conference on Real Numbers and Computers*, pages 65–80, 2003. [154]

72. J. L. Brown. Systems of coordinates associated with points scattered in the plane. *Comput. Aided Design*, 14:547–559, 1997. [242]

73. C. Burnikel, S. Funke, K. Mehlhorn, S. Schirra, and S. Schmitt. A separation bound for real algebraic expressions. Technical Report ECG-TR-123101-02, MPI Saarbrücken, 2002. [139, 140]

74. C. Burnikel, S. Funke, and M. Seel. Exact geometric computation using cascading. *Internat. J. Comput. Geom. Appl.*, 11:245–266, 2001. [122, 123]

75. C. Burnikel, K. Mehlhorn, and S. Schirra. How to compute the Voronoi diagram of line segments: Theoretical and experimental results. In *Proc. 2nd Annu. European Sympos. Algorithms*, volume 855 of *Lecture Notes Comput. Sci.*, pages 227–239. Springer-Verlag, 1994. [140]

76. C. Burnikel, K. Mehlhorn, and S. Schirra. The LEDA class real number. Technical Report MPI-I-96-1-001, Max-Planck Institut Inform., Saarbrücken, Germany, Jan. 1996. [155]

77. J. Canny. *The Complexity of Robot Motion Planning*. ACM – MIT Press Doctoral Dissertation Award Series. MIT Press, Cambridge, MA, 1987. [140]

78. H. Carr, J. Snoeyink, and U. Axen. Computing contour trees in all dimensions. *Computational Geometry*, 24:75–94, 2003. [309]

79. J. Carr, R. Beatson, J. Cherrie, T. Mitchell, T. Fright, B. McCallum, and T. Evans. Reconstruction and representation of 3d objects with radial basis functions. In *Siggraph*, pages 67–76. ACM, 2001. [232]

80. H. Cartan. *Differential Calculus*. Houghton Mifflin Co, 1971. [169]

81. F. Cazals, J. Giesen, M. Pauly, and A. Zomorodian. Conformal alpha shapes. In *Symposium on Point Based Graphics*, 2005. [269]

82. R. Chaine. A convection geometric-based approach to surface reconstruction. In *Symp. Geometry Processing*, pages 218–229, 2003. [266, 276]

83. F. Chazal and A. Lieutier. The lambda medial axis. *Graphical Models*, 67: 304–331, 2005. [113]

84. B. Chazelle. An optimal convex hull algorithm in any fixed dimension. *Discrete Comput. Geom.*, 10:377–409, 1993. [77]

85. B. Chazelle, H. Edelsbrunner, L. J. Guibas, and M. Sharir. A singly-exponential stratification scheme for real semi-algebraic varieties and its applications. *Theoret. Comput. Sci.*, 84:77–105, 1991. [42, 47]

86. B. Chazelle et al. Application challenges to computational geometry: CG impact task force report. Technical Report TR-521-96, Princeton Univ., Apr. 1996. [313]

87. B. Chazelle et al. Application challenges to computational geometry: CG impact task force report. In B. Chazelle, J. E. Goodman, and R. Pollack, editors,

Advances in Discrete and Computational Geometry, volume 223 of *Contemporary Mathematics*, pages 407–463. American Mathematical Society, Providence, 1999. [313]

88. H.-L. Cheng, T. K. Dey, H. Edelsbrunner, and J. Sullivan. Dynamic skin triangulation. *Discrete Comput. Geom.*, 25:525–568, 2001. [248]

89. S.-W. Cheng, T. K. Dey, H. Edelsbrunner, M. A. Facello, and S.-H. Teng. Sliver exudation. In *Proc. 15th Ann. Sympos. Comput. Geom.*, pages 1–13, 1999. [235]

90. S.-W. Cheng, T. K. Dey, E. A. Ramos, and T. Ray. Sampling and meshing a surface with guaranteed topology and geometry. In *Proc. 20th Ann. Sympos. Comput. Geom.*, pages 280–289, 2004. [187, 209, 211, 222]

91. L. P. Chew. Guaranteed-quality mesh generation for curved surfaces. In *Proc. 9th Ann. Sympos. Comput. Geom.*, pages 274–280, 1993. [201]

92. Y.-J. Chiang, T. Lenz, X. Lu, and G. Rote. Simple and optimal output-sensitive construction of contour trees using monotone paths. *Computational Geometry, Theory and Applications*, 30:165–195, 2005. [309]

93. D. Cohen-Steiner. *Topics in Surface Discretization*. PhD thesis, Ecole Polytechnique, 2004. [178]

94. D. Cohen-Steiner and F. Da. A greedy Delaunay based surface reconstruction algorithm. *The Visual Computer*, 20:4–16, 2004. [255, 256, 272, 274]

95. D. Cohen-Steiner and J.-M. Morvan. Approximation of normal cycles. Research Report 4723, INRIA, 2003. [169, 173]

96. D. Cohen-Steiner and J.-M. Morvan. Approximation of the curvature measures of a smooth surface endowed with a mesh. Research Report 4867, INRIA, 2003. [169, 173]

97. D. Cohen-Steiner and J.-M. Morvan. Approximation of the second fundamental form of a hypersurface of a riemannian manifold. Research Report 4868, INRIA, 2003. [169, 173]

98. D. Cohen-Steiner and J.-M. Morvan. Restricted Delaunay triangulations and normal cycles. In *Proc. 19th Ann. Symposium on Computational Geometry*, pages 237–246, San Diego, CA, June 2003. [175]

99. K. Cole-McLaughlin, H. Edelsbrunner, J. Harer, V. Natarajan, and V. Pascucci. Loops in Reeb graphs of 2-manifolds. *Discrete Comput. Geom.*, 32(2):231–244, 2004. [310]

100. E. Collins. Quantifier elimination by cylindrical algebraic decomposition—twenty years of progress. In B. Caviness and J. Johnson, editors, *Quantifier Elimination and Cylindrical Algebraic Decomposition*, pages 8–23. Springer Verlag, 1998. [5]

101. G. E. Collins. Quantifier elimination for real closed fields by cylindrical algebraic decomposition. In *Proc. 2nd GI Conference on Automata Theory and Formal Languages*, volume 33 of *Lecture Notes Comput. Sci.*, pages 134–183. Springer-Verlag, 1975. [5, 47]

102. G. E. Collins and J. R. Johnson. Quantifier elimination and the sign variation method for real root isolation. In *ISSAC '89: Proceedings of the ACM-SIGSAM 1989 international symposium on Symbolic and algebraic computation*, pages 264–271, New York, NY, USA, 1989. ACM Press. [135]

103. M. L. Connolly. Analytical molecular surface calculation. *J. Appl. Cryst.*, 16:548–558, 1983. [63]

104. M. L. Connolly. Molecular surfaces: A review, 1996. http://www.netsci.org/Science/Compchem/feature14.html. [64]

105. R. Corless, P. Gianni, and B. Trager. A reordered Schur factorization method for zero-dimensional polynomial systems with multiple roots. In *Proc. International Conference on Symbolic and Algebraic Computation*, pages 133–140. ACM Press, 1997. [141]

106. M. Coste. An introduction to semi-algebraic geometry. RAAG network school, 2002. [145]

107. D. Cox, J. Little, and D. O'Shea. *Ideals, Varieties, and Algorithms: An Introduction to Computational Algebraic Geometry and Commutative Algebra.* Undergraduate Texts in Mathematics. Springer Verlag, New York, 1992. [141, 146]

108. H. S. M. Coxeter. *Introduction to Geometry.* John Wiley & Sons, New York, 2nd edition, 1969. [85]

109. M. de Berg, L. J. Guibas, and D. Halperin. Vertical decompositions for triangles in 3-space. *Discrete Comput. Geom.*, 15:35–61, 1996. [42, 43]

110. M. de Berg, M. van Kreveld, M. Overmars, and O. Schwarzkopf. *Computational Geometry: Algorithms and Applications.* Springer-Verlag, Berlin, 1997. [69]

111. M. de Berg, M. van Kreveld, M. Overmars, and O. Schwarzkopf. *Computational Geometry: Algorithms and Applications.* Springer-Verlag, Berlin, Germany, 2nd edition, 2000. [6, 9, 21, 23]

112. M. de Carmo. *Differential Geometry of Curves and Surfaces.* Prentice Hall, Englewood Cliffs, NJ, 1976. [158]

113. C. J. A. Delfinado and H. Edelsbrunner. An incremental algorithm for Betti numbers of simplicial complexes on the 3-sphere. *Comput. Aided Geom. Des.*, 12(7):771–784, 1995. [291]

114. R. Descartes. *Géométrie*, volume 90-31 of *A source book in Mathematics*. Havard University press, 1636. [131]

115. O. Devillers. The Delaunay hierarchy. *Internat. J. Found. Comput. Sci.*, 13:163–180, 2002. [107, 109]

116. O. Devillers, A. Fronville, B. Mourrain, and M. Teillaud. Algebraic methods and arithmetic filtering for exact predicates on circle arcs. *Comput. Geom. Theory Appl.*, 22:119–142, 2002. [4, 40, 136]

117. O. Devillers and S. Pion. Efficient exact geometric predicates for Delaunay triangulations. In *Proc. 5th Workshop Algorithm Eng. Exper.*, pages 37–44, 2003. [122, 123]

118. T. Dey, H. Woo, and W. Zhao. Approximate medial axis for CAD models. In *Proc. 8th ACM symposium on Solid modeling and applications*, pages 280–285, 2003. [116]

119. T. K. Dey, H. Edelsbrunner, and S. Guha. Computational topology. In B. Chazelle, J. E. Goodman, and R. Pollack, editors, *Advances in Discrete and Computational Geometry - Proc. 1996 AMS-IMS-SIAM Joint Summer Research Conf. Discrete and Computational Geometry: Ten Years Later*, number 223 in Contemporary Mathematics, pages 109–143. American Mathematical Society, 1999. [277]

120. T. K. Dey and J. Giesen. Detecting undersampling in surface reconstruction. In *Proc. 16th Ann. Symposium on Computational Geometry*, pages 257–263. ACM Press, 2001. [253]

121. T. K. Dey, J. Giesen, and J. Hudson. Delaunay based shape reconstruction from large data. In *IEEE Symposium in Parallel and Large Data Visualization and Graphics*, pages 19–27, 2001. [260]

122. T. K. Dey, J. Giesen, E. A. Ramos, and B. Sadri. Critical points of the distance to an epsilon-sampling on a surface and flow-complex-based surface reconstruction. In *Proc. 21st Ann. Symposium on Computational Geometry*, pages 218–227. ACM Press, 2005. [253, 266]

123. T. K. Dey and S. Goswami. Tight cocone: A water-tight surface reconstructor. *Journal of Computing and Information Science in Engineering*, 3:302–307, 2003. [260, 272]

124. T. K. Dey and S. Goswami. Provable surface reconstruction from noisy samples. In *Proc. 20th Ann. Symposium on Computational Geometry*, pages 330–339. ACM Press, 2004. [261, 272]

125. T. K. Dey and P. Kumar. A simple provable algorithm for curve reconstruction. In *SODA '99: Proceedings of the Tenth Annual ACM-SIAM Symposium on Discrete Algorithms*, pages 893–894, Philadelphia, PA, USA, 1999. Society for Industrial and Applied Mathematics. [253]

126. T. K. Dey, G. Li, and T. Ray. Polygonal surface remeshing with Delaunay refinement. In *Proc. 14th Internat. Meshing Roundtable*, pages 343–361, 2005. [213]

127. T. K. Dey, K. Mehlhorn, and E. A. Ramos. Curve reconstruction: Connecting dots with good reason. *Comput. Geom. Theory Appl.*, 15:229–244, 2000. [228, 253]

128. L. Dupont, D. Lazard, S. Lazard, and S. Petitjean. Near-optimal parameterization of the intersection of quadrics. In *Proc. 19th Ann. Sympos. Comput. Geom.*, pages 246–255, 2003. [49]

129. R. A. Dwyer. Higher-dimensional Voronoi diagrams in linear expected time. *Discrete Comput. Geom.*, 6:343–367, 1991. [77]

130. H. Edelsbrunner. *Algorithms in Combinatorial Geometry*, volume 10 of *EATCS Monographs on Theoretical Computer Science*. Springer-Verlag, Heidelberg, West Germany, 1987. [2, 66]

131. H. Edelsbrunner. Weighted alpha shapes. Technical Report UIUCDCS-R-92-1760, Dept. Comput. Sci., Univ. Illinois, Urbana, IL, 1992. [238]

132. H. Edelsbrunner. Surface reconstruction by wrapping finite point sets in space. In B. Aronov, S. Basu, J. Pach, and M. Sharir, editors, *Ricky Pollack and Eli Goodman Festschrift*, pages 379–404. Springer-Verlag, 2003. [243, 264, 272]

133. H. Edelsbrunner. Biological applications of computational topology. In J. Goodman and J. O'Rourke, editors, *CRC Handbook of Discrete and Computational Geometry*, chapter 63, pages 1395–1412. Chapman & Hall/CRC, 2004. [277]

134. H. Edelsbrunner, J. Harer, V. Natarajan, and V. Pascucci. Morse-Smale complexes for piecewise linear 3-manifolds. In *Proc. 19th Ann. Sympos. Comput. Geom.*, pages 361–370, 2003. [306]

135. H. Edelsbrunner, J. Harer, and A. Zomorodian. Hierarchical Morse-Smale complexes for piecewise linear 2-manifolds. *Discrete Comput. Geom.*, 30(1): 87–107, 2003. [306]

136. H. Edelsbrunner and E. P. Mücke. Simulation of simplicity: A technique to cope with degenerate cases in geometric algorithms. *ACM Trans. Graph.*, 9(1):66–104, 1990. [210, 235]

137. H. Edelsbrunner and E. P. Mücke. Three-dimensional alpha shapes. *ACM Trans. Graph.*, 13(1):43–72, Jan. 1994. [238]

138. H. Edelsbrunner and N. R. Shah. Triangulating topological spaces. *Int. J. on Comp. Geom.*, 7:365–378, 1997. [84, 209, 228, 241]

139. A. Eigenwillig. Exact arrangement computation for cubic curves. M.Sc. thesis, Universität des Saarlandes, Saarbrücken, Germany, 2003. [20]

140. A. Eigenwillig, L. Kettner, E. Schömer, and N. Wolpert. Complete, exact, and efficient computations with cubic curves. In *Proc. 20th Annual Symposium on Computational Geometry*, pages 409–418, 2004. accepted for Computational Geometry: Theory and Applications. [4, 11, 18, 36, 38]

141. D. Eisenbud. *Commutative Algebra with a view toward Algebraic Geometry*, volume 150 of *Graduate Texts in Math.* Berlin, Springer-Verlag, 1994. [146]

142. M. Elkadi and B. Mourrain. *Introduction à la résolution des systèmes d'équations algébriques*, 2003. Notes de cours, Univ. de Nice (310 p.). [142, 148, 149]

143. M. Elkadi and B. Mourrain. Symbolic-numeric tools for solving polynomial equations and applications. In A. Dickenstein and I. Emiris, editors, *Solving Polynomial Equations: Foundations, Algorithms, and Applications.*, volume 14 of *Algorithms and Computation in Mathematics*, pages 125–168. Springer, 2005. [141]

144. I. Emiris and E. P. Tsigaridas. Computing with real algebraic numbers of small degree. In *Proc. 12th European Symposium on Algorithms, LNCS 3221*, pages 652–663. Springer-Verlag, 2004. [4, 40]

145. I. Z. Emiris, A. Kakargias, S. Pion, M. Teillaud, and E. P. Tsigaridas. Towards an open curved kernel. In *Proc. 20th Ann. Sympos. Comput. Geom.*, pages 438–446, 2004. [4, 11, 39]

146. I. Z. Emiris and M. I. Karavelas. The predicates of the apollonius diagram: algorithmic analysis and implementation. *Computational Geometry: Theory and Applications*, 33:18–57, 2006. [109]

147. I. Z. Emiris, B. Mourrain, and E. P. Tsigaridas. Real algebraic numbers: Complexity analysis and experimentations. Research Report 5897, INRIA, Avril 2006. [135, 136, 138]

148. I. Z. Emiris and E. P. Tsigaridas. Comparison of fourth-degree algebraic numbers and applications to geometric predicates. Technical Report ECG-TR-302206-03, INRIA Sophia-Antipolis, 2003. [136, 137, 138, 139]

149. I. Z. Emiris and E. P. Tsigaridas. Methods to compare real roots of polynomials of small degree. Technical Report ECG-TR-242200-01, INRIA Sophia-Antipolis, 2003. [136]

150. J. Erickson. Nice point sets can have nasty Delaunay triangulations. *Discrete Comput. Geom.*, 30(1):109–132, 2003. [206]

151. J. Erickson. Dense point sets have sparse delaunay triangulations. *Discrete Comput. Geom.*, 33:85–115, 2005. [77]

152. E. Eyal and D. Halperin. Improved implementation of controlled perturbation for arrangements of spheres. Technical Report ECG-TR-363208-01, Tel-Aviv University, 2004. [64]

153. E. Eyal and D. Halperin. Dynamic maintenance of molecular surfaces under conformational changes. In *Proc. 21st Ann. Symposium on Computational Geometry*, pages 45–54, 2005. [50, 64]

154. E. Eyal and D. Halperin. Improved maintenance of molecular surfaces using dynamic graph connectivity. In *Proc. 5th Workshop on Algorithms in Bioinformatics - WABI 2005*, volume 3692 of *LNCS*, pages 401–413. Springer-Verlag, 2005. [50, 53, 64]

155. A. Fabri, G.-J. Giezeman, L. Kettner, S. Schirra, and S. Schönherr. The CGAL kernel: A basis for geometric computation. In M. C. Lin and D. Manocha, editors, *Proc. 1st ACM Workshop on Appl. Comput. Geom.*, volume 1148 of *Lecture Notes Comput. Sci.*, pages 191–202. Springer-Verlag, 1996. [318]

156. A. Fabri, G.-J. Giezeman, L. Kettner, S. Schirra, and S. Schönherr. On the design of CGAL a computational geometry algorithms library. *Softw. - Pract. Exp.*, 30(11):1167–1202, 2000. [3, 36, 316]

157. G. Farin. *Curves and surfaces for computer aided geometric design: a practical guide*. Comp. science and sci. computing. Acad. Press, 1990. [132, 136]

158. G. Farin. An SSI bibliography. In *Geometry Processing for Design and Manufacturing*, pages 205–207. SIAM, Philadelphia, 1992. [117]

159. R. Farouki and R. Ramamurthy. Voronoi diagram and medial axis algorithm for planar domains with curved boundaries i. theoretical foundations. *Journal of Computational and Applied Mathematics*, 102(1):119–141, 1999. [110]

160. O. Faugeras. *Three-Dimensional Computer Vision: A Geometric Viewpoint*. MIT Press, Cambridge, MA, 1993. [116]

161. H. Federer. Curvature measures. *Trans. Amer. Math. Soc.*, 93:418–491, 1959. [162, 169]

162. H. Federer. *Geometric Measure Theory*. Springer-Verlag, 1970. [162, 174]

163. E. Flato, D. Halperin, I. Hanniel, and O. Nechushtan. The design and implementation of planar maps in CGAL. In *Abstracts 15th European Workshop Comput. Geom.*, pages 169–172. INRIA Sophia-Antipolis, 1999. [26]

164. E. Flato, D. Halperin, I. Hanniel, O. Nechushtan, and E. Ezra. The design and implementation of planar maps in CGAL. *The ACM Journal of Experimental Algorithmics*, 5:1–23, 2000. [23]

165. E. Fogel, D. Halperin, R. Wein, S. Pion, M. Teillaud, I. Emiris, A. Kakargias, E. Tsigaridas, E. Berberich, A. Eigenwillig, M. Hemmer, L. Kettner, K. Mehlhorn, E. Schomer, and N. Wolpert. An empirical comparison of software for constructing arrangements of curved arcs (preliminary version). Technical Report ECG-TR-361200-01, Tel-Aviv University, INRIA Sophia-Antipolis, MPI Saarbrücken, 2004. [4]

166. E. Fogel, D. Halperin, R. Wein, M. Teillaud, E. Berberich, A. Eigenwillig, S. Hert, and L. Kettner. Specification of the traits classes for CGAL arrangements of curves. Technical Report ECG-TR-241200-01, INRIA Sophia-Antipolis, 2003. [4]

167. E. Fogel, R. Wein, and D. Halperin. Code flexibility and program efficiency by genericity: Improving CGAL's arrangements. In *Proc. 12th Annual European Symposium on Algorithms*, pages 664–676. Springer-Verlag, 2004. [4, 25, 26]

168. S. Fortune and C. J. Van Wyk. Static analysis yields efficient exact integer arithmetic for computational geometry. *ACM Trans. Graph.*, 15(3):223–248, July 1996. [121, 122]

169. S. Fortune and C. V. Wyk. *LN User Manual*. AT&T Bell Laboratories, 1993. [123]

170. J. Fu. Convergence of curvatures in secant approximations. *Journal of Differential Geometry*, 37:177–190, 1993. [163, 169, 171]

171. J. Fu. Curvature of singular spaces via the normal cycle. *Amer. Math. Soc.*, 116:819–880, 1994. [171]

172. S. Funke, C. Klein, K. Mehlhorn, and S. Schmitt. Controlled perturbations for Delaunay triangulations. In *Proc. 16th ACM-SIAM Sympos. Discrete Algorithms (SODA)*, pages 1047–1056, 2005. [50, 53]

173. S. Funke, K. Mehlhorn, and S. Schmitt. The LEDA class real number – extended version. Technical Report ECG-TR-363110-01, MPI Saarbrücken, 2004. [155]

174. S. Funke and E. A. Ramos. Smooth-surface reconstruction in near-linear time. In *ACM SODA'02*, pages 781 – 790, 2002. [260, 273]

175. E. Gamma, R. Helm, R. Johnson, and J. Vlissides. *Design Patterns*. Addison-Wesley, Reading, MA, 1995. [27, 32]

176. T. Garrity and J. Warren. Geometric continuity. *Computer Aided Geometric Design*, 8:51–65, 1991. [117]

177. G. Gatellier, A. Labrouzy, B. Mourrain, and J.-P. Técourt. Computing the topology of three-dimensional algebraic curves. In T. Dokken and B. Jüttler, editors, *Computational Methods for Algebraic Spline Surfaces (COMPASS)*, pages 27–44. Springer-Verlag, 2005. [222]

178. N. Geismann, M. Hemmer, and E. Schömer. Computing a 3-dimensional cell in an arrangement of quadrics: Exactly and actually! In *Proc. 17th Ann. Sympos. Comput. Geom.*, pages 264–273, 2001. [48]

179. B. Gerkey. Pursuit-evasion with teams of robots. http://robotics.stanford.edu/~gerkey/research/pe/index.html. [65]

180. B. Gerkey, S. Thrun, and G. J. Gordon. Visibility-based pursuit-evasion with limited field of view. In *Proc. National Conference on Artificial Intelligence (AAAI)*, pages 20–27, 2004. [64]

181. P. Giblin and B. Kimia. A formal classification of 3d medial axis points and their local geometry. *IEEE Transactions on Pattern Analysis and Machine Intelligence*, 26(2):238–251, 2004. [245, 253]

182. C. Gibson. *Elementary Geometry of Algebraic Curves*. Cambridge University Press, 1998. [19]

183. J. Giesen and M. John. Surface reconstruction based on a dynamical system. In *Proceedings of the 23rd Annual Conference of the European Association for Computer Graphics (Eurographics), Computer Graphics Forum 21*, pages 363–371, 2002. [266]

184. J. Giesen and M. John. The flow complex: A data structure for geometric modeling. In *ACM SODA*, pages 285 – 294, 2003. [243, 266]

185. L. González-Vega and I. Necula. Efficient topology determination of implicitly defined algebraic plane curves. *Comput. Aided Geom. Design*, 19(9):719–743, 2002. [145, 152, 214]

186. L. González-Vega, F. Rouillier, and M.-F. Roy. Symbolic recipes for polynomial system solving. In A. M. Cohen, H. Cuypers, and H. Sterk, editors, *Some Tapas of Computer Algebra*, volume 4 of *Algorithms and Computation in Mathematics*, chapter 2, pages 34–65. Springer, 1999. [142]

187. M. T. Goodrich and R. Tamassia. Dynamic trees and dynamic point location. *SIAM J. Comput.*, 28:612–636, 1998. [44]

188. M. Gopi, S. Krishnan, and C. Silva. Surface reconstruction based on lower dimensional localized Delaunay triangulation. In *Eurographics*, 2000. [253]

189. M. Granados, P. Hachenberger, S. Hert, L. Kettner, K. Mehlhorn, and M. Seel. Boolean operations on 3d selective nef complexes: Data structure, algorithms, and implementation. In *Proc. 11th European Symposium on Algorithms*, pages 174–186, 2003. [55]

190. T. A. Grandine. Applications of contouring. *SIAM Review*, 42:297–316, 2000. [117]

191. T. A. Grandine and F. W. Klein. A new approach to the surface intersection problem. *Computer Aided Geometric Design*, 14:111–134, 1997. [117, 152]

192. G.-M. Greuel and G. Pfister. *A Singular introduction to commutative algebra.* Springer-Verlag, Berlin, 2002. With contributions by Olaf Bachmann, Christoph Lossen and Hans Schönemann. [147]

193. A. Griewank. *Evaluating derivatives: principles and techniques of algorithmic differentiation.* Society for Industrial and Applied Mathematics, Philadelphia, PA, USA, 2000. [191]

194. V. Guillemin and A. Pollack. *Differential Topology.* Prentice Hall, Englewood Cliffs, NJ, 1974. [304]

195. P. Hachenberger and L. Kettner. Boolean operations on 3D selective Nef complexes: Optimized implementation and experiments. In *Proc. of 2005 ACM Symposium on Solid and Physical Modeling (SPM'05)*, pages 163–174, Cambridge, MA, June 2005. [55]

196. D. Halperin. Arrangements. In J. E. Goodman and J. O'Rourke, editors, *Handbook of Discrete and Computational Geometry*, chapter 24, pages 529–562. Chapman & Hall/CRC, 2nd edition, 2004. [2, 30, 42, 66]

197. D. Halperin, L. E. Kavraki, and J.-C. Latombe. Robotics. In J. E. Goodman and J. O'Rourke, editors, *Handbook of Discrete and Computational Geometry*, chapter 41, pages 755–778. CRC Press LLC, Boca Raton, FL, 1997. [116]

198. D. Halperin and E. Leiserowitz. Controlled perturbation for arrangements of circles. *International Journal of Computational Geometry and Applications*, 14(4 & 5):277–310, 2004. [50, 52, 53]

199. D. Halperin and M. H. Overmars. Spheres, molecules, and hidden surface removal. *Computational Geometry: Theory and Applications*, 11(2):83–102, 1998. [63, 64]

200. D. Halperin and C. R. Shelton. A perturbation scheme for spherical arrangements with application to molecular modeling. *Comput. Geom. Theory Appl.*, 10:273–287, 1998. [50, 64]

201. I. Haran and D. Halperin. An experimental study of point location in general planar arrangements. In *Proc. ALENEX 2006*, 2006. To appear. [23, 31]

202. J. Harris. *Algebraic Geometry, a First Course*, volume 133 of *Graduate Texts in Math.* New-York, Springer-Verlag, 1992. [146]

203. A. Hatcher. *Algebraic Topology.* Cambridge University Press, 2002. [283]

204. S. Hert, M. Hoffmann, L. Kettner, S. Pion, and M. Seel. An adaptable and extensible geometry kernel. In *Proc. Workshop on Algorithm Engineering*, volume 2141 of *Lecture Notes Comput. Sci.*, pages 79–90. Springer-Verlag, 2001. [26, 39]

205. D. Hilbert. *Foundations of Geometry (Grundlagen der Geometrie).* Open Court, 1971. [118]

206. D. Hilbert and S. Cohn-Vossen. *Geometry and the Imagination.* Reprint from the American Mathematical Society, 1999. [177]

207. K. Hildebrandt, K. Polthier, and M. Wardetzky. On the convergence of metric and geometric properties of polyhedral surfaces. *To appear in Geom. Dedicata.* [158]

208. M. W. Hirsch. *Differential Topology.* Springer-Verlag, New York, NY, 1976. [184, 249]

209. S. Hirsch and D. Halperin. Hybrid motion planning: Coordinating two discs moving among polygonal obstacles in the plane. In *Proc. 5th Workshop on Algorithmic Foundations of Robotics*, pages 225–241, 2002. [58]

210. H. Hiyoshi and K. Sugihara. Improving continuity of Voronoi-based interpolation over Delaunay spheres. *Comput. Geom.*, 22(1-3), 2002. [242]

211. S. Ho, S. Sarma, and Y. Adachi. Real-time interference analysis between a tool and an environment. *Computer-Aided Design*, 33(13):935–947, 2001. [59]

212. *IEEE Standard for binary floating point arithmetic, ANSI/IEEE Std 754 –* 1985. New York, NY, 1985. Reprinted in SIGPLAN Notices, 22(2):9–25, 1987. [119]

213. O. Ilushin, G. Elber, D. Halperin, R. Wein, and M.-S. Kim. Precise global collision detection in multi-axis machining. *Computer-Aided Design*, 37(9): 909–920, Aug 2005. [59]

214. J. Jost. *Riemannian Geometry and Geometric Analysis*. Universitext. Springer-Verlag, 2002. [305]

215. M. Karavelas. A robust and efficient implementation for the segment voronoi diagram. In *Proc. International Symposium on Voronoi Diagrams in Science and Engineering*, pages 51–62, 2004. [115]

216. M. Karavelas and M. Yvinec. Dynamic additively weighted voronoi diagrams in 2d. In *Proc. 10th European Symposium on Algorithms*, pages 586–598, 2002. [115]

217. M. Karavelas and M. Yvinec. The Voronoi diagram of convex objects in the plane. In *Proc. 11th European Symposium on Algorithms*, pages 337–348, 2003. [106, 107, 108, 109]

218. M. I. Karavelas and I. Z. Emiris. Predicates for the planar additively weighted Voronoi diagram. Technical Report ECG-TR-122201-01, INRIA Sophia-Antipolis, 2002. [109]

219. M. I. Karavelas and I. Z. Emiris. Root comparison techniques applied to computing the additively weighted Voronoi diagram. In *Proc. 14th ACM-SIAM Sympos. Discrete Algorithms (SODA)*, pages 320–329, 2003. [109, 136]

220. L. E. Kavraki, P. Švestka, J.-C. Latombe, and M. H. Overmars. Probabilistic roadmaps for path planning in high dimensional configuration spaces. *IEEE Trans. Robot. Autom.*, 12:566–580, 1996. [58]

221. L. Kettner, K. Mehlhorn, S. Pion, S. Schirra, and C. Yap. Classroom examples of robustness problems in geometric computations. In *Proc. 12th European Symposium on Algorithms*, volume 3221 of *Lecture Notes Comput. Sci.*, pages 702–713. Springer-Verlag, 2004. [317]

222. L. Kettner and S. Näher. Two computational geometry libraries: LEDA and CGAL. In J. E. Goodman and J. O'Rourke, editors, *Handbook of Discrete and Computational Geometry*, chapter 65, pages 1435–1463. CRC Press LLC, Boca Raton, FL, second edition, 2004. [3, 66, 316]

223. L. Kettner, J. Rossignac, and J. Snoeyink. The Safari interface for visualizing time-dependent volume data using iso-surfaces and contour spectra. *Computational Geometry: Theory and Applications*, 25:97–116, 2003. [308]

224. L. Kettner and J. Snoeyink. A prototype system for visualizing time-dependent volume data. In *Proc. 17th Ann. Symp. Computational Geometry*, pages 327–328. ACM Press, 2001. [308]

225. J. Keyser, T. Culver, M. Foskey, S. Krishnan, and D. Manocha. ESOLID: A system for exact boundary evaluation. *Computer-Aided Design*, 36(2):175–193, 2004. [4]

226. J. Keyser, T. Culver, D. Manocha, and S. Krishnan. Efficient and exact manipulation of algebraic points and curves. *Computer-Aided Design*, 32(11): 649–662, 2000. [4]

227. D.-S. Kim, C.-H. Cho, Y. Cho, C. I. Won, and D. Kim. Pocket recognition on a protein using Euclidean Voronoi diagrams of atoms. In *Proc. 3rd International Conference on Computational Science and its Applications*, volume 1, pages 707–715, 2005. [116]

228. D.-S. Kim, D. Kim, Y. Cho, J. Ryu, C.-H. Cho, J. Y. Park, and H.-C. Lee. Visualization and analysis of protein structures using Euclidean Voronoi diagrams of atoms. In *Proc. 3rd International Conference on Computational Science and its Applications*, volume 1, pages 993–1002, 2005. [116]

229. D. A. Klain and G.-C. Rota. *Introduction to Geometric Probability*. Cambridge University Press, 1997. [158]

230. R. Klein. *Concrete and Abstract Voronoi Diagrams*, volume 400 of *Lecture Notes Comput. Sci.* Springer-Verlag, 1989. [92, 93]

231. R. Klein, K. Mehlhorn, and S. Meiser. Randomized incremental construction of abstract Voronoi diagrams. *Comput. Geom. Theory Appl.*, 3(3):157–184, 1993. [102, 109]

232. W. Krandick and K. Mehlhorn. New bounds for the descartes method. *J. of Symb. Comp.*, 41(1), Jan 2006. [135]

233. S. Krishnan and D. Manocha. An efficient intersection algorithm based on lower dimensional formulation. *ACM Transactions on Computer Graphics*, 16:74–106, 1997. [117]

234. F. Labelle and J. Shewchuk. Anisotropic voronoi diagrams and guaranteed-quality anisotropic mesh generation. In *Proc. 19th Ann. Symposium on Computational Geometry*, pages 191–200. ACM Press, 2003. [86]

235. J. M. Lane and R. F. Riesenfeld. Bounds on a polynomial. *BIT*, 21(1):112–117, 1981. [135]

236. S. Lang. *Algebra*. Addison-Wesley, 1980. [125, 148]

237. J.-C. Latombe. *Robot Motion Planning*. Kluwer Academic Publishers, Boston, 1991. [57, 116]

238. S. Lazard, L. M. Peñaranda, and S. Petitjean. Intersecting quadrics: An efficient and exact implementation. In *Proc. 20th Ann. Sympos. Comput. Geom.*, pages 419–428, 2004. [50]

239. G. Leibon and D. Letscher. Delaunay triangulations and Voronoi diagrams for Riemannian manifolds. In *Proc. 16th Ann. Sympos. Comput. Geom.*, pages 341–349, 2000. [69]

240. E. Leiserowitz and S. Hirsch. Exact construction of Minkowski sums of polygons and a disc with application to motion planning. Technical Report ECG-TR-181205-01, Tel-Aviv University, 2002. [57, 58]

241. J. Levin. Algorithm for drawing pictures of solid objects composed of quadratic surfaces. *Commun. ACM*, 19(10):555–563, Oct. 1976. [50]

242. C. Li and C. Yap. A new constructive root bound for algebraic expressions. In *Proc. 12th ACM-SIAM Symposium on Discrete Algorithms*, pages 496–505, 2001. [35, 140]

243. A. Lieutier. Any open bounded subset of ∇^n has the same homotopy type than its medial axis. *Computer-Aided Design*, 11(36):1029–1046, 2004. [114]

244. W. E. Lorensen and H. E. Cline. Marching cubes: A high resolution 3d surface construction algorithm. *SIGGRAPH Comput. Graph.*, 21(4):163–169, 1987. [188, 189]

245. V. Luchnikov, M. Gavrilova, and N. Medvedev. A new development of the Voronoi-Delaunay technique for analysis of pores in packings of non-spherical

objects and in packings confined in containers. In *Proc. of the 21st Int. Conference on Applied Physics*, volume 1, pages 273–275, 2001. [116]

246. V. Luchnikov, N. Medvedev, and M. Gavrilova. The Voronoi-Delaunay approach for modeling the packing of balls in a cylindrical container. In *Proc. Int. Conf. Computational Science*, volume 1 of *Lecture Notes in Computer Science*, pages 748–752. Springer, 2001. [116]

247. F. S. Macaulay. On the resolution of a given modular system into primary systems including some properties of Hilbert numbers. *Math. Ann.*, 74(1): 66–121, 1913. [142]

248. A. Mantler and J. Snoeyink. Intersecting red and blue line segments in optimal time and precision. In J. Akiyama, M. Kano, and M. Urabe, editors, *Discrete and Computational Geometry, Japanese Conference, JCDCG 2000, Tokyo, Japan, November, 22-25, 2000, Revised Papers*, volume 2098 of *Lecture Notes in Computer Science*, pages 244–251. Springer, 2001. [54]

249. J. Matoušek. *Lectures on Discrete Geometry*, volume 212 of *Graduate Texts in Mathematics*. Springer-Verlag, 2002. [2, 66]

250. Y. Matsumoto. *An Introduction to Morse Theory*, volume 208 of *Translations of Mathematical Monographs*. American Mathematical Society, 2002. [295]

251. K. Mehlhorn and S. Näher. *LEDA: A Platform for Combinatorial and Geometric Computing*. Cambridge University Press, Cambridge, UK, 2000. [4, 9, 11, 38, 50, 55, 316]

252. G. Melquiond and S. Pion. Formally certified floating-point filters for homogenous geometric predicates. *Special issue on REAL NUMBERS of Theoretical Informatics and Applications*, 2006. to appear. [123, 154]

253. P. G. Mezey. Molecular surfaces. In K. B. Lipkowitz and D. B. Boyd, editors, *Reviews in Computational Chemistry*, volume 1. VCH Publishers, 1990. [64]

254. M. Mignotte. *Mathematics for Computer Algebra*. Springer-Verlag, 1992. [140]

255. J. W. Milnor. *Morse Theory*. Princeton University Press, Princeton, NJ, 1963. [295]

256. J. S. B. Mitchell. Shortest paths and networks. In J. E. Goodman and J. O'Rourke, editors, *Handbook of Discrete and Computational Geometry*, chapter 24, pages 445–466. CRC Press LLC, Boca Raton, FL, 1997. [116]

257. R. E. Moore. *Interval Analysis*. Prentice Hall, Englewood Cliffs, NJ, 1966. [120]

258. F. Morgan. Minimal surfaces, crystals, and norms on R^n. In *Proc. 7th Ann. Sympos. Comput. Geom.*, pages 204–213, 1991. [161, 171]

259. J.-M. Morvan and B. Thibert. Smooth surface and triangular mesh: Comparison of the area, the normals and the unfolding. In *ACM Symposium on Solid Modeling and Applications*, pages 147–158, 2002. [252]

260. J.-M. Morvan and B. Thibert. Approximation of the normal vector field and the area of a smooth surface. *Discrete & Computational Geometry*, 32(3): 383–400, 2004. [162, 163, 164]

261. B. Mourrain. Computing isolated polynomial roots by matrix methods. *J. of Symbolic Computation, Special Issue on Symbolic-Numeric Algebra for Polynomials*, 26(6):715–738, Dec. 1998. [141]

262. B. Mourrain, F. Rouillier, and M.-F. Roy. Bernstein's basis and real root isolation. In J. E. Goodman, J. Pach, and E. Welzl, editors, *Combinatorial and Computational Geometry*, Mathematical Sciences Research Institute Publications, pages 459–478. Cambridge University Press, 2005. [135, 136]

263. B. Mourrain and J.-P. Técourt. Isotopic meshing of a real algebraic surface. Technical Report RR-5508, INRIA-Sophia Antipolis, France, Feb. 2005. 21 pp. [187, 213, 217, 220, 222]

264. B. Mourrain, J.-P. Técourt, and M. Teillaud. Sweeping an arrangement of quadrics in 3d. In *Proc. 19th European Workshop on Computational Geometry*, pages 31–34, 2003. [46]

265. B. Mourrain, J.-P. Técourt, and M. Teillaud. On the computation of an arrangement of quadrics in 3d. *Computational Geometry: Theory and Applications*, 30:145–164, 2005. [46, 47]

266. B. Mourrain and P. Trébuchet. Algebraic methods for numerical solving. In *Proc. of the 3rd International Workshop on Symbolic and Numeric Algorithms for Scientific Computing'01 (Timisoara, Romania)*, pages 42–57, 2002. [141]

267. B. Mourrain and P. Trébuchet. Generalised normal forms and polynomial system solving. In M. Kauers, editor, *Proc. Intern. Symp. on Symbolic and Algebraic Computation*, pages 253–260. New-York, ACM Press., 2005. [141]

268. B. Mourrain, M. Vrahatis, and J. Yakoubsohn. On the complexity of isolating real roots and computing with certainty the topological degree. *J. of Complexity*, 18(2):612–640, 2002. [135]

269. K. Mulmuley. A fast planar partition algorithm, I. *J. Symbolic Comput.*, 10(3-4):253–280, 1990. [23]

270. K. Mulmuley. A fast planar partition algorithm, II. *J. ACM*, 38:74–103, 1991. [32]

271. D. A. Musser and A. A. Stepanov. Generic programming. In *Proc. Intern. Symp. on Symbolic and Algebraic Computation, LNCS 358*, pages 13–25. Springer-Verlag, 1988. [314]

272. N. Myers. Traits: A new and useful template technique. *C++ Report*, 7(5): 32–35, 1995. [314]

273. A. Nanevski, G. Blelloch, and R. Harper. Automatic generation of staged geometric predicates. *Higher-Order and Symbolic Computation*, 16(4):379–400, Dec. 2003. [123]

274. W. Nef. *Beiträge zur Theorie der Polyeder*. Herbert Lang, Bern, 1978. [55]

275. C. Ó'Dúnlaing and C. K. Yap. A "retraction" method for planning the motion of a disk. *J. Algorithms*, 6:104–111, 1985. [57]

276. A. Okabe, B. Boots, and K. Sugihara. *Spatial Tessellations: Concepts and Applications of Voronoi Diagrams*. John Wiley & Sons, Chichester, UK, 1992. [69]

277. G. J. Olling, B. K. Choi, and R. B. Jerard. *Machining Impossible Shapes*. Kluwer Academic Publishers, 1998. [59]

278. J. Owen and A. Rockwood. Intersection of general implicit surfaces. In *Geometric Modeling: Algorithms and New Trends*, pages 335–345. SIAM, Philadelphia, 1987. [117]

279. E. Packer. Finite-precision approximation techniques for planar arrangements of line segments. M.Sc. thesis, Tel Aviv University, Tel Aviv, Israel, 2002. [50]

280. M. P. Patrikalakis and T. Maekawa. *Shape Interrogation for Computer Aided Design and Manufacturing*. Springer Verlag, 2002. [117]

281. S. Petitjean and E. Boyer. Regular and non-regular point sets: Properties and reconstruction. *Comput. Geom. Theory Appl.*, 19:101–126, 2001. [269]

282. S. Pion. *De la géométrie algorithmique au calcul géométrique*. Thèse de doctorat en sciences, Université de Nice-Sophia Antipolis, France, 1999. TU-0619 http://www.inria.fr/rrrt/tu-0619.html. [154]

283. S. Pion. Interval arithmetic: An efficient implementation and an application to computational geometry. In *Workshop on Applications of Interval Analysis to systems and Control*, pages 99–110, 1999. [154]

284. S. Pion and M. Teillaud. Towards a CGAL-like kernel for curves. Technical Report ECG-TR-302206-01, MPI Saarbrücken, INRIA Sophia-Antipolis, 2003. [39]

285. S. Pion and C. K. Yap. Constructive root bound for k-ary rational input numbers. In *Proc. 19th Ann. Sympos. Comput. Geom.*, pages 256–263, 2003. [35, 140, 155]

286. S. Plantinga and G. Vegter. Isotopic approximation of implicit curves and surfaces. In *SGP '04: Proceedings of the 2004 Eurographics/ACM SIGGRAPH symposium on Geometry processing*, pages 245–254, New York, NY, USA, 2004. ACM Press. [187, 190, 196, 198, 200, 227]

287. H. Pottmann and J. Wallner. *Computational Line Geometry*. Springer-Verlag, Berlin, Heidelberg, 2001. [2]

288. F. P. Preparata and M. I. Shamos. *Computational Geometry: An Introduction*. Springer-Verlag, New York, NY, 1985. [317]

289. M. A. Price and C. G. Armstrong. Hexahedral mesh generation by medial surface subdivision: Part II, solids with flat and concave edges. *International Journal for Numerical Methods in Engineering*, 40:111–136, 1997. [116]

290. M. A. Price, C. G. Armstrong, and M. A. Sabin. Hexahedral mesh generation by medial surface subdivision: Part I, solids with convex edges. *International Journal for Numerical Methods in Engineering*, 38(19):3335–3359, 1995. [116]

291. W. Pugh. Skip lists: a probabilistic alternative to balanced trees. *Commun. ACM*, 33(6):668–676, 1990. [107]

292. S. Raab. Controlled perturbation for arrangements of polyhedral surfaces with application to swept volumes. In *Proc. 15th Ann. Symposium on Computational Geometry*, pages 163–172, 1999. [50]

293. G. Reeb. Sur les points singuliers d'une forme de Pfaff complètement intégrable ou d'une fonction numérique. *Comptes Rendus Acad. Sciences Paris*, 222: 847–849, 1946. [307]

294. F. M. Richards. Areas, volumes, packing, and protein structure. *Annu. Rev. Biophys. Bioeng.*, 6:151–176, 1977. [63]

295. J. Risler. *Méthodes mathématiques pour la CAO*. Masson, 1991. [132]

296. V. Rogol. Maximizing the area of an axially-symmetric polygon inscribed by a simple polygon. Master's thesis, The Technion, Haifa, Israel, 2003. `ftp://ftp.cs.technion.ac.il/pub/barequet/theses/rogol-msc-thesis.pdf.gz`. [63]

297. J. Rossignac and M. O'Connor. Sgc: A dimension-independent model for pointsets with internal structures and incomplete boundaries. In M. Wozny, J. Turner, and K. Preiss, editors, *Geometric Modeling for Product Engineering*. North-Holland, 1989. [55]

298. F. Rouillier. Solving zero-dimensional polynomial systems throuhg Rational Univariate Representation. *App. Alg. in Eng. Com. Comp.*, 9(5):433–461, 1999. [142]

299. F. Rouillier and P. Zimmermann. Efficient isolation of a polynomial real roots. *Journal of Computational and Applied Mathematics*, 162(1):33–50, 2003. [135, 136]

300. M. Roy. Basic algorithms in real algebraic geometry: from Sturm theorem to the existential theory of reals. In *Lectures on Real Geometry in memoriam of Mario Raimondo*, volume 23 of *Exposition in Mathematics*, pages 1–67, 1996. [138]

301. J. Ruppert. A Delaunay refinement algorithm for quality 2-dimensional mesh generation. *J. Algorithms*, 18:548–585, 1995. [182]

302. T. Sakkalis and T. J. Peters. Ambient isotopic approximations for surface reconstruction and interval solids. In *SM '03: Proceedings of the eighth ACM symposium on Solid modeling and applications*, pages 176–184, New York, NY, USA, 2003. ACM Press. [185]

303. M. F. Sanner, A. J. Olson, and J.-C. Spehner. Fast and robust computation of molecular surfaces. In *Proc. 11th Ann. Sympos. Comput. Geom.*, pages C6–C7, 1995. [63]

304. E. Scheinerman. When close enough is close enough. *American Mathematical Monthly*, 107:489–499, 2000. [140]

305. S. Schirra. Robustness and precision issues in geometric computation. In J.-R. Sack and J. Urrutia, editors, *Handbook of Computational Geometry*, chapter 14, pages 597–632. Elsevier Science Publishers B.V. North-Holland, Amsterdam, 2000. [317]

306. S. Schmitt. Improved separation bounds for the diamond operator. Technical Report ECG-TR-363108-01, MPI Saarbrücken, 2004. [140, 155]

307. S. Schmitt. The diamond operator – implementation of exact real algebraic numbers. In *Proc. 8th Internat. Workshop on Computer Algebra in Scient. Comput. (CASC 2005)*, volume 3718 of *Lecture Notes in Computer Science*, pages 355–366. Springer, 2005. http://www.mpi-sb.mpg.de/projects/EXACUS/leda_extension/. [38]

308. M. Seel. Implementation of planar Nef polyhedra. Research Report MPI-I-2001-1-003, Max-Planck-Institut für Informatik, Stuhlsatzenhausweg 85, 66123 Saarbrücken, Germany, August 2001. [55]

309. M. Seel. *Planar Nef Polyhedra and Generic High-dimensional Geometry*. PhD thesis, Universität des Saarlandes, September 2001. [55]

310. R. Seidel and N. Wolpert. On the exact computation of the topology of real algebraic curves. In *Proc. 21th Annual Symposium on Computational Geometry*, pages 107–115, 2005. [4]

311. M. Sharir. Almost tight upper bounds for lower envelopes in higher dimensions. *Discrete Comput. Geom.*, 12:327–345, 1994. [70]

312. M. Sharir and P. K. Agarwal. *Davenport-Schinzel Sequences and Their Geometric Applications*. Cambridge University Press, New York, 1995. [2, 23, 42, 66, 70]

313. V. Sharma and C. Yap. Sharp amortized bounds for descartes and de casteljau's methods for real root isolation. www.cs.nyu.edu/yap/papers, Oct. 2005. [135]

314. H. Shaul. Improved output-sensitive construction of vertical decompositions of triangles in three-dimensional space. M.Sc. thesis, School of Computer Science, Tel Aviv University, Tel Aviv, Israel, 2001. [43, 44]

315. H. Shaul and D. Halperin. Improved construction of vertical decompositions of three-dimensional arrangements. In *Proc. 18th Ann. Sympos. Comput. Geom.*, pages 283–292, 2002. [41, 43, 44, 47]

316. A. Sheffer and M. Bercovier. Hexahedral meshing of non-linear volumes using Voronoi faces and edges. *Numerical Methods in Engineering*, 49(1):329–351, 2000. [116]

317. J. R. Shewchuk. Delaunay refinement algorithms for triangular mesh generation. *Computational Geometry: Theory and Applications*, 22:21–74, 2002. [182]

318. R. Sibson. A brief description of natural neighbour interpolation. In V. Barnet, editor, *Interpreting Multivariate Data*, pages 21–36. John Wiley & Sons, Chichester, 1981. [242]

319. R. B. Simpson. Anisotropic mesh transformations and optimal error control. In *Proceedings of the third ARO workshop on Adaptive methods for partial differential equations*, pages 183–198, New York, NY, USA, 1994. Elsevier North-Holland, Inc. [179]

320. S. Smale. On gradient dynamical systems. *Ann. of Math.*, 74:199–206, 1961. [306]

321. J. M. Snyder. *Generative modeling for computer graphics and CAD: symbolic shape design using interval analysis*. Academic Press, 1992. [187, 190, 194]

322. J. M. Snyder. Interval analysis for computer graphics. *SIGGRAPH Comput. Graph.*, 26(2):121–130, 1992. [187, 190, 194]

323. M. Spivak. *Analysis on Manifolds*. Perseus Book Publishing, 1965. [299]

324. B. T. Stander and J. C. Hart. Guaranteeing the topology of an implicit surface polygonization for interactive modeling. *Computer Graphics*, 31(Annual Conference Series):279–286, 1997. [187, 223]

325. J. Stillwell. *Classical Topology and Combinatorial Group Theory*. Springer-Verlag, New York, 1993. [281]

326. B. Tagansky. A new technique for analyzing substructures in arrangements of piecewise linear surfaces. *Discrete Comput. Geom.*, 16:455–479, 1996. [42]

327. J.-P. Técourt. *Sur le calcul effectif de la topologie de courbes et surfaces implicites*. Thèse de doctorat en sciences, Université de Nice–Sophia Antipolis, France, Dec. 2005. [187, 213, 217, 220, 222]

328. P. Trébuchet. *Vers une résolution stable et rapide des équations algébriques*. PhD thesis, Université Pierre et Marie Curie, 2002. [141]

329. G. Vegter. Computational topology. In J. Goodman and J. O'Rourke, editors, *CRC Handbook of Discrete and Computational Geometry*, chapter 32, pages 719–742. Chapman & Hall/CRC, 2004. [277, 295]

330. J. von zur Gathen and J. Gerhard. *Modern computer algebra*. Cambridge University Press, New York, 1999. [130, 149]

331. R. Wein. High-level filtering for arrangements of conic arcs. In *Proc. 10th European Symposium on Algorithms*, volume 2461 of *Lecture Notes Comput. Sci.*, pages 884–895, 2002. [3, 11, 35]

332. R. Wein. High-level filtering for arrangements of conic arcs. M.Sc. thesis, School of Computer Science, Tel Aviv University, Tel Aviv, Israel, 2002. [35]

333. R. Wein, E. Fogel, B. Zukerman, and D. Halperin. Advanced programming techniques applied to CGAL's arrangements. In *Proc. Workshop on Library-Centric Software Design (LCSD 2005), at the Object-Oriented Programming, Systems, Languages and Applications (OOPSLA) Conference.*, October 2005. [4, 26]

334. R. Wein and D. Halperin. Generic implementation of the construction of lower envelopes of planar curves. Technical Report ECG-TR-361100-01, Tel-Aviv University, 2004. [61]

335. R. Wein, O. Ilushin, G. Elber, and D. Halperin. Continuous path verification in multi-axis nc-machining. In *Proc. 20th Annual Symposium on Computational Geometry*, pages 86–95, 2004. [60, 61]

336. R. Wein, J. P. van den Berg, and D. Halperin. The visibility–Voronoi complex and its applications. In *Proc. 21th Annual Symposium on Computational Geometry*, pages 63–72, 2005. [65]

337. R. Wein and B. Zukerman. Exact and efficient construction of planar arrangements of circular arcs and line segments with applications. Technical report, Tel-Aviv University, 2006. [56]

338. H. Whitney. *Complex analytic varieties.* Addison-Wesley Publishing Co., Reading, Mass.-London-Don Mills, Ont., 1972. [149]

339. H.-M. Will. Fast and efficient computation of additively weighted Voronoi cells for applications in molecular biology. In *Proc. 6th Scand. Workshop Algorithm Theory*, volume 1432 of *Lecture Notes Comput. Sci.*, pages 310–321. Springer-Verlag, 1998. [116]

340. N. Wolpert. *An Exact and Efficient Approach for Computing a Cell in an Arrangement of Quadrics.* Ph.D. thesis, Universität des Saarlandes, Saarbrücken, Germany, 2002. [20, 48]

341. N. Wolpert. Jacobi curves: Computing the exact topology of non-singular algebraic curves. In G. D. Battista and U. Zwick, editors, *Proc. 11th European Symposium on Algorithms*, Lecture Notes Comput. Sci., pages 532–543, 2003. [4, 20]

342. G. Wyvill, C. McPheeters, and B. Wyvill. Data structure for soft objects. *The Visual Computer*, 2(4):227–234, February 1986. [188]

343. C. Yap. Towards exact geometric computation. *Comput. Geom. Theory Appl.*, 7(1):3–23, 1997. [140]

344. C. K. Yap. Symbolic treatment of geometric degeneracies. *J. Symbolic Comput.*, 10:349–370, 1990. [210]

345. C. K. Yap. *Fundamental Problems in Algorithmic Algebra.* Princeton University Press, 1993. [140]

346. C. K. Yap. Robust geomtric computation. In J. E. Goodman and J. O'Rourke, editors, *Handbook of Discrete and Computational Geometry*, chapter 41, pages 927–952. Chapman & Hall/CRC, 2nd edition, 2004. [50, 318]

347. Y. Yomdin. On the general structure of a generic central set. *Compositio Math.*, 43:225–238, 1981. [245]

348. L. Yuan-Shin and C. Tien-Chien. 2-phase approach to global tool interference avoidance in 5-axis machining. *Computer-Aided Design*, 27(10):715–729, 1995. [59]

349. M. Zähle. Integral and current representations of federer's curvature measures. *Arch. Math. (Basel)*, 46:557–567, 1986. [171]

350. H. K. Zhao, S. Osher, and R. Fedkiw. Fast surface reconstruction using the level set method. In *Proc. IEEE Workshop on Variational and Level Set Methods in Computer Vision*, page 194, 2001. [232, 266]

351. A. Zomorodian. *Topology for Computing*, volume 16 of *Cambridge Monographs on Applied and Computational Mathematics.* Cambridge University Press, 2005. [277]

Index